# BIOTECHNOLOGY

# BIOTECHNOLOGY

*Editor*

**Carsten Weeber**

scitus
academics

**Biotechnology**
Edited by **Carsten Weeber**

This Book is Printed on Acid-Free Paper. ♾



ISBN: 9781681179087
Library of Congress Control Number: 2017948936


SCITUS Academics LLC,
616, Corporate Way, Suite 2, 4766,
Valley Cottage, NY 10989

www.scitusacademics.com

**Notice**



Printed in United States of America.

# Preface

For about 10,000 years, farmers have been improving wild plants and animals throughout the selection and breeding of enviable characteristics. This breeding has resulted in the domesticated plants and animals that are normally used in crop and livestock agriculture. In the twentieth century, breeding became more sophisticated, as the traits that breeders select for include increased yield, disease and pest resistance, drought resistance and enhanced flavor. Agricultural biotechnology deals with the scientific techniques used to improve plants, animals and microorganisms. DNA is the key to biotechnology and based on its understanding scientists have developed solutions to increase agricultural productivity. Starting from the ability to identify genes that may confer advantages on certain crops, and the ability to work with such characteristics very precisely, biotechnology enhances breeders' ability to make improvements in crops and livestock. Traits are passed from one generation to the next through genes, which are made of DNA. All living things—including the fruits, vegetables and meat that we eat—contain genes that tell cells how to function. Recently, scientists have learned enough to begin to identify and work with the genes (DNA) that are responsible for traits. It encompasses genetic engineering, inclusive of enzyme and protein engineering plant and animal tissue culture technology, biosensors for biological monitoring, bioprocess and fermentation technology.. it is concerned with upgradation of quality and also utilization of livestock and resources for the well being of both animals and plants. New approaches in biotechnology can develop high yielding and more nutritious crop varieties, improve resistance to disease and also reduce the need for fertilizer and other expensive agricultural chemicals. It could also improve forestry and its products, fibre crops and chemical feedstocks.

This book, a collaborative endeavor involving contributions from renowned academics around the globe, discusses the techniques of plant tissue culture, the fundamental basis for the development of innovative crop improvement strategies. It brings together the principles and contemporary agricultural biotechnology. The information provided in this book should be of interest for students, academic researchers and for agriculturalists.

# Table of Contents

# CHAPTER 1

# Biotechnology and Agriculture

**Akinruli Samuel Ayobami, Azevedo Valesca, Barbosa Francisco Vidal, Azevedo Vasco***

*Universidade Federal De Minas Gerais (UFMG), Brazil*

## ABSTRACT

Here, we provide a comprehensive insight into current advances in the use of modern biotechnology in agriculture. Considering that biotechnology involves the use of scientific and engineering principles in the processing of materials by biological agents to provide goods and services. This study does not only take a historical retrace of biotechnology into account but considers the pros and cons of the use of biotechnology in attending the need for global food production as well as the challenges there of.

**Keywords:** Biotecnology; Agriculture; Cropping systems; Agricultural Production Systems

## 1. INTRODUCTION

Agriculture could be seen as the use of cultivation toward achieving life sustenance. However, cultivation includes that of animals, plants, fungi and other life forms for food, fiber, and other products. The study of agriculture is known as agricultural science, and also observed in certain species of ant and termite. It could be holistically refer to human activities as it was the key factor in the rise of sedentary human civilization, whereby farming of domesticated species created food surpluses that nurtured the development of civilization.

Biotechnology, on the other hand, is the use of scientific and engineering principles in the biological materials processing of agents in order to provide goods and services for human use. There has been a strong tie between its existence and societal development. In other words, history has it that biotechnology has been principally associated with food, addressing such provision, and as such, acting as solutions to solving problems related malnutrition and famine in societies. Nonetheless, Biotechnology has recently been coupled with the development of vital drugs.

Globally, the systems of agriculture have advanced tremendously owing to the advancement of technological knowhow. More so, recent centuries marked the peak of the agricultural embracement with the use of modern Biotechnology. Although, agricultural practices such as irrigation, crop rotation, fertilizers, and pesticides were developed long ago

The history of agriculture has played a major role in human history, as agricultural progress has been a crucial factor in worldwide socioeconomic change and evident ideologically in cultural advancement, and social organization such as hunter-gathering, settlements, arts, architecture, law etc. Furthermore, human and social science researcher of various specialties have long argued civilization emerged from agricultural development and as such its attainment owed to agricultural advancement.

# 2. AGRICULTURAL PRODUCTION SYSTEMS

Agricultural Activities or production can be divided into two major systems called 'Cropping system and the livestock production system.

## 2.1 The cropping systems

The Cropping systems are of different kinds and the farm's choice is often determined by considering the constraints and availability of some motivating factors. These could be traced to regional and geographical structure, climate, government policies, socio-economic and political factors etc. However, cropping system is also referred to as cultivation system which includes the shifting cultivation system and annual cultivation system. The shifting cultivation system involves slashing and burning of the forests to bring about the release of nutrients on the soil for annual and perennial crops cultivation. The annual crop cultivation does not necessarily require the fallow period, while it appears preferable in the perennial cultivation whereby the tampered land is sequentially left fallow for forest to germinate. During this period, famers move to another piece of land to return after ten to twenty years. When necessary, farmers may choose to reduce the fallow period. However, this requires a boost in nutrient using fertilizer or manure and pest control which is also applicable to monocultures and polycultures at large scale or large acreage where nutrient tends to be relatively low as the biodiversity therein.

**AGRICULTURAL PRACTICES:** Agricultural practices are the methods through which agriculture production processes is performed. These practices include;

**TILLAGE:** Tillage this involves plowing soil in preparation for planting, incorporation of nutrient or pest control. The intensity of the tillage practice varies from conventional to no-till and may improve productivity through soil warming, weeds controlling and incorporating fertilizer, provoking organic matter decomposition and consequently releasing $CO_2$ which brings about reduction in the abundance and diversity of soil organisms. However, soil may be more exposed to erosion [1,2].

**PEST CONTROL:** Pest control involves the management of weeds, insects/mites, and diseases. Here, chemical (pesticides), biological (bio-control), mechanical (tillage), and cultural practices are employed. Cultural practices include crop rotation, culling, cover crops, intercropping, composting, avoidance, and resistance. "Integrated pest management attempts to use all of these methods to keep pest populations below the number which would cause economic loss, and recommends pesticides as a last resort" [3].

**NUTRIENT MANAGEMENT:** Nutrient management includes both the source of nutrient inputs for crop and livestock production, and the method of utilization of manure produced by livestock. Nutrient inputs can be chemical inorganic fertilizers, manure, green manure, compost and mined minerals [4]. Crop nutrient use may also be managed using cultural techniques such as crop rotation or a fallow period [5,6]. Manure is used either by holding livestock where the feed crop is growing, such as in managed intensive rotational grazing, or by spreading either dry or liquid formulations of manure on cropland or pastures.

**WATER MANAGEMENT**: Water management is where rainfall is insufficient or variable, which occurs to some degree in most regions of the world [7]. Some farmers use irrigation to supplement rainfall. In other areas such as the Great Plains in the U.S. and Canada, farmers use a fallow year to conserve soil moisture to use for growing a crop in the following year [8]. Agriculture represents 70% of freshwater use worldwide.

## 2.2 Livestock production systems

Livestock production systems involves animal husbandry which includes the breeding and raising of species either for work and companionship or for production of food like meat, milk, eggs, or wool. These animals include cattle, horses, mules, oxen, camels, llamas, alpacas, and dogs, goats, sheep etc. Animal husbandry can be classified by the methods of feeding which include grassland based, mixed, and landless livestock production system.

**GRASSLAND BASED LIVESTOCK PRODUCTION:** Grassland based livestock production depends on plant material such as shrub land, rangeland, and pastures for feeding ruminant animals. Outside nutrient inputs may be used, however manure is returned directly to the grassland as a major nutrient source. This system is particularly important in areas where crop production is not feasible because of climate or soil, representing 30-40 million pastoralists [7]. Mixed production systems use grassland, fodder crops and grain feed crops as feed for ruminant and monogastic (one stomach; mainly chickens and pigs) livestock. Manure is typically recycled in mixed systems as a fertilizer for crops. Approximately 68% of all agricultural land is permanent pastures used in the production of livestock. Landless systems rely upon feed from outside the farm, representing the de-linking of crop and livestock production found more prevalently in OECD member countries. In the U.S., 70% of the grain grown is fed to animals on feedlots." [7]

**Biotechnology and Agriculture:**Recent centuries witnessed high level of specialized and concentrated agriculture based on the use of machines, agricultural chemical technologies like fertilizers and pesticides, and plant breeding (hybrids and GMOs). In recent decades, sustainable agriculture became the most sought-after means to attain development coupling ideas of socio-economic justice, resource conservation and environment in the farming system [9,10]. This triggered solutions to problems of conventional agriculture approaches and the growth of agricultural diversification bringing about organic agriculture, community supported agriculture, urban agriculture, ecological or biological agriculture, holistic management and integrated farming system.

Since time immemorial, farming practices have embraced genetic modification of grown crops through selection of specie with characteristics like fast growth rate, larger seeds or sweeter fruits which make the features of domesticated plant species different from the wild relatives. It may therefore be affirmed that many of our modern crops were developed by farmers who are considered to have lacked an understanding of the scientific basis of plant breeding. Most of who are unrecognized or unknown to the academic environment.

Record has it that in ancient time, there exist flakes of what is today known as Biotechnology. The ancient Egyptians invented scientific processes such as the fermentation and breeding practices. Other early records include; 1750 BC- The Sumerians brew beer; 500 BC-The Chinese use moldy soybean curds as an antibiotic to treat boils; 250 BC- The Greeks practice crop rotation to maximize soil fertility; 100 BC-Powdered chrysanthemum is used in China as an insecticide etc. (Timeline of Biotechnology above).

# 3. PLANT BREEDING

Early civilization was strongly base on agriculture and as such a the need for alteration of crop which has been a longstanding practice which is traceable to 2500-2000 BC as recorded that in Africa the ancient Egyptians, apart from breeding of animals like geese and cattle for dietary and nutritional use in their society, produced wine using fermentation techniques with the aid of the microbiological processes in the absence of oxygen. The fermentation technology was equally used in bakery to provoke dough rise which consequently brought about 50 varieties of bread over 4,000 years ago.

In the 17$^{th}$ century, advance plants study and the understanding were relatively poor. Nonetheless, breeding of plant was a popular practice when the study of Gregor Mendel, "the father of genetics" who was a plant breeder, on inheritable traits of peas spore a better understanding of genetic inheritance bringing about a new terminology called crossbreeding which is today referred to as hybridization. Furthermore, he published a paper in 1865 called *Versuche über Pflanzenhybriden (Experiments on Plant Hybridization*) at two meetings of the Natural History Society of Brünn in Moravia which was criticized, but seen as merely useful to hybridization rather than inheritance and as such ignored. In this paper distinctively demonstrated how dominant and recessive alleles produce their notable traits and possibly transfer this to offspring.

In 1900, the work was rediscovered by a group of scientists who were working on breeding problems and made Medel's findings more popular. Thereafter, major advances in plant breeding were achieved with reference to Mendel study and new understanding of genetics found tangential points to the traditional self-pollination and crosspollination techniques.

The identification of traits and their incorporating is crucial in plant breeding. Therefore breeders toured around foreign lands in search of plants with specific desirable traits. Occasionally, these traits emerge through the mutation process which was rather considered too slow and unreliable for the desired rate of production of diverse plants. This gesture of domestication of plants further became the most important means of increasing yield, improving disease resistance and drought tolerance, ameliorated nutrition and taste of crop as well as easing their harvests for centuries. Remarkably, *The Green Revolution* played a vital role in popularizing the use of conventional hybridization to multiply yields into many folds through "high-yielding varieties" and has a huge impact on the world's crop production.

# 4. MODERN BIOTECHNOLOGY AND AGRICULTURE (MODERN GMO)

A Genetically Modified Organism (GMO) is that whose genetic material is altered using genetic engineering techniques called recombinant DNA technology. Modern Genetic engineering has provided a wide range of genes which are valuable in the creation of desired germilines for new crops. A good account of this progress is traceable to the use of genetic engineering to better tomatoes so they can be more resistant to mechanical handling which was stimulated by the emergence of mechanical tomato-harvester in the early 1960s. Genetic engineering has also helped immensely in recent creation of crops with various beneficial and development traits like flavors and aromas, sizes, harvest durability, colors, tastes, resistance etc.

# 5. HERBICIDE-TOLERANT GMO CROPS

One of the major problems of cropping systems is the attack of pests and unwanted and sabotaging weeds. A long combating solution to this is the herbicide use in the farms. Herbicides (weed killers) are pesticides applicable to unwanted plants and may be used to target specific kind of plants. According to the United State Department of Agriculture Natural Resources Conservation Service, 70% of the overall agriculture pesticide use and landscape turf management are based on herbicides [11]. Herbicides have proven to be the most probable solution, but recently, a deeper study into Genomic Engineering tend to bring more affordable proposals of seeds that are herbicide tolerant.

An example of the herbicide tolerant seed is The Roundup Ready seed. It is today grown around the world, possesses an herbicide resistant gene, built with modern genetic engineering, in its genome which enables plants to tolerate exposure to glyphosate. The Roundup Ready technology permits growing of a crop that can withstand direct contact with herbicides especially glyphosate, while combating weeds. In other words, with RR seed one can spray the farm with glyphosate to curb the weed without damaging the genetically modified herbicidetolerant crop. A suitable sample is RR soybean which accounts for 92% of the US soybeans acreage.

Notwithstanding, there are record of cases that some weeds have developed resistance to glyphosate amongst other crop deficiencies complaints, leaving farmers with little or no option than switching to other herbicides.

## 6. INSECT-RESISTANT GMO CROPS

The Insect-Resistant Plant is other kind of grown genetic engineered crops which possesses a gene from soil *Bacillus thuringiensis* (Bt) bacterium, that releases a specific toxin against insects protecting the plant against attacks by insects. This kind of technology already is being used in the starlink crops for example. However, it has also been argued that similar or better pest-resistance traits can be achieved by traditional breeding practices and hybridization or cross-pollination with wild species can bring about plant resistance to various pests. Furthermore, some wild species seldom present the primary source of resistance traits. A good example is some tomato cultivar with wild species traits that have gained resistance to over 19 diseases and have done so by crossing with wild populations of tomatoes.

## 7. GENETIC ENGINEERING AND GM CROPS

For the past 30 years, agricultural biotech has strived rapidly owing to the better understanding of DNA as the chemical double-helix code to genes creation. It is one of the most modern tools in agriculture based on recombinant DNA technology. This involves the use of tools and enzyme to cut out, insert, and modify pieces of DNA that present interesting genes. However, the main distinction between the mutation in the traditional plant breeding and genetic engineering is in the ability to manipulate individual gene and transfer gene between species which cannot be derive naturally in the wild. More so, in the traditional or conventional plant breeding, there are slim chances of obtaining any particular desired gene combination from the millions of crosses generated in the process. There can be transfer of undesired genes with the desired genes or loss of some desired genes from both parents due to the random re-assorting of their genes by the offspring, limiting the chances of improvements (Figure 1).

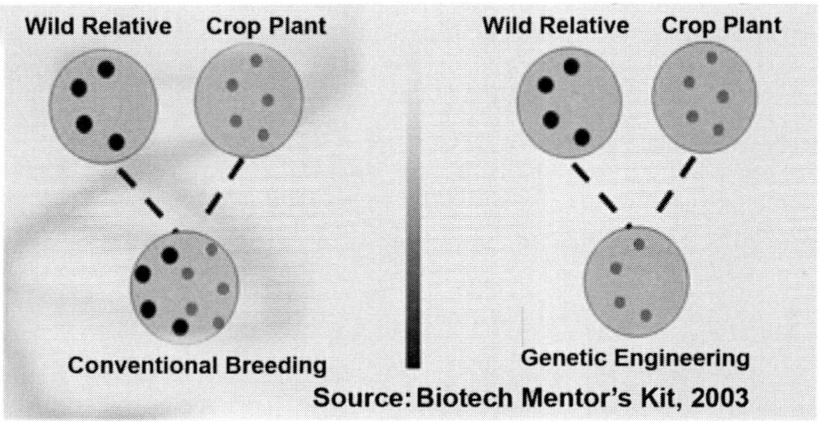


**Figure 1:** Conventional vs. Genetic engineering (Source Biotech Mentor's Kit, 2003)

## 7.1 Process of Genetic Engineering

Genetic engineering on the other hand enables the direct transfer of one or specific set of genes between closely or distantly related organisms. Nevertheless, genetic engineering doesn't only involve inserting DNA from other organisms, but may also involve reduction or switching off specific genes, thereby controlling the undesired genes.

## 7.2 Application of genetic engineering in crop production

Genetic engineering techniques are applicable when all related techniques are assumed exhausted and when:

a) The trait to be introduced is not present in the germplasm of the crop;

b) The trait is very difficult to improve by conventional breeding methods; and

c) It will take a very long time to introduce and/or improve such trait in the crop by conventional breeding methods (Figure 2)

It is worth saying that modern plant breeding has paved way for multi-disciplinary process where a wide range of tools and elements of conventional breeding techniques and bioinformatics, molecular biology, molecular genetics, genetic engineering, biochemistry, biochemistry are integrated.

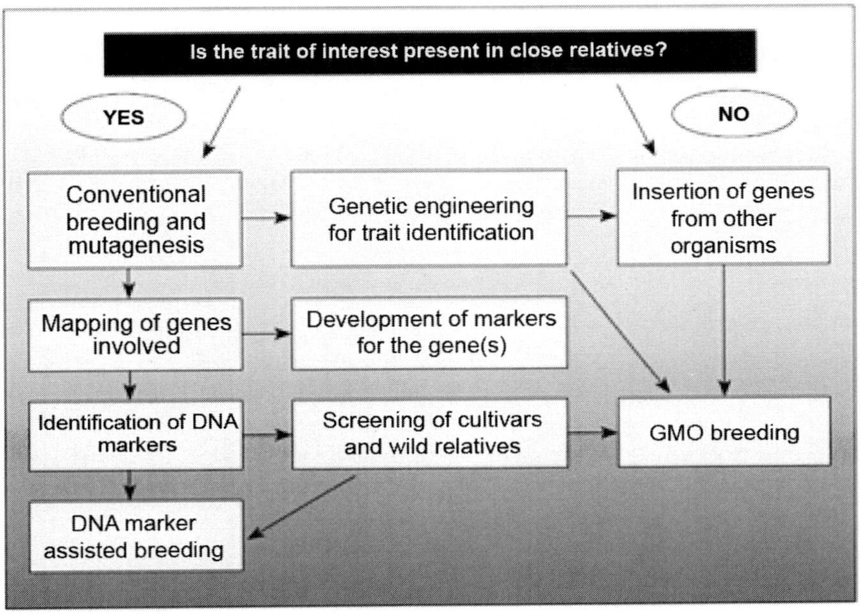


**Figure 2:** Genetic Engineering Process

## 7.3 Development of transgenic crops

Genetic engineering has its reasonable and basic principles despite the diverse and complex techniques involved. However, it is crucially important to identify the biochemical and physiological mechanisms of action, regulation of gene expression and safety of gene and gene product to be utilized. The process of genetic engineering requires the successful completion of a series of six steps.

**STEP 1-NUCLEIC ACID (DNA/RNA) EXTRACTION:** Nucleic acid extraction, either DNA or ribonucleic acid (RNA) is the first step in the genetic engineering process. It is therefore important that reliable methods are available for isolating these components from the cell. In any isolation procedure, the initial step is the disruption of the cell of the desired organism, which may be viral, bacterial or plant cells, in order to extract the nucleic acid. After a series of chemical and biochemical steps, the extracted nucleic acid can be precipitated to form thread-like pellets of DNA/RNA.

**STEP 2-THE SECOND STEP IS GENE CLONING:** There are basically four stages in any cloning experiment: generation of DNA fragments, joining to a vector, propagation in a host cell, and selection of the required sequence. In DNA extraction, all DNA from the desired organism is extracted. This genomic DNA is treated with specific enzymes called restriction enzymes cutting it into smaller fragments with defined ends to allow it to be cloned into bacterial vectors. Copies of the vector will then harbor many different inserts of the genome. These vectors are transformed into bacterial cells and thousands of copies are produced (Figure 3).

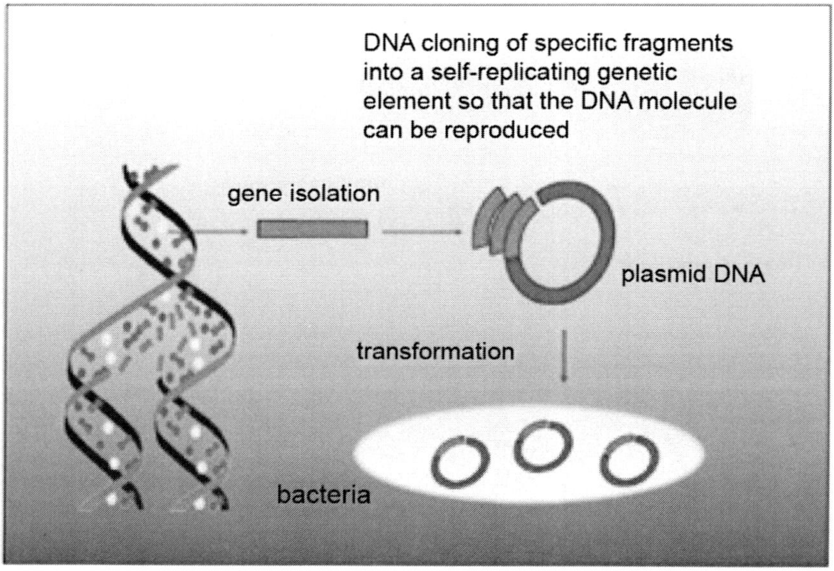


**Figure 3:** DNA cloning of specific fragments into a self-replicating genetic element so that the DNA molecule can be reproduced.

**STEP 3-GENE DESIGN AND PACKAGING:** Once the gene of interest has been cloned, it has to be linked to pieces of DNA that will control its expression inside the plant cell (Figure 9). These pieces of DNA will switch on (promoter) and off (terminator) the expression of the gene inserted. Gene designing/packaging can be done by replacing an existing promoter with a new one, incorporating a selectable marker gene and reporter gene, adding gene enhancer fragments, introns, and organelle-localizing sequences, among others.

**PROMOTERS:** Promoters allow differential expression of genes. For instance some promoters cause the inserted genes to be expressed all the time, in all parts of the plant (constitutive) whereas others allow expression only at certain stages of plant growth, in certain plant tissues, or in response to external environmental signals. The amount of the gene product to be expressed is also controlled by the promoter. Some promoters are weak, whereas others are strong. Controlling the gene expression is an advantage in developing GM plants."

# 8. CURRENT STATISTICS AND THE FUTURE OF BIOTECHNOLOGY AND AGRICULTURE

## 8.1 Biotech crops surge over 1 billion hectares

Developing nations drive growth at adoption rates exceeding industrialized countries. According to Clive James, 2010, "the 1 billionth hectare was planted in 2010 by one of the 15.4 million farmers in 29 countries who now benefit from the technology. For comparison, 1 billion hectares is roughly equivalent to the vast land area of China, or of the United States. With an unprecedented 87-fold increase between 1996 and 2010, biotech crops are the fastest-adopted crop technology in the history of modern agriculture. Growth remains strong, with biotech hectarage increasing 14 million hectares -- or 10 percent – between 2009 and 2010" (Figure 4).

For the first time, in 2010, the ten largest biotech crop growing countries all had more than 1 million hectares in production, providing a broad and stable base for future growth. In hectarage rank order, they include: USA (66.8 million), Brazil (25.4 million), Argentina (22.9 million), India (9.4 million), Canada (8.8 million), China (3.5 million), Paraguay (2.6 million), Pakistan (2.4 million), South Africa (2.2 million) and Uruguay (1.1 million) (Figure 5).

Clive James report further portrays that "for the second consecutive year, Brazil had the world's largest year-over-year increase in absolute biotech crop plantings, adding 4 million hectares in 2010 -- a 19 percent increase to grow a total of 25.4 million hectares. Only the United States leads Brazil in total cropland devoted to biotech crops. Australia, which recovered from a multi-year drought, saw the largest proportional yearon- year increase in biotech crop plantings at 184 percent. Burkina Faso followed at 126 percent growth with 80,000 farmers planting 260,000 hectares, a 65 percent adoption rate (Figure 6).

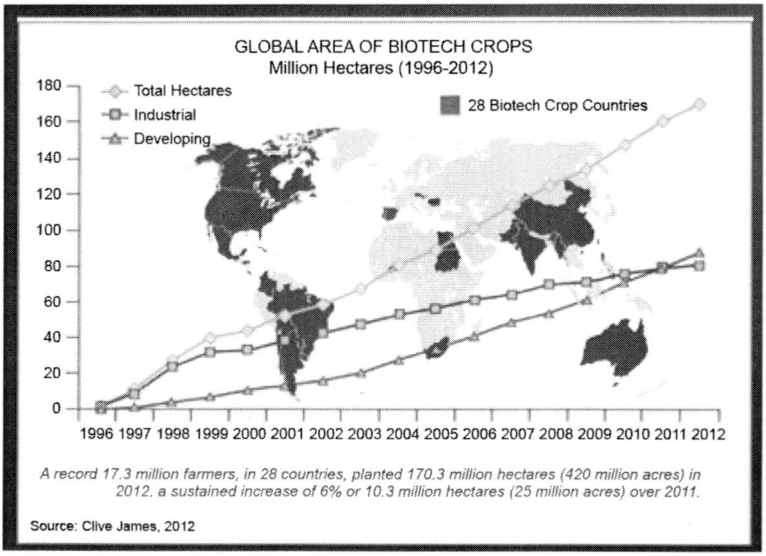


**Figure 4:** Global area of Biotech crops-Million Hectares (1996-2012) (Source: Cleve James, 2012)

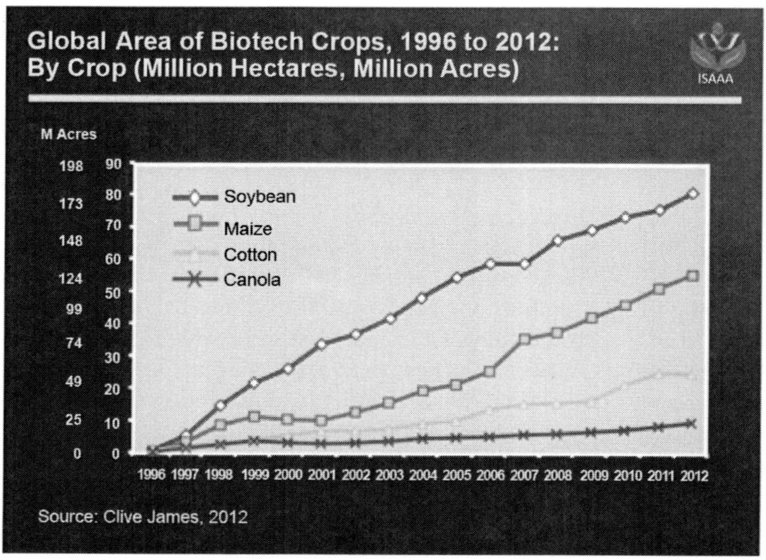


**Figure 5:** Global area of Biotech Crops, 1996 to 2012: by Crop (Million Hectares, Million Acres) (Source: Cleve James, 2012)

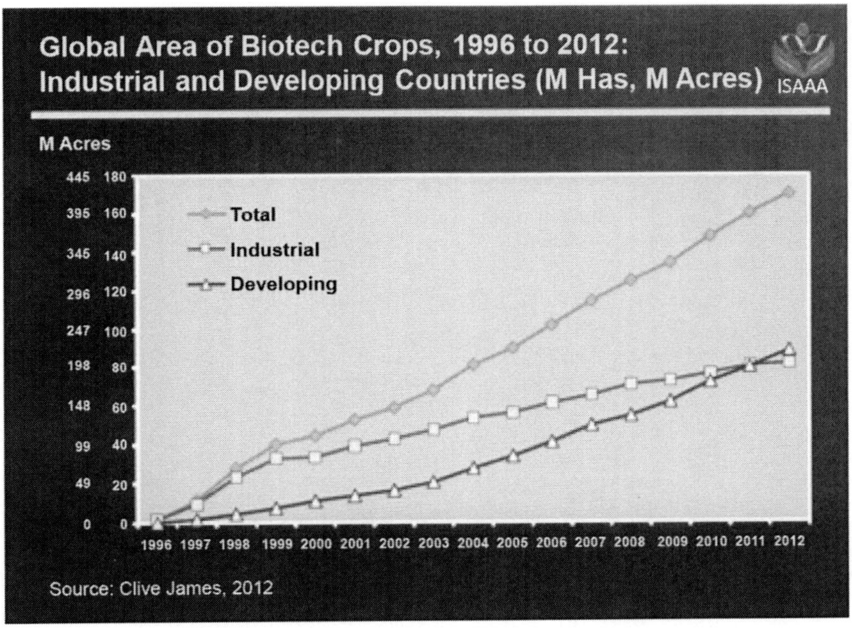


**Figure 6:** Global area of Biotech crops, 1996 to 2012: Industrial and Developing countries (M Has, M Acres) (Source: Cleve James, 2012)

Brazil, after expediting approvals of biotech crops (a total of 27, and 8 in 2010 alone) and securing export trade agreements, now plants 17 percent of the world's biotech crops. Productivity increases attributed to biotech crops helped fuel Brazil's ability to double its annual grain production since 1990 while increasing cropland by only 27 percent. The benefits from biotech crops are spurring strong political will and substantial new R & D investments in biotech crops, with speed and effectiveness increasing access to technology, Gomes noted. With an ability to bring up to 100 million more hectares of cropland, with water, into production, Brazil will continue to be a driving force in the global adoption of biotech crops and is investing in infrastructure to support that growth (Figure 7).

The five principal developing countries growing biotech crops – China, India, Brazil, Argentina and South Africa – planted 63 million hectares of biotech crops in 2010, equivalent to 43 percent of the global total. All told, 19 of the 29 countries that have adopted biotech crops are developing nations, which grew at a rate of 17 percent or 10.2 million hectares over 2009 compared to only 5 percent growth or 3.8 million hectares in industrialized countries" (Figure 9).

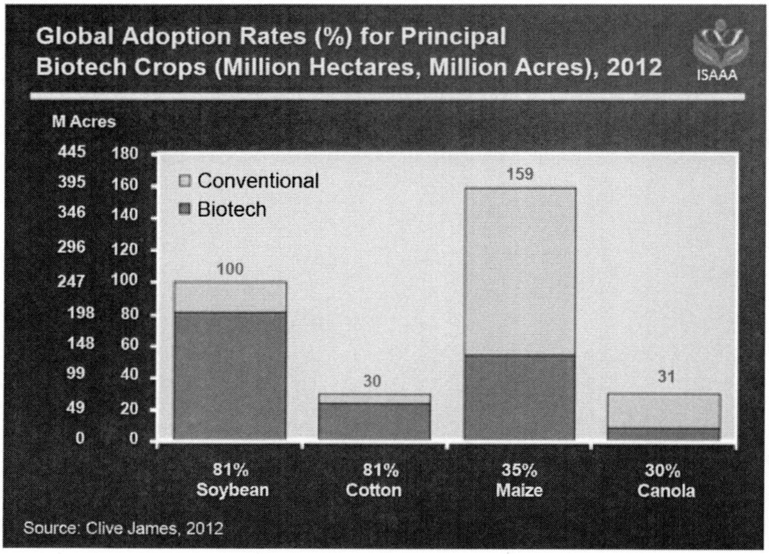


**Figure 7:** Global Adoption Rates (%) for principal biotech crops (Million Hectare, Million Acres), 2012 (Source: Cleve James, 2012)

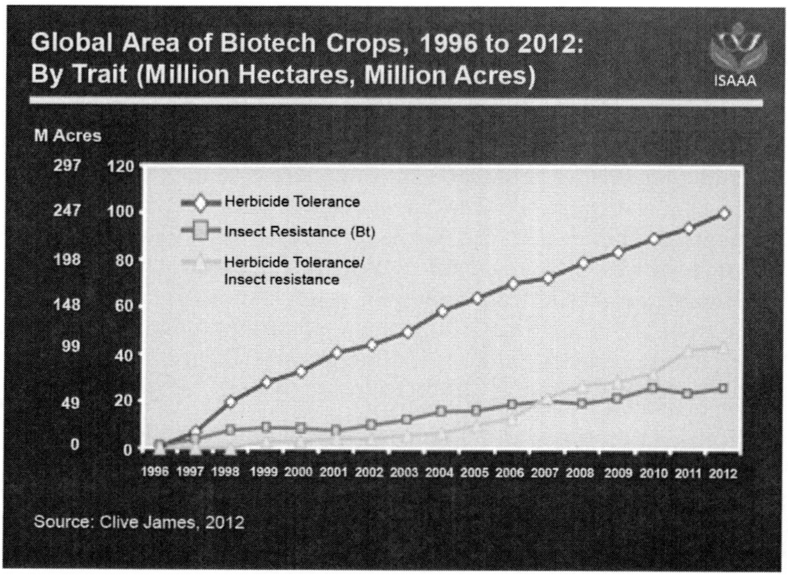


**Figure 8:** Global area of biotech crops, 1996 to 2012: by Trait (Million Hectares, Million Acres) (Source: Cleve James, 2012)

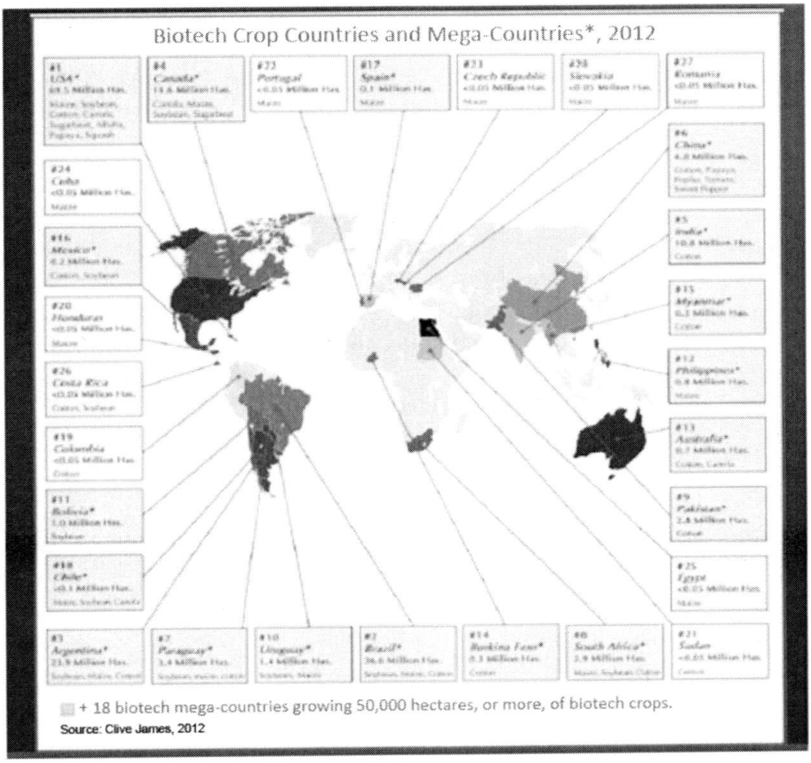


*18 Biotech mega-countries growing 50, 000 hectares, or more, of biotech crops.

**Figure 9:** Biotech Crops Countries and Mega-Countries* (Source: Cleve James, 2012)

Developing countries grew 48 percent of global biotech crops in 2010 and will exceed industrialized nations in their plantings of biotech crops by 2015. Clearly, the countries of Latin America and Asia will drive the most dramatic increases in global hectares planted to biotech crops during the remainder of the technology's second decade of commercialization (Figure 8).

## 8.2 Merits of modern biotechnology and agriculture

The merits of modern Biotech and agriculture include:

**SAFETY:** Agriculture as an industry works hand in hand with the government agencies and some other organizations to guarantee farmers' access to necessary technologies and ensuring modern agriculture system. Farmers can be aided by education and certification programs that ensure careful application of agricultural practices and avoid malpractices.

**SUSTAINABILITY:** this advancement in technology provides farmers with useful tools and resources to improve sustainable farming and limit environmental footprint. It has also incentivized innovations like conservation tillage as prevention against erosion, water pollution and promotes carbon sequestration.

**AFFORDABILITY:** Modern technology system promotes affordable supply of food to attend to the demand of the world's growing population. Modern agriculture practice also permits large production utilizing smaller pieces of land in order to reduce cost of production.

## 8.3 Demerits of modern biotechnology and agriculture

- The British Medical Association has reported that antibiotic resistant marker genes inserted into specific crops could provoke microbes that cause diseases in the humans' or other animals' gut as it would lead to the microbes becoming antibiotics resistant thereby compounding public health problem of antibiotic resistance.
- Foreign genes are known to alter nutritional value of foods in unpredicted manner. It may reduce some nutrients values while increasing the level of some others and there are fewer findings into the effect of changes in nutrient composition of food plant and animals.
- Genetic modification of food plants could transfer allergenic properties of the donor source into the recipient plant or animal. In addition, many genetically engineered foods use microorganisms as donors whose allergenic potential are either unknown or untested.
- Genetically modified crops having herbicide and insect resistance could cross-pollinate with wild species, and unintentionally create hard to-eradicate super-weeds especially in small farm fields surrounded by wild plants. This unintentional gene transfer, although hard to substantiate, can have consequences that are not yet known.
- Genetic modification could inadvertently enhance natural plant toxins by switching on a gene that has both the desired effect and capacity to pump out a poison.
- Religious concerns are also voiced as some of the reasons for opposing genetic engineering of foods, while some people object to bio-engineered foods for personal, ethical, cultural, and esthetic reasons, as

well as infringement on consumer choice, and inability to distinguish GM foods from non-GM counterparts.

- Critics of genetic modification of foods fear that commercialization of transgenic crops
- Will pose a new threat to crop genetic diversity already endangered by current agricultural practices that favor the worldwide adoption of a few crop varieties.
- Plants engineered to contain virus particles as part of a strategy to enhance resistance could facilitate the creation of new viruses in the environment.
- Some critics of genetic modification argue that patenting allows corporations to have monopoly control of genetically altered plants or animals.
- Animal rights groups strongly oppose any form of cloning or genetic engineering involving animals, or use of animals in research, and have sometimes resorted to vandalizing animal research.
- Many critics are concerned that GM foods are not labeled. They insist that labeling can help the consumer trace unintended consequences to a certain consumed GM food.
- Consumers also have a genuine 'fear of the unknown' in that deadly microorganisms or super plants might be released during field testing or field trials, and accidents in biotech labs might lead to release of toxic agents, poisons, or biological toxins which will threaten human and animal populations.
- Organic farmers fear that GM foods would obscure organic foods because of lack of labeling, and they feel that the biotech revolution could make it difficult for people to locate non-GM crops.

Using information relating to specific molecular marker sequences and the desired phenotype, the vector harboring the desired sequence is detected, selected, isolated and clones are produced. Restriction enzymes are again utilized to determine if the desired gene insert was cloned completely and correctly.

## 9. CONCLUSION

New nations are around the world opt for the planting of biotech crops. ISAAA 2010 report shows that approximately 600,000 farmers in Pakistan and 375,000 farmers in Myanmar, planted insect-resistant Bt cotton, and

Sweden have started planting a new biotech high-quality starch potato approved for industrial and feed use. Germany also planted the same biotech potatoes in 2010, resuming its place among the eight EU nations now growing either biotech maize or potatoes.

Many more companies are expected to adopt biotech crops in the nearest future as the number of growers and global hectarage of biotech crops are also expected to increase tremendously. It has also been stimulated that at least three more countries from each of the three regions of Asia, West Africa, East/Southern Africa and some few others from Latin/Central America, and Western/Eastern Europe will grow more of biotech crops according to the ISAAA 2010 publication. In 2010, Mexico which is notable for its variety of maize has also embarked some field trials of Bt and herbicide tolerant maize which was reportedly successful.

In 2010, ISAAA further emphasized the potential for increasing the biotech adoption for four main and current large hecterage biotech crops which include maize, soybeans, cotton and canola which also are reported to have represented almost 150 million hectares in 2010 being a half way attained capacity of the hectarage around the world. It was also demonstrated in the same research that the 29 countries which planted biotech crops in 2010 represented 59 percent of the world population, and as such argued that its 2015 Millennium Development Goals could find solution to food security and poverty problems and would be made popular around the world. However, there being to side to a coin, it is advisable to have a well scrutinized and checkmated biotechnology practice as well as related agriculture practices. Therefore, every national and international agency as well as the governing bodies should ensure control and protection against its abuse.

## REFERENCES

1. Brady NC, Weil RR (2002) Practical Nutrient Management. Elements of the Nature and Properties of Soils, Pearson Prentice Hall, Upper Saddle River, NJ, USA.
2. Acquaah G (2002) Land Preparation and Farm Energy. Principles of Crop Production, Theories, Techniques and Technology. Prentice Hall, Upper Saddle River, NJ, USA.
3. Acquaah G (2002) Pesticide Use in U.S. Crop Production. Principles of Crop Production, Theories, Techniques and Technology. Prentice Hall, Upper Saddle River, NJ, USA.
4. Acquaah, G (2002) Soil and Land. Principles of Crop Production, Theories, Techniques and Technology. Prentice Hall, Upper Saddle River, NJ, USA.
5. Chrispeels MJ Sadava DE (1994) Farming Systems: Development, Productivity, and Sustainability. Plants, Genes, and Agriculture. Jones and Bartlett, Boston, USA.
6. Brady NC, Weil RR (2002) Elements of the Nature and Properties of Soils. Pearson Prentice Hall, Upper Saddle River, NJ.
7. Chrispeels MJ Sadava DE (1994) Nutrition from the Soil. Plants, Genes, and Agriculture. Jones and Bartlett, Boston, MA, USA.

8. Acquaah G (2002) Plants and Soil Water. Principles of Crop Production, Theories, Techniques and Technology. Prentice Hall, Upper Saddle River, NJ.
9. Gold MV (1999) Sustainable Agriculture: Definitions and Terms. USDA National Agriculture Library. Beltsville, MD.
10. Earles R, Williams P (2005) ATTRA National Sustainable Agriculture Information Service.
11. Kellogg RL, Nehring R, Grube A, Goss DW, Plotkin S (2000) Environmental indicators of pesticide leaching and runoff from farm fields. United States Department of Agriculture Natural Resources Conservation Service.

# CHAPTER 2

# Gene Revolution in Agriculture: 20 Years of Controversy

**Tatjana Brankov[1]*, Koviljko Lovre[1], Bozidar Popovic[2] and Vladimir Bozovic[2]**

*1 University of Novi Sad, Faculty of Economics, Subotica, Serbia*

*2 University of Montenegro, Faculty of Mathematics and Natural Sciences, Podgorica, Montenegro*

## ABSTRACT

In the spirit of the general debate over genetically modified (GM) food which is not slowing down, we wanted to answer some questions, such as: Is Gene Revolution an answer to world hunger? Do GM crops with more complex transformation contribute to the enrichment of multinationals? Why U.S. increases food aids? To this end, we firstly describe the diffusion of GM crops around the world during the previous 20 years. Starting from 1996, we present global progress with adoption of biotech crops, its distribution in developed and developing countries, global area by trait, adoption rate and global value of biotech crops. The findings reveal 10 countries, four crops, and two traits domination. The findings of this study clarify the failure of transgenic technology to eradicate hunger. In addition, the results have shown statistically significant correlation between stacked trait and global market value of biotech crops as well as between raising production of biotech crops in U.S. and an increase in U.S. food aid through World Food Program (WFP).

**Keywords**: gene revolution, famine, stacked traits, global market value, U.S. food aid

# 1. INTRODUCTION

Although a significant amount of food, which is obtained from genetically modified (GM) plants, is involved in a food chain, public debate on the issue is not slowing down. There is a large expert disagreement about environmental and health effect as well as about socioeconomic implications [1–4].

The consumer willingness to use and pay GM products varies among the nations [5–8]. From the very beginning, the U.S. supported biotechnology industry. The same policy that considers GM and conventional foods substantially equivalent carried out all administrations, from R. Reagan to B. Obama administration. At present, the total U.S. revenues from GM systems reached at least $350 billion, the equivalent of approximately 2.5% of GDP [9]. Directly or indirectly numerous international organizations have contributed to the spread of GM products [10]. Among them, the most important are World Bank (WB), World Trade Organization (WTO), Consultative Group on International Agricultural Research (CGIAR), Food and Agriculture Organization (FAO), various international foundations, research centers, and universities. The WTO's Agreement on trade-related intellectual property rights (TRIPS) negotiated at the end of the Uruguay Round obliges member states to patenting biotech inventions (products and processes) and plant varieties and for the first time provides a legal means for the protection of intellectual property rights. Patent protection in this field gives the corporations unprecedented control over research and development as well as over whole food chain [11]. This strongly accelerated already started transformation process of the commercial seed industry from a sector composed primarily of family-owned small firms to a small number corporate domination [12]. After many acquisitions and joint ventures formation, the top three seed firms currently control 85% of transgenic corn patent in the U.S. [13]. Global recession has contributed to the further strengthening of the monopoly power of the biggest biotech companies [14]. Because overall industry data are not shared publicly, Farm Journal magazine has reported market shares by interviewing consultants, industry executives, and other sources to assemble the data [15]. According to this source in 2014, DuPont, Pioneer, and Monsanto dominate, accounting for 70% of the corn seed business and 60% of the soybean seed business.

While majority of farmers worldwide still rely on saving their own seeds, the prevalence of this practice, especially in developed countries, is declining rapidly [16]. In the leading GMOs country, the U.S., the rate of saving soybean seed fell to 10% in 2001 from 63% in 1960 [17], while the rate of saving corn seed fell to less than 5% [18]. Ignoring the interests of the poor countries can be observed from the fact that multinational corporations from North America and Europe take advantage of natural wealth located mainly in the Global South [10, 19]. The best example is maybe the neem tree, Azadirachta indica, which has religious and cultural significance throughout India. U.S. Patent Office issued Patent No. 5,124,349 to W.R. Grace & Co. on the particular derivative of the neem tree. Long before any official discovery, farmers in India had been using a neem, among many other purposes, as a powerful insecticide that is not harmful to humans [20].

## 2. RESEARCH METHODOLOGY

In the spirit of the general debate, which is not slowing down in this chapter, we wanted to answer some questions, such as Is Gene Revolution an answer to world hunger? Do GM crops with more complex transformation contribute to the enrichment of multinationals? Why U.S. increases food aids? To this end, we have set up the following research hypotheses:

1) GM food has failed to feed the hungry.
2) There is a statistically significant correlation between stacked trait and global market value of biotech crops.
3) There is a correlation between raising production of biotech crops in the U.S. and an increase in the U.S. food aid through World Food Program (WFP).

For the purpose of general understanding of the issue at the beginning, we present global progress with adoption of biotech crops, its distribution in developed and developing countries, global area by trait, adoption rate, and global value of biotech crops. The first hypothesis "GM food has failed to feed the hungry" we set because agribusiness segment has often promoted transgenic technology as the next hunger-quenching Green revolution. "There is a statistically significant correlation between stacked trait and global market value of biotech crops" hypothesis has been settled because the multinationals claim that this complex transformation can allow farmers to address multiple old problems with two or more traits in the same seed and are more effective at meeting the needs of farmers and consumers than the traditional, mono-trait seed varieties [21]. Finally, our last hypothesis "There is a correlation between raising production of biotech crops in the U.S. and an increase in the U.S. food aid through WFP" was tested due to the controversy over the food aid. Excluding the many benefits of emergency humanitarian aid, recent finding shows that an increase in U.S. food aid increases the incidence of armed civil conflict in recipient countries by prolonging existing conflicts [22]. In addition to this aspect, a disagreement over the food aid in terms of GM food arose in 2000s when several African countries and countries in transition (such as Serbia and Montenegro) refused to accept U.S. aid due to GM ingredients presence. "A fiery" debate continues via WTO Doha Round in 2003 and 2004, when the other countries have put pressure on the U.S., as the principal donor country, to move away from in-kind food aid. The great power of lobby groups in America could be visible due to the U.S. Congress rejection of Bush administration proposal that one-quarter of U.S. food aid should be cash-based [11]. The principal reason why the U.S. continues to insist on giving its food aid in-kind may be the inability to find export market for its GM maize [23].

Data required for the analysis were collected from relevant sources: ISAAA, FAO, USDA, Earth Policy Institute. The arithmetic means of the confidence interval as a measure of variability in the descriptive statistics have been used. The method of parametric statistics (t test, ANOVA) was applied in processing the results, while Pearson's correlation coefficient and simple linear regression were used as a measure of association. To detect trends in the variables considered, the collected data were compared by linear regression analysis using the factor time as independent variable. Data were analyzed using Microsoft Excel and R Studio software.

# 3. GLOBAL DIFFUSION

## 3.1. International diffusion of biotech crops

In the period 1996–2015, up to approximately 18 million farmers grew biotech crops annually. Global biotech hectares increased more than 100-fold, from 1.7 million hectares in 1996 to 179.7 million hectares in 2015. In other words, GM crops are grown on approximately 3.7% of the world's total agricultural land and 13% of arable land and by <1% of the world's farming population [24]. In the 19 years period 1996–2014, the growth rate of area under GM was continually growing. Total crop plantings were decreased for the first time in 2015 comparing with 2014 (-1%). A clear linear trend of the total biotech hectarage growth can be expressed by equation:

$$y=9.919x-5.781 \ (R2=0.993) \qquad (1)$$

As represented in Figure 1, linear trend of GM crops diffusion is obvious in both developing and developed countries. Linear trend equation for developing and developed countries, respectively, is as follows:

$$y=5.749x-16.79\ (R2=0.969) \tag{2}$$

$$y=4.161x+11.07\ (R2=0.946) \tag{3}$$

For the first time in 2012, developing countries planted more biotech crops than industrial countries, 51.2 and 48.8%, respectively. Higher participation of developing countries was also observed in subsequent years. Latin American, Asian, and African farmers collectively grew 53.6% of the global biotech hectares in 2013, 52.9% in 2014, and 54% in 2015.

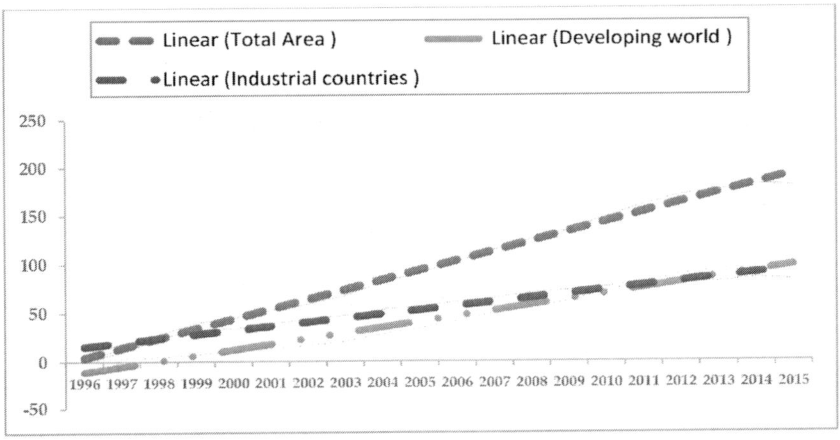


**Figure 1.**Total area under GM, distribution in developing and developed countries (million hectares). Source: Author's calculations based on data from James (1996–2015).

The number of producer countries also grew by time, starting from 6 in 1996 to 28 in 2015 (Figure 2). The first countries that engaged themselves in the transgenic production were U.S., China, Argentina, Canada, Australia, and Mexico. In 1998, South Africa, Spain, and France grew transgenic crops for the first time. Portugal, Rumania, and Ukraine joined them in 1999 and Bulgaria, Uruguay, Indonesia, and Germany in 2000. India planted GM crops for the first time in 2002. In the same year, Colombia and Honduras started their production. Two countries, Brazil and the Philippines, approved planting of GM crops in 2003. In 2004, for the first time, Paraguay reported its cultivation. New producing countries become Iran and Czech Republic in 2005; Slovakia in 2006; Chile and Poland in 2007; Burkina Faso, Egypt, and Bolivia in 2008; Costa Rica joined them in 2009; Pakistan, Myanmar, and Sweden in 2010; Sudan and Cuba in 2012; Bangladesh in 2014; Vietnam in 2015. From Graph 2, it can be seen a significant increase in the producing countries in numbers. For example, the number of developing countries increases from 2 in 1996 to 20 in 2015 [21, 25–43]. It is very important to stress that total number of producing countries has not increased since 2010. This is an indication of problems in the geographical diffusion of GMOs.

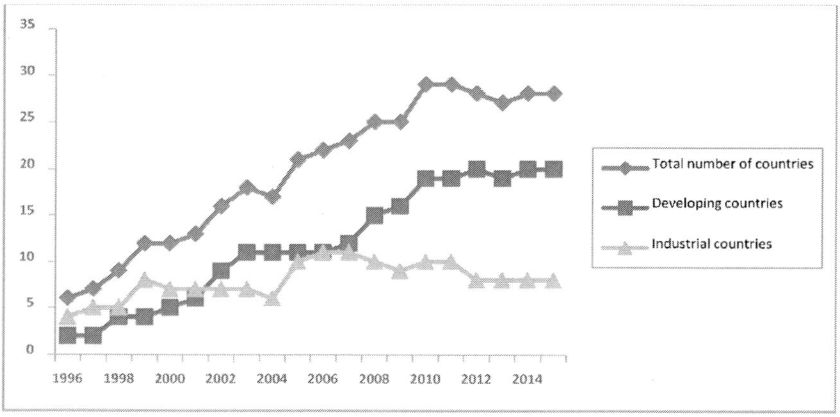


**Figure 2**. Producing countries developing and developed. Author's calculations based on data [21, 25–43].

From the very beginning, U.S. is absolute leader in the production, but U.S. involvement has decreased with the inclusion of the other countries. For example, U.S. accounted for 64% in the total area under GM in 1997, while their share in 2015 was 39.4%. Brazil, the second largest producer, started relatively late with the production but at present accounts for 24.6% in the total areas. Argentina, the third biggest producer, participates in the total acreage of 13.6%. India and Canada areas reached 6.4 and 6.1% of global hectarage, respectively. China share in the total area is reduced significantly, from 16% in 1997 to 2% in 2015. Australia produces GM on just 0.7 million hectares (0.4% of total area under GM crops), which is a just slight increase from 1997 (0.1 million hectares). In 2015, there were 28 countries that grew GM crops, but in reality, a large majority of the area planted to GM crops still remains in just a few countries. Top 10 countries accounted for 98% of the total global GM hectarage. The list of countries that grew GM crops in 2015 includes Bangladesh, which grew as little as 25 hectares of GM crops in 2015, and Costa Rica, which grew 38 hectares in 2014 [21, 25–43]. After the initial years of high growth rates, the global growth of GM crops is, in fact, slowing down, as the amount of land under GM crops in the few countries that are large-scale adopters has become saturated [24]. Annual global growth rates were between 3 and 10% in the past seven years, with the global GM area in 2013 and 2014 approximately 3% higher than the years before [43].

At first glance, this diffusion is impressive but certainly is not without problems. Despite the great support to genetically modified organisms (GMOs) in the several countries of Latin America, there is a resistance on this continent. Venezuela with over 30 million inhabitants approved a new law on December 23, 2015, that imposes one of the world's toughest regulations on GMOs. The Seed Law seeks to consolidate national food sovereignty, regulates the production of hybrid seed, rejects the production, distribution, and import of GMO seeds, and also bans transgenic seed research [44]. Similarly, in the Middle East, Iran, country which initiated its rice biotech activities in 2005 with several hundred farmers growing 4000 hectares of Bt rice on their farms and cloned the first GM animal, sheep in 2006 [45] very soon, has decided to drop the commercialization of GMOs. Currently, government is preparing new law to facilitate GMO production. At the same, time the resistance of certain groups strengthens [46].

Apart from Russia where situation is more than clear and where the food safety regulator Rosselkhoznadzor bans GMO production and import [47] and Prime Minister has described as amounting to little more than a form of biological warfare weapon [48], the biggest challenge in terms of adoption and acceptance of GMOs remains Africa and Europe. For a long time, South Africa was the only country on the continent of Africa to commercialize biotech crops. GMO proponents have expressed their pleasure because of broad geographical coverage in Africa in 2008 when Egypt as represent of North Africa and Burkina Faso from West Africa become GM producer countries [37]. Historically, important event was also the involvement of Sudan in 2012. But, all commercial production in Egypt is currently stalled due to Ministerial Decree 378 (MD378), 2012, rescinding the registration of Ajeeb-YG corn variety for its commercial use [49]. Bt cotton remains the only crop commercialized in Burkina Faso [50]. Sofitex, Burkina Faso's major cotton production association, on April 5, 2015, decided to switch its production to 100 percent conventional methods and has announced a rejection of GM cotton in the next three years due to the disappointing yields and poor quality fiber [51]. South Africa, the largest producer country in Africa, has experienced the greatest reduction—23% less for its GM maize crop hectarage in 2015 comparing to 2014. Sudan, GM cotton producer country, seized a shipment of GM soybeans reportedly imported from the U.S. [52]. Nigeria, one country being targeted for expansion, has recently submitted a written petition (more than 100 groups representing at least 5 million Nigerians) to the country's authorities in hopes of impeding GMO entrance [53].

Cultivation in the EU fell 10.4% in 2015 to only 128.103 hectares, almost all of that insect-resistant corn grown in Spain (93.7%). Share of GM corn in total corn area is just 1.3%. Commercial cultivation of GE crops is minimal in other EU producing countries: Portugal, Czech Republic, Romania, Slovakia, France, Germany, and Poland. According to USDA Foreign Agriculture Services, EU member states can be divided into three categories depending on GMOs view. The adopters include the GM producer countries as well as countries where the government and industries mostly favor biotechnology (Denmark, Estonia, Finland, Flanders in northern Belgium, the Netherlands, and the United Kingdom). Conflicted members are those states where scientific community, farmers, and the feed industry are willing to adopt the technology, but consumers and governments, influenced by activist Green parties and NGOs, reject it (France, Germany, Poland, Southern Belgium (Wallonia), Bulgaria, Ireland, Lithuania Sweden, Germany). In the opposed member states, most stakeholders reject the technology and government generally supports organic agriculture and geographical indications (Austria, Croatia, Cyprus, Greece, Hungary, Italy, Malta, Slovenia, and Latvia) [54]. One of the leading opposed countries is Hungary which initiated a joint alliance of EU member states rejecting the use of GE crops with the objective to make the entire EU free from GM crops.

## 3.2. Four crops domination

Between 1986 and 1995, 56 GM crops were field tested around the world the majority of which were on eight crops [25]. Besides expiration that there are more than 85 potential new products in the pipeline now being field tested [43], 20 years later, only 10 crops are grown commercially (soybean, corn, cotton, canola, alfalfa, sugar beet, papaya, squash, potato, and eggplant/brinjal). There are no commercialized GM varieties of a number of key global staple crops, including wheat, rice, barley, millet, sorghum, cassava, and yam. The first generations of GM potato that resist blight were planted on 162 hectares in 2015 after FDA approval [55]. In significant quantities (about 99%), only four crops soybean, corn, cotton, and canola have been grown over the past 20 years. Principal four crops are generally used in food system as ingredients in processed food and in animal feed. Exceptions used as whole foods are some fruit and vegetables GM sweet corn, squash, papaya, eggplant/brinjal, potato which together with sugar beet and alfalfa collectively account for only 1% of global GM crop hectares. Area under all four crops, soybean, maize, cotton, and canola, depicts linear trend of growth, respectively (Figure 3):

$$y=4.740x+1.742\ (R2=0.986) \tag{4}$$

$$y=3.265x-5.836\ (R2=0.959) \tag{5}$$

$$y=1.43x-2.2\ (R2=0.964) \tag{6}$$

$$y=0.445x+0.367\ (R2=0.954) \tag{7}$$

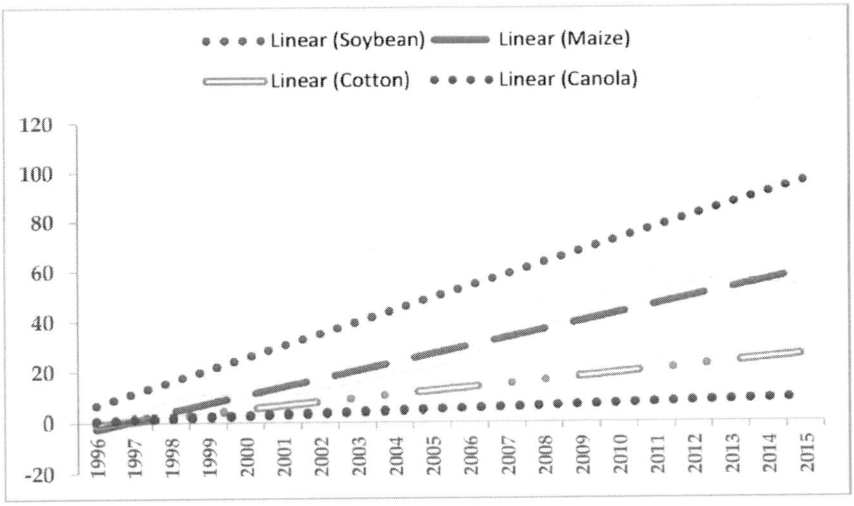


**Figure 3.** Linear growth trend for four principal GM crops (million hectares/year). Author's calculation based on Refs. [21, 25–43].

As could be seen in Figure 4, maximum share of soybeans in total areas was in 2001 (63.3%), and overtime, it decreases to 51.2% in 2015. Share of corn stabilized at the level of about 30% in 2015 such as it was in 1998. Share of cotton increases from 9% in 1996 to 13.3% in 2015. The biggest drop is recorded in the share of canola in the total GM area, from 8.6% in 1998 to 4.7% in 2015. GM sugar beet production started in 2008, but until now areas under this crop are about 0.5 million hectares, which means there is no enlargement as it was the case with previous crops. Regarding global adoption rate, it can be noticed from Figure 5 that only soybean rate continuously was growing but with a sharp slowdown after 2009. Cotton adoption rate reached its maximum value in 2011 (82%) and then fell to 75% in 2015. Similarly, maximum corn value was 35% in 2012, after slipped to 29% in 2015. Canola global adoption rate slipped from max 30 to 24% in 2012 and 2015, respectively.

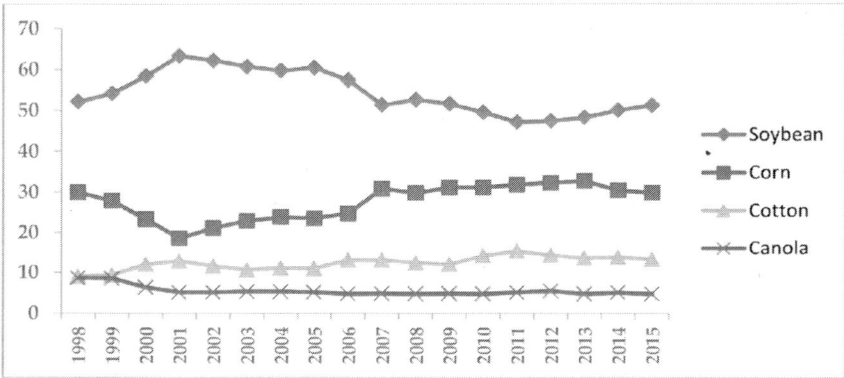


**Figure 4**.Percentage share of the most important crops in the total area under GMOs (1996–2015). Author's calculation based on Refs. [21, 25–43].

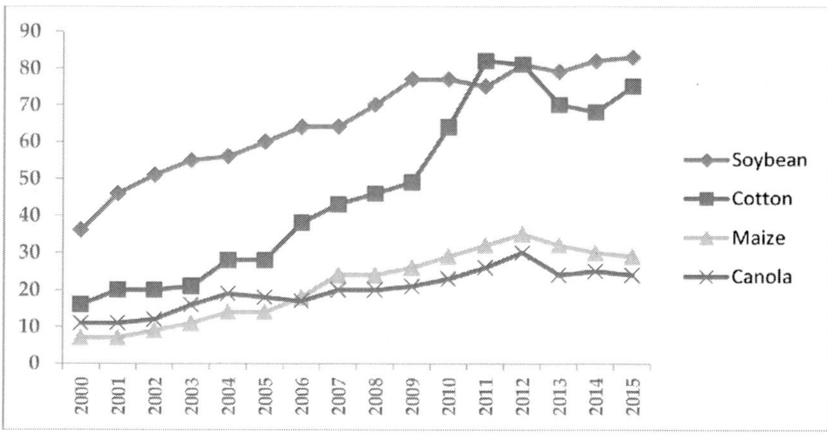


**Figure 5.**Global adoption rate of the principal GM crops (2000–2015). Author's calculation based on Refs. [21, 25–43].

One look at the slower growth of plantation area is due to the saturation of adoption rate in countries with mature GM models. For example, adoption rate of GM cotton in Australia is 99%, in Canada 95% of canola is GM, 95% cotton in India is GM, 99% cotton and 100% soybean in Argentina are GM, 95% of soybean in Paraguay is GM and over 90% of all four principal GM crops in U.S. is already GM [56]. Another view indicates that the resistance becoming stronger by opposition from consumer and environmental groups, regulatory hurdles, and, in some cases, scientific obstacles. There are opinions that the problems with diffusion of GM canola lay in the fact that canola is worst candidate crop species for practical segregation of GE and non-GE because it is inherently promiscuous [57]. Also, there are reports that state "Hundreds of thousands of farmers have died in India, after having been allegedly forced to grow GM cotton instead of traditional crops. The seeds are so expensive and demand so much more maintenance that farmers often go bankrupt and kill themselves" [58]. One might ask whether the refusal of consumers around the world to accept GM sugar is the reason for the stagnation of GM sugar beets production. Anyway, the fact is that the U.S. corn and soybeans more and more are channeled into biofuels production. For example, in the marketing year MY Aug/Sept 2015/2016, 38.8% of the total corn disappearance was ethanol share, 10.1% of total use was directed to total food, seed, and industrial use (not including ethanol), feed and residual use accounted for 38.8%, while export accounted for 12.2% of the total corn disappearance. The situation was very different in MY Aug/Sept 1995/1996: 4.6% of the total use was ethanol use, 14.4% was total food, seed, and industrial use (not including ethanol), 54.9% was dedicated to feed and residual use, and 26.1% for export [59]. In the period 1995–2015, U.S. almost doubled its corn production, from $7.4 \times 10^9$ to $13.6 \times 10^9$ bushels [60]. Taking into consideration problems with exports market, it is logical routing of surplus into ethanol production. Share of soybean oil used for biodiesel increased from 0.3% in 2000/2001 to 23% in 2014/2015. In the same period of time, soybean oil production has increased for just 11.7% (from $18.4 \times 10^9$ to $20.6 \times 10^9$ pounds). It can be observed that share of the U.S. soybean oil export in total use significantly declined from 17.6% in 2009/2010 to 10% in 2014/2015 [59]. The reasons for these trends may lay in the efforts of the European retailers and major manufacturers (including REWE Group, Lidl, Edeke in Germany, Carrefour in France, and many of the retailers in Austria, along with Waitrose in the United Kingdom) to support the transition from GMO to non-GMO for animal production. The most prominent initiative Danube Soy was launched in 2012 as a "mainstream" vehicle for providing access to non-GMO soy for EU markets. It is undergoing rapid growth as a source of sustainable and non-GMO soy for Europe, and soya production in the Danube area (excluding Ukraine) has increased from 560,000 hectares in 2011 to 960,000 hectares in 2015 [61]. Signing this declaration means that EU wants to increase self-sufficiently in soy production. As of 2011, the EU imported 72% of its protein feed needs (majority coming from the U.S.,

Argentina, and Brazil—GM producing countries), so the plan is to provide at least 50% soybean needs until 2025 from non-GMO Danube region.

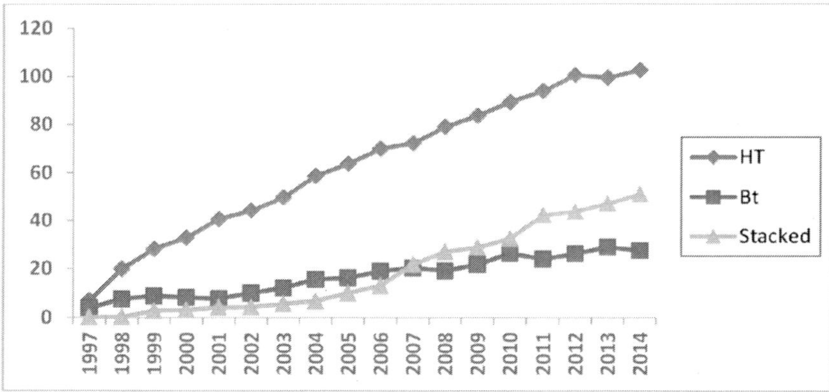


**Figure 6**. GM plants, by traits 1997–2014 (million hectares) [21, 26–42].

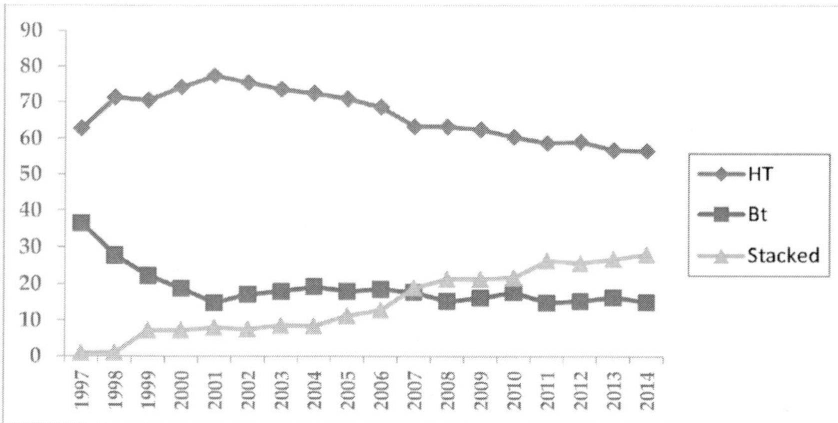


**Figure 7**. Percentage share of the most important traits in the total area under GMOs (1997–2014). Author's calculation based on Refs. [21, 26–42].

## 3.3. Herbicide tolerance and insect resistance domination

GMOs diffusion is uneven if we are speaking about traits also. The most widely used commercial GM traits are herbicide tolerance (HT) and insect resistance (IR). HT crops are developed to survive application of specific herbicides, which would otherwise kill the crop plants. "Roundup Ready" is the most common HT crops today. These crops have been created by Monsanto in order to tolerate applications of the company's glyphosate-based herbicide "Roundup." IR crops are engineered with a gene from the bacteria Bacillus thuringiensis (Bt), which is toxic to some insects. When certain insects ingest the protein produced by Bt, the function of their digestive systems is disrupted, producing slow growth and, ultimately, death. In recent years, various combinations of these two properties are used in creation of stacked plants. Gene stacking refers to the process of combining two or more genes of interest into a single plant; the combined traits resulting from this process are called stacked traits. As represented in Figure 6, HT plants occupied from the very beginning the most of area under GMOs. From 2007, areas under stacked crops are higher than the areas under Bt crops. In 2014, 56.4% of the GM crops worldwide were engineered to be HT, 15.1% were engineered to be Bt, and 28.1% were stacked with both HT and Bt traits (Figure 7). These numbers are significantly different in 2001 when 77.2% of crops were HT, 14.8% were Bt, and 8% were stacked. It is important to stress that referred data do not mean that the area under HT crops was reduced. Contrarily, it was increased. When we consider stacked and single trait crops together in 2014, 84.5% of all GM crops in the world are engineered to be tolerant to herbicides. Companies that have patent protection over GM seeds in the same time are herbicides producers. This means that multinationals in the same time sell both seeds and pesticides. HT and Bt crops account for almost all the GM crops grown commercially over the past 20 years. Other traits, virus resistance and drought tolerance, are very important for poor small-scale farmers in developing countries, collectively account for <1% of global GM crop hectares [43].

## 3.4. Global market value of biotech crops

Global market value of transgenic crops has increased overtime 165 times in terms of current prices and 115 times in terms of constant prices (2002 = 100) (Figure 8). Linear trend of growth in current prices can be expressed through equation:

$$y=899.0x-2028 \ (R2=0.969) \qquad (8)$$

The trend growth in constant prices is expressed as:

$$y=691.3x-940.7 \ (R2=0.978) \qquad (9)$$

We have also obtained the simple linear trend for transgenic crops global market value per hectare in current prices and polynomial trend of transgenic crops global market value per hectare in constant prices (Figure 9):

$$y=2.029x+47.20\ (R2=0.919) \quad (10)$$

$$y=0.050x2-0.477x+59.36\ (R2=0.561) \quad (11)$$

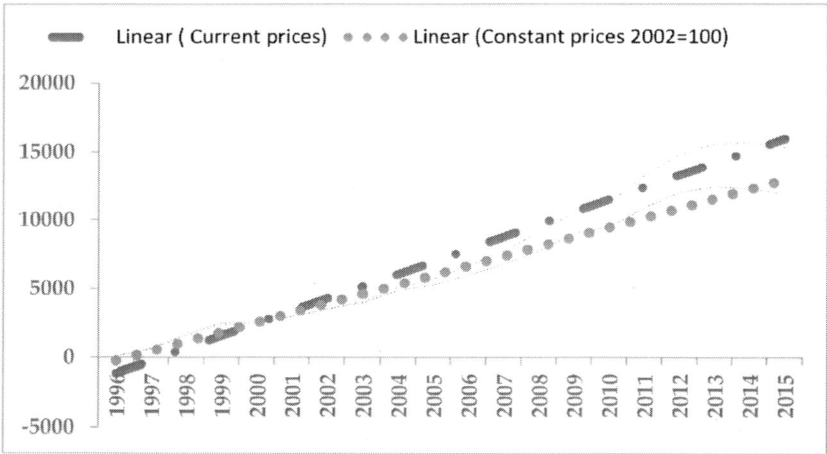


**Figure 8**. Global market value of biotech crops (million US$). Authors calculation based on Refs. [21, 25–43].

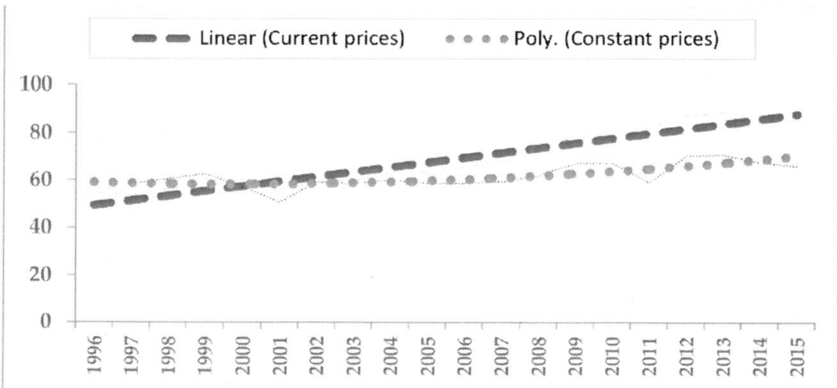


**Figure 9**. Transgenic crops global market value per hectare in current and constant prices. Source: Author's calculation based on Ref. [21, 25–43].

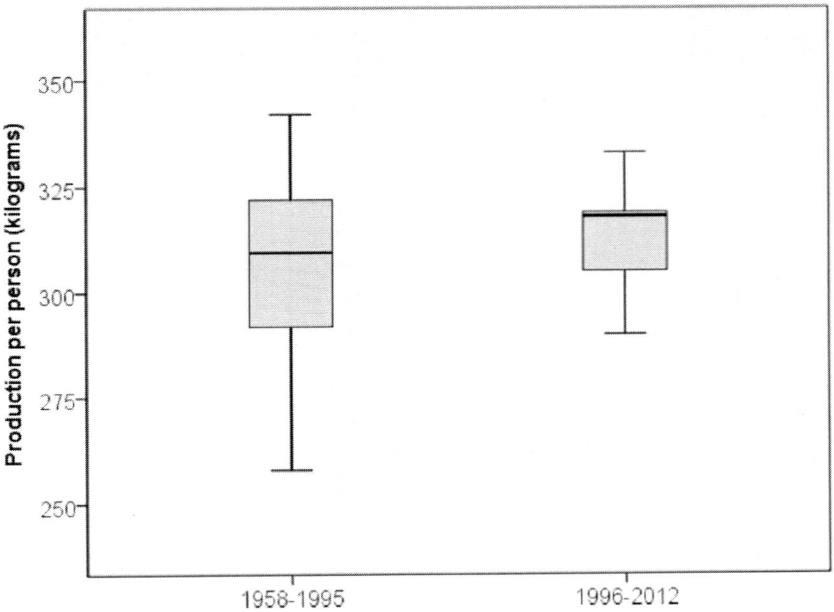


**Figure 10**. Cereals production (1958–1995, 1996–2012). Authors' calculation based on Ref. [62].

# 4. RESULTS OF HYPOTHESES TESTING

In order to test our first hypothesis "GM food has failed to feed the hungry," we calculate the significance of the difference between per capita grain production and surplus/deficit cereal productions in the two periods, period before and after GM commercialization, 1958–1995 and 1996–2012. We found that there is no statistically significant per capita grain production differences between the two observed periods ($t = -1.481$; $df = 53$; $p = 0.145$) (Figure 10). Average cereals production level in the period 1958–1995 was 305.13 kg/person/year (CI 297.57–312.69) and 313.94 in the period 1996–2012 (CI 307.54–320.34). Transgenic technology has not contributed to the improvement in food balance, since there is no statistically significant difference in surplus/deficit cereal production between the two observed periods $t = 0.941$; $df = 51$; $p = 0.51$) (Figure 11). Average value of this parameter was 7.91 × 106 tons (CI −5.02 to 20.86) in the period before transgenic plant cultivation and $-4.0 \times 10^6$ tons (CI −30.48 to 22.48) in the period after GMOs cultivation. These results clearly demonstrated that transgenic technology has failed to contribute to hunger and famine reduction. Unfortunately, food insecurity and world hunger are still a recurring problem in a significant number of countries.

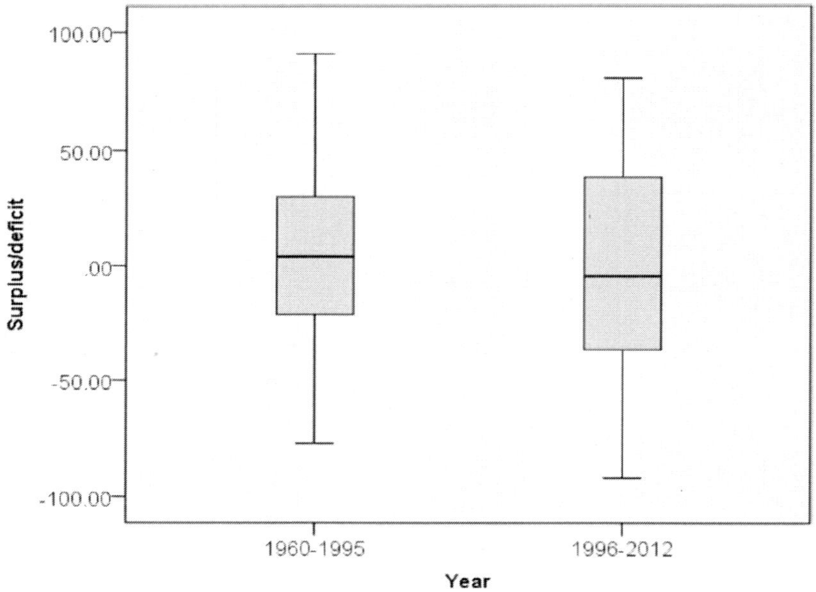


**Figure 11**. Cereal surplus/deficit before and after GM cultivation. Author's calculations based on Ref. [62].

The second hypotheses "There is a statistically significant correlation between stacked trait and global market value of biotech crops" have been confirmed as presented in Figure 12. A statistically significant correlation exists between global market value (G) (const price, 2002 = 100) of biotech crops and areas under stacked crops (S) ($r = 0.984$; $p < 0.001$). The regression analysis suggests that an increase in 1 million hectares under stacked crops causes an increase in global market value by 198.3 million in terms of 2002 US$ (Table 1). This can be represented by the following equation:

$$G = 198.3xS + 243.834 \tag{12}$$

The results can lead us to conclusion that stacked traits obtained through more complex transformations potentially leading to a further enrichment of multinational companies.

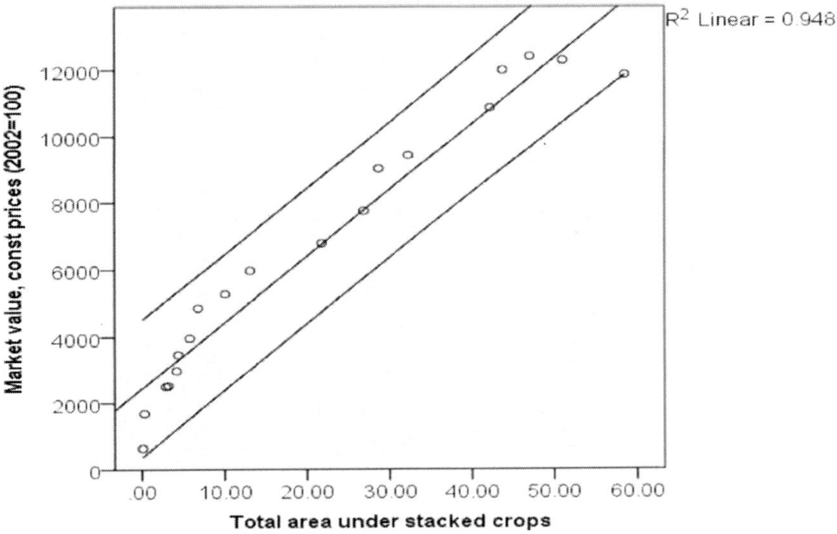


**Figure 12.**Correlation between global market value of transgenic crops and stacked crops. Authors' calculations based on Refs. [21, 25–43].

**Table 1**. The regression analyses (stacked crops/global market value).

| | | Coefficients | | *t* | *p* |
|---|---|---|---|---|---|
| | | *B* | Std. Error | | |
| 1 | Constant | 2438.834 | 320.191 | 7.517 | 0.000 |
| | S | 198.335 | 11.247 | 17.734 | 0.000 |

Finally, we pay attention to U.S. food aid which is in all forms procured by the U.S. Department of Agriculture (USDA) and administered by either the USDA or the U.S. Agency for International Development (USAID). A substantial portion of the U.S. food aid is channeled through the WFP where the U.S. as the major donor provided over the last 18 years around 43% of total WFP contributions [63]. With some fluctuation, U.S. food aid increased 3.2 times in const prices and 4.5 times in current prices from 1996 to 2015 (Figure 13). Maximum value reached in 2008 in terms of 2002 US$ prices. Polynomial trend line equations in current and constant prices are as follows:

$$y= 0.009x6-0.556x5+11.99x4-122.8x3+604.4x2-1169x+1183 \ (R2=0.826) \quad (13)$$

y= 0.008x6−0.482x5+10.5x4−108.8x3+538.9x2−103.4x+1147 (R2=0.721) (14)

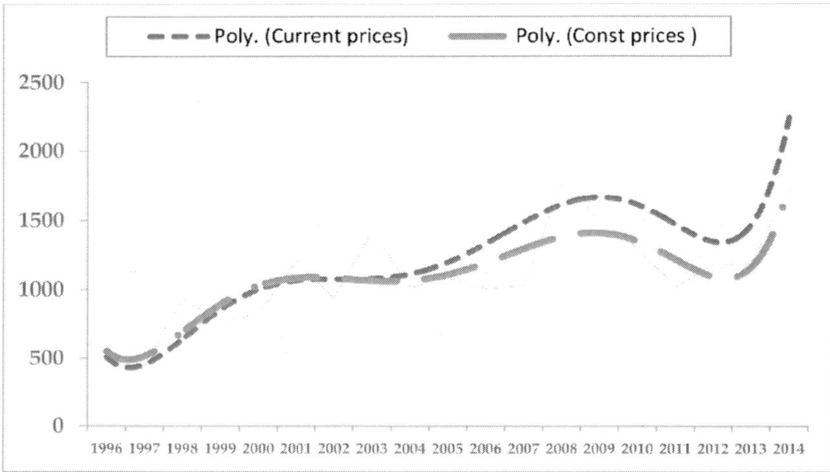


**Figure 13**. The U.S. food aid (1996–2014). Author's calculations based on Refs. [63, 64].

Taking into consideration that the U.S. increases in food emergencies in the last two decades, we have settled our third, last research hypothesis: "There is a correlation between raising production of biotech crops in the U.S. and an increase of the U.S. food aid through WFP." The results have shown significant correlation between U.S. GMOs areas (G) and the U.S. aid (A) ($r = 0.72$; $p = 0.001$) (Figure 14). The regression analysis (Table 2) suggests that an increase in 1 million hectares under biotech crops causes an increase in U.S. aid for 11.1 million US$ (const price, 2002 = 100) as expressed by the following equation:

$$A=11093.17xG+574475.15 \quad (15)$$

**Table 2**. The regression analyses (GMOs/U.S. aid).

| | | Coefficients | | *t* | *p* |
|---|---|---|---|---|---|
| | | *B* | Std. Error | | |
| 1 | Constant | 574475.15 | 133289 | 4.310 | 0.009 |
| | G | 11093.17 | 2669.74 | 4.156 | 0.000 |

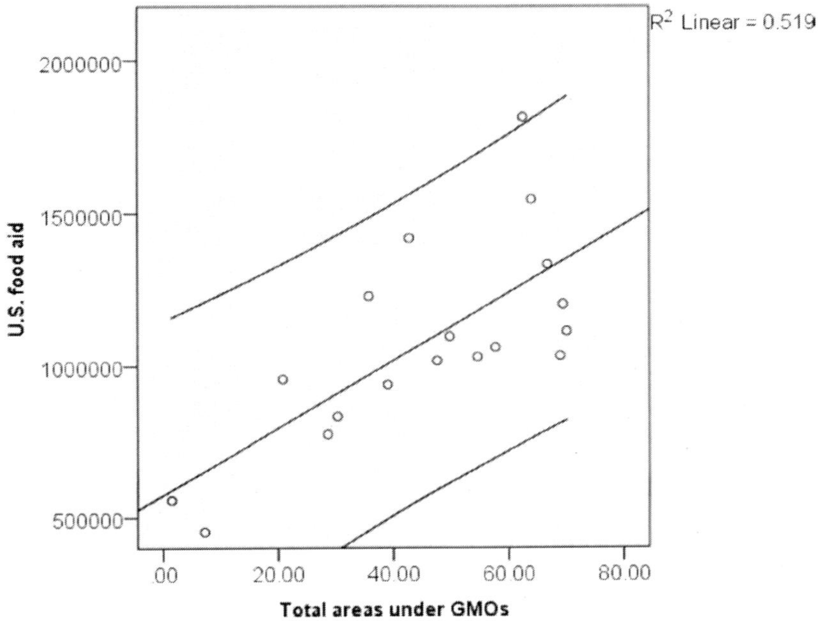


**Figure 14**. Correlation between U.S. GMOs areas and U.S. food aid. Authors' calculation based on Refs. [21, 25–43, 63, 64].

## 5. CONCLUSION

Unlike the Green revolution, in which public research institutes developed technologies and freely disseminated them around the world, Gene revolution is led by multinational corporations, which place on the market GM crops protected by patents. Despite the claims of GMOs impressive diffusion, unequal distribution is present in all segments. Herbicide tolerance and insect resistance are the main GM traits that are currently under commercial cultivation, and the main crops are soybean, maize, canola, and cotton. Soybean and maize share in the total acreage was 81% in 2015. The reason for such a large share is their widespread use as food, feed, for industrial purposes and for biofuels production. These two plant species are broadly present as ingredients of processed foods. For example, soya components are in margarine, cooking fat, mayonnaise, biscuits, coffee creams, fried products, ice creams, baked goods, chocolate, smoothies, tofu, sauces, as well as in meat, meat products, milk, and eggs (as 75% of soy worldwide is used for animal feed, especially for poultry and pigs) [65].

GM crops are grown by <1% of the world's farming population on 13% arable land mostly located in Americas and in some part of Asia. Transgenic technology is facing problems on all continents. Russia has learned a lesson from the past when even a strong nuclear power failed to provide food self-sufficiency. That's why the modern tendency of the Russian Federation is to ensure food security. In case, it comes to fruition that Russia will be a key player in reshaping tomorrow's global food system. Because the U.S. is powerless to oppose the Russian ban on GMOs and because of strong possibility of further strengthening of the European opposition, the U.S. intensifies the pressure on week, poor countries or countries in transition to accept transgenic technology. Apart from several African countries (Cameroon, Egypt, Ghana, Kenya, Malawi, Nigeria, and Uganda) where multinationals conducted in recent years field trials on GM crops as the penultimate step prior to its approval, in the Balkan region Serbia faces the greatest pressure to accept this technology. The reason lays in the facts that Serbia is leading agricultural country in the region, among the 10 largest world exporter of corn (non-GM), one of the biggest European producing countries of non-GM soya exporting soybean products into five continents and because the U.S. has lost export market for soybean meal. So far, 127 municipalities in Serbia have signed "declaration against GMOs." On May 29, 2009, National Parliament adopted new law on GMOs that fully prohibit the possibility of commercial growing of GMOs, or trade and products derived from GMOs. On the other hand, corrupt politicians and livestock agribusiness sector are pressuring to amend the Law on GMOs. Although multinationals conduct court disputes with farmers who sown GM seed without paying from the countries that allow the GM cultivation, in Serbia as a part of its expansion, strategy does not require protection of patent rights for illegally planted seeds.

Transgenic technology has failed to contribute to hunger and famine reduction since there is no statistically significant per capita grain production differences between the two periods, period before GM cultivation and period after it. In order to ensure food security for a growing population, it is necessary to launch a new revolution in food production or revoke the patent protection of transgenic technology and develop crops and traits vitally important for poor farmers and developing countries. At present, generally speaking, the main winners are multinationals who as time passes more and more strengthening its financial position. It can be seen, inter alia, on the basis of a correlation between the stacked trait and global market value of biotech crops. U.S. food aid will still remain a controversial issue as there is a significant correlation between raising production of biotech crops in the U.S. and an increase in the U.S. food aid through WFP.

Finally, it is important that governments around the world provide a transparent source of information for consumers on GM foods, which clearly listed of what GM foods are on the market. That way, consumer rights to informed choice will be satisfied.

# REFERENCES


1. Seralini G.E., Mesnage R., Defrage N., Gress S., Hennequin D., Claie E., et al. Answers to critics: why there is a long term toxicity due to a Roundap-tolerant genetically modified maize and a Roundap herbicide. Food and Chemical Toxicology. 2013;53:476–483.
2. Domingo J.L. Toxicity studies of genetically modified plants: a review of the published literature. Critical Review of Food Science and Nutrition. 2007;47(8):721–733.
3. Brookes G., Barfoot P. GM crops: the global socioeconomic and environmental impact-the first nine years 1996-2004. AgBioForum. 2005;8(2 and 3):187–196.
4. Brookes G. Genetically modified crops: how long before Africa benefits? In: Heap B., Bennet D., editors. Africa's Future...Can Bioscience Contribute? Cambridge UK: Banson/B4FA; 2014. pp. 50–57.
5. Brankov T.P., Sibalija T., Lovre K., Cvijanovic D., Subic J. The impact of biotechnology knowledge on the acceptance of genetically modified food in Serbia. Romanian Biotechnological Letters. 2013;18(3):8295–8306.
6. Chern W.S., Rickertsen K. Consumer acceptance of GMO: survey results from Japan, Norway, Taiwan and the United States. Taiwanese Economic Review. 2001;7(1):1–28.
7. Burton M., Rigbi D., Young T., James S. Consumer attitudes to genetically modified organisms in food in the UK. European Review of Agricultural Economics. 2001;28(4):479–498.
8. Costa-Font M., Gil J.M., Traill W.B. Consumer acceptance, valuation of and attitudes towards genetically modified food: review and implication for food policy. Food Policy. 2008;33(2):99–111.
9. Carlson R. The U.S. Bioeconomy in 2012 reached $350 billion in revenues, or about 2.5% of GDP [Internet]. January 6, 2014 [Updated: Jun 18 2016]. Available from: http://www.synthesis.cc/2014/01/the-us-bioeconomy-in-2012.html
10. Paul H., Streinbrecher R., Michaels L., Kuyek D. Hungry Corporation: Transnational Biotech Companies Colonise the Food Chain. London UK: Zed Books; 2004. 256 p.
11. Brankov T.P., Lovre K. WTO law and genetically modified products. In: Tomic D., Sevarlic M., Lovre K., Zekic S., editors. 135th EAAE Seminar: Challenges for the Global Agriculture Trade Regime after Doha; 28–30 August; Belgrade. Belgrade: Serbian Association of Agricultural Economists Faculty of Economics Subotica; 2013. pp. 99–113.
12. Fernandez-Cornejo J., Just R.E. Research ability of modern agricultural input markets and growing concentration. America Journal of Agricultural Economics. 2007;89:1269–1275.
13. Glenna L., Cahoy D.R. Agribusiness concentration, intellectual property, and the prospects for rural economic benefits from the emerging biofuel economy. Southern Rural Sociology. 2009;24:111–129.
14. Brankov T.P., Lovre K. Implikacije svetske ekonomske krize na biotehnolosku industriju. Ekonomika Poljoprivrede. 2010;3:369–376.
15. Farm Journal Magazine. Estimated U.S. seed market shares released [Internet]. July 29, 2015 [Updated: June 10, 2015]. Available from: http://www.agrimarketing.com/s/97703
16. Howard P.H. Visualizing consolidation in the global seed industry: 1996–2008. Sustainability. 2009;1(4):1266–1287.
17. Mascarenhas M., Bush L. Seeds of change: intellectual property rights, genetically modified soybean and seed saving in the United States. Sociologia Ruralis. 2006;46:122–138.
18. Fernandez-Cornejo J. The Seed Industry in U.S. Agriculture: An Exploration of Data and Information on Crop Seed Markets, Regulation, Industry Structure, and Research and Development. Washington, DC: USDA Economic Research Service; 2004. 71 p.

19. Eaton D., Louwaars N. Intellectual Property Rights in the International Seed Sector and Options for Resource-Poor Farmers, Report 09-019. Hague: LEI Wageningen UR; 2009.
20. Marden E. The neem tree patent: international conflict over the commodification. International and Comparative Law Review. 1999;22(2):279–295.
21. James C. Global Status of Commercialized Biotech/GM Crops ISAAA Brief No. 42. Ithaca, NY: ISAAA; 2010.
22. Nathan N., Qian N. US food aid and civil conflict. The American Economic Review. 2014;104(6):1630–1666.
23. Clapp J. The political economy of food aid in an era of agricultural biotechnology. Global Governance: A Review of Multilateralism and International Organizations. 2005;11(4):467–485.
24. Canadian Biotechnology Action Network (CBAN). Where in the World Are GM Crops and Foods? Report 1. Ottava, Ontario: CBAN; 2015. 30 p.
25. James C., Krattinger A.F. Global Review of the Field Testing and Commercialization of Transgenic Plants, No. 1, 1986–1995. The First Decade of Crop Biotechnology. Ithaca, NY: ISAAA; 1996.
26. James C. Global Status of Transgenic Crops in 1997. ISAAA Brief No. 5. Ithaca, NY: ISAAA; 1997.
27. James C. Global Review of Commercialized Transgenic Crops: 1998. ISAAA Brief No. 8. Ithaca, NY: ISAAA; 1998.
28. James C. Global Status of Commercialized Transgenic Crops 1999. ISAAA Brief No. 12: Preview. Ithaca, NY: ISAAA; 1999.
29. James C. Global Status of Commercialized Transgenic Crops 2000. ISAAA Brief No. 21. Ithaca, NY: ISAAA; 2000.
30. James C. Preview: Global Status of Commercialized Transgenic Crops 2001. ISAAA Brief No. 24. Ithaca, NY: ISAAA; 2001.
31. James C. Preview: Global Status of Commercialized Transgenic Crops 2002. ISAAA Brief No. 27. Ithaca, NY: ISAAA; 2002.
32. James C. Preview: Global Status of Commercialized Transgenic Crops 2003. ISAAA Brief No. 30. Ithaca, NY: ISAAA; 2003.
33. James C. Preview: Global Status of Commercialized Transgenic Crops 2004. ISAAA Brief No. 32. Ithaca, NY: ISAAA; 2004.
34. James C. Global Status of Commercialized Transgenic Crops 2005. ISAAA Brief No. 34. Ithaca, NY: ISAAA; 2005.
35. James C. Global Status of Commercialized Transgenic Crops 2006. ISAAA Brief No. 35. Ithaca, NY: ISAAA; 2006.
36. James C. Global Status of Commercialized Transgenic Crops 2007. ISAAA Brief No. 37. Ithaca, NY: ISAAA; 2007.
37. James C. Global Status of Commercialized Transgenic Crops 2008. ISAAA Brief No. 39. Ithaca, NY: ISAAA; 2008.
38. James C. Global Status of Commercialized Transgenic Crops. ISAAA Brief No. 41. Ithaca, NY: ISAAA; 2009.
39. James C. Global Status of Commercialized Transgenic Crops 2011. ISAAA Brief No. 43. Ithaca, NY: ISAAA; 2011.
40. James C. Global Status of Commercialized Transgenic Crops 2012. ISAAA Brief No. 44. Ithaca, NY: ISAAA; 2012.
41. James C. Global Status of Commercialized Transgenic Crops 2013. ISAAA Brief No. 46. Ithaca, NY: ISAAA; 2013.
42. James C. Global Status of Commercialized Transgenic Crops 2014. ISAAA Brief No. 49. Ithaca, NY: ISAAA; 2014.
43. James C. Global Status of Commercialized Transgenic Crops 2015. ISAAA Brief No. 51. Ithaca, NY: ISAAA; 2015.
44. GMWATCH. Venezuela passes national anti-GMO and anti-patent seed law [Internet]. December 15, 2015. Available from: http://www.gmwatch.org/news/latest-news/16602-venezuela-passes-national-anti-gmo-and-anti-patent-seed-law [Accessed: July 15, 2016].

45. PRESSTV. GM crops growing in Iran amid lax laws [Internet]. July 6, 2015. Available from: http://presstv.com/Detail/2015/07/06/418990/iran-gm-crops-agriculture-imports [Accessed: July 10, 2016].
46. JIHAD WATCH. Iranian scientist: genetically modified food a "Zionist plot" [Internet]. February 16, 2016. Available from: http://www.jihadwatch.org/2016/02/ iranian-scientist-genetically-modified-foods-a-zionist-plot [Accessed: July 1, 2016].
47. NEO. Russia Bans US GMO Imports [Internet]. February 26, 2016. Available from: http://journal-neo.org/2016/02/26/russia-bans-us-gmo-imports/ [Accessed: June 11, 2016].
48. Todhunter C. India: selling out to Monsanto. GMOs and the Bigger Picture [Internet]. August 8, 2014. Available from: http://www.globalresearch.ca/independent- india-selling-out-to-monsanto-gmos-and-the-bigger-picture/5395187 [Accessed: June 1, 2016].
49. Global Agricultural Information Network (GAIN). Egypt Agricultural Biotechnology Annual 2015 [Internet]. July 16, 2015. Available from: http://www.gain. fas.usda.gov/Recent%20GAIN%20Publications/Agricultural%20Biotechnology%20Annual_Cairo_Egypt_7-16-2015.pdf [Accessed: May 30, 2016].
50. Global Agricultural Information Network (GAIN). Senegal Agricultural Biotechnology Annual 2015. Francophone West Africa Biotechnology Report [Internet]. August 19, 2015. Available from: http://gain.fas.usda.gov/Recent%20GAIN% 20Publications/Agricultural%20Biotechnology%20Annual_Dakar_Senegal_8-19-2015.pdf [Accessed: May 15, 2016].
51. GMWATCH. Burkina Faso dumps GM Bt cotton [Internet]. June 9, 2015. Available from: http://www.gmwatch.org/news/latest-news/87-news/archive/2015/16219-burkina-faso-dumps-gm-bt-cotton [Accessed: June 9, 2016].
52. BUZZ. Sudan blocks import of GM soybeans considers total ban on GM foodstuffs [Internet]. Available from: http://www.buzz.naturalnews.com/ 001110-Sudan-GM_food-total_ban.html [Accessed: July 1, 2016].
53. GMWATCH. 5 Million Nigerians oppose Monsanto's plants to introduce GMO cotton and corn [Internet]. March 29, 2016. Available from: http://www.ecowatch.com/2016/03/29/monsantoes-gmo-cotton-corn-nigeria/ [Accessed: June 6, 2016].
54. Global Agriculture Information Network (GAIN). EU-28 Agricultural biotechnology annual 2015 [Internet]. July 23, 2015. Available from: http://gain.fas.usda.gov/Recent%20GAIN%20Publications/Agricultural%20Biotechnology%20Annual_Paris_EU-28_7-23-2015.pdf [Accessed: June 18, 2016].
55. Staropoli N. FDA approves GMO potato that resists blight that caused Irish potato famine [Internet]. January 14, 2016. Available from: https://www.geneticliteracyproject.org/2016/01/14/fda-approves-gmo-potato-resist-blight-caused-irish-potato-famine/ [Accessed: June 18, 2016].
56. Starke L, editor. The Trends That Are Sharping Our Future. Vital Signs. Volume 22. Washington, DC: Island Press; 2015. 133 p.
57. GMWATCH. Scientist warns of problems with GM canola [Internet]. February 5, 2008. Available from: http://www.gmwatch.org/latest-listing/47-2008/ 6171-scientist-warns-of- problems-with-gm-canola-522008 [Accessed: July 2, 2016].
58. Reuters. GMO that kills: GM-cotton problems drive Indian farmers to suicide [Internet]. May 24, 2015. Available from: https://www.rt.com/news/ 261673-india- gmo-cotton-suicides/ [Accessed: July 7, 2016].
59. USDA. World agricultural supply and demand estimates [Internet]. Available from: https://www.nass.usda.gov [Accessed: June 1, 2016].
60. National Corn Growers Association. U.S. corn production 1935–2015 [Internet]. January 12, 2016. Available from: http://www.worldofcorn/#us-corn-production [Accessed: May 6, 2016].
61. Kron M., Bittner U. Danube soya-improving European GM-free soya supply for food and feed. Oilseeds and Fets, Crops and Lipids. 2015;22(5):10.
62. Earth Policy Institute. Food and Agriculture [Internet]. January 17, 2013. [Accessed: May 2, 2016].

63. Hanrahan C.E., Canada C. International food aid: U.S. and other donor contributions [Internet]. November 12, 2013. Available from: https://www.fas.org/sgp/crs/misc/RS21279.pdf [Accessed: May 10, 2016].
64. World Food Programme (WFP). Contributions by year [Internet]. Available from: http://www.wfp.org/funding/year [Accessed: June 1, 2016].
65. WWF Global. Soy is everywhere [Internet]. Available from: http://wwf.panda.org/what_we_do/footprint/agriculture/soy/facts/ [Accessed: June 2, 2016].

# CHAPTER 3

# The Use of Lactic Acid Bacteria in the Fermentation of Fruits and Vegetables — Technological and Functional Properties

**Dalia Urbonaviciene[1, 2], Pranas Viskelis[2], Elena Bartkiene[3], Grazina Juodeikiene[1] and Daiva Vidmantiene[1]**

*[1] Department of Food Science and Technology, Kaunas University of Technology, Lithuania*

*[2] Institute of Horticulture, Lithuanian Research Centre for Agriculture and Forestry, Lithuania*

*[3] Veterinary Academy, Lithuanian University of Health Sciences, Lithuania*

## 1. INTRODUCTION

The relationship between food and health has been investigated for many years, and therefore, the development of foods that promote health and well-being is a key research priority of the food industry [1]. Fruits and vegetables are an essential part of human nutrition. Unfortunately, the daily intake of fruits and vegetables is estimated to be lower than the recommendation of the World Health Organization (WHO) [2], who suggest a dietary intake of 450 and 500 g of fruits and vegetables, respectively. Vegetables are strongly recommended in the human diet because they are rich in antioxidants, vitamins, dietary fibres and minerals. The majority of vegetables consumed in the human diet are fresh, minimally processed, pasteurised or cooked by boiling in water or microwaving, and vegetables can be canned, dried, or juiced or made into pastes, salads, sauces, or soups. Fresh vegetables or those that have been minimally processed have a particularly short shelf-life because they are subjected to rapid microbial spoilage. In addition, the above cooking processes can cause a number of potentially undesirable changes in physical characteristics and chemical composition [3,4].

Therefore, these drawbacks could be reduced by novel technologies, such as new packaging systems, high-hydrostatic pressure processing, ionisation radiation and pulsed electric fields [5-7]. The use of natural antimicrobial preservatives is considered to be the simplest and most valuable biological technique to keep and/or enhance the safety, nutrition, palatability and shelf-life of fruits and vegetables [5]. Lactic acid fermentation of vegetables, currently used as the bio-preservation method for the manufacture of finished and half-finished foods, is an important biotechnology for maintaining and/or improving safety, nutritional, sensory and shelf-life properties of vegetables. Three technology options are usually considered for lactic acid fermentation of vegetable: spontaneous fermentation by autochthonous lactic acid bacteria, fermentation by starter cultures that are added into raw vegetables, and fermentation of mild heat-treated vegetables by starter cultures [18]. For thousands of years, microorganisms have been used to produce and preserve foods through the process of fermentation. Fermented foods have been adopted in various ways depending on the properties of the available raw materials and the desired features of the final products [8-10]. Food produced by traditional methods has become popular among consumers who know that their food is manufactured from high quality raw materials, without preservatives and other synthetic additives that are characterised by unique flavour values [11].

## 2. FERMENTATION FROM A BIOCHEMICAL POINT OF VIEW

Bourdichon et al. [12] describe the fermentation process as "a metabolic process of deriving energy from organic compounds without the involvement of an exogenous oxidising agent". Fermented foods are subjected to the actions of microorganisms or enzymes. Fermentation plays different roles in food processing, such that desirable biochemical changes have occurred [13]. The fermentation process is very important in the improvement of technological properties of preservation, such as a relative cost-effectiveness and low energy requirements, which are essential for ensuring the shelf-life and microbiological safety of the product [8]. The major roles of fermentation are considered to be the following:

1) preservation of food: the formation of inhibitory metabolites, such as organic acid (lactic acid, acetic acid, formic acid, propionic acid), ethanol, bacteriocins, etc., often in combination with a decrease in water activity (by drying or the use of salt) [14-15].
2) improving food safety through the inhibition of pathogens [16,17] or the removal of toxic compounds [18].
3) improving nutritional value: biological enrichment of food substrates with proteins, essential amino acids, essential fatty acids and vitamins [19,20].
4) organoleptic food quality: enrichment of the diet through the development of a diversity of flavours, aromas, and textures in food substrates [21-24].
5) decrease in cooking times and fuel requirements [25].

Interest in the biopreservation of food has created a demand for more natural and minimally processed food, with particular interest in naturally produced antimicrobial agents [26].

# 3. LACTIC ACID BACTERIA (LAB) IN FOOD FERMENTATION AND NEW NATURAL ANTIMICROBIAL COMPOUNDS

LAB have traditionally been associated with food fermentation. LAB are generally considered beneficial microorganisms, with some strains even considered to promote good health (probiotic), and their extensive historical use contributes to their acceptance as being GRAS (generally recognised as safe) for human consumption [27]. LAB are used as natural or selected starters in food fermentation and exert health benefits through the antimicrobial effect produced from different metabolic processes (lactose metabolism, proteolytic enzymes, citrate uptake, bacteriophage resistance, bacteriocin production, polysaccharide biosynthesis, metal-ion resistance and antibiotic resistance) [28,29,9]. Spontaneous fermentation typically results from the competitive activity of a variety of autochthonous and contaminating microorganisms, which may lead to a high risk for failure. Both from a hygiene and safety perspective, the use of starter cultures is recommended, as it leads to rapid inhibition of spoilage and pathogenic bacteria while yielding processed fruit with consistent sensory and nutritional quality [30].

Interest in the biopreservation of food has prompted the quest for novel antimicrobial compounds from different natural origins. The LAB of genera such as *lactobacilli* and *lactococcus* are amongst the most important known members that have probiotic activity [31]; these bacteria produce antimicrobial peptides most frequently referred to as bacteriocins [32,33]. Bacteriocins ensure the stability of fermented plant products, reduce microbial contamination during fermentation, inhibit the growth of moulds and delay microbiological spoilage of baked goods [34].

LAB have strong inhibitory effects on the growth and toxin production of other bacteria. This activity can occur due to the following factors: competition for available nutrients; decrease in redox potential; production of lactic acid and acetic acid and the resulting decrease in pH; production of other inhibitory primary metabolites, such as hydrogen peroxide, carbon dioxide or diacetyl; and production of special antimicrobial compounds, such as bacteriocins and antibiotics [35].

Each of these properties, particularly when combined, can be used to extend the shelf-life and safety of food products [36].

Amongst the various technologies, lactic acid fermentation may be thought of as a simple and valuable biotechnology for maintaining and/or improving the safety, nutritional, sensory and shelf-life properties of fruits and vegetables [37,38]. Overall, LAB are a small part (2–4 $\log_{10}$ CFU $g^{-1}$) of the autochthonous microbiota of raw vegetables, and their cell density is mainly influenced by the vegetable species, temperature and harvesting conditions [37].

Interest in the use of LAB fermentation of vegetable products stems largely from the nutritional, physiological and hygienic aspects of the process and their corresponding implementation and production costs [39].

LAB fermentation represents the easiest and most suitable way to increase the daily consumption of nearly fresh fruits and vegetables.

# 4. UNIQUE FEATURES OF FERMENTED FRUITS AND VEGETABLES

Buckenhuskes and colleagues [40] generally agreed that fermented plant products are the "food of the future". The following factors support this idea: products can be marked as "natural" or "biological"; desirable flavour compounds are enhanced while negative flavour compounds (for example, glucosinolates) are destroyed; handling and storage (without cooling) is simple; easy methods exist for the pre-handling of raw material before further processing; desired metabolites (lactic acid, amino acids) are enriched; and the process results in the detoxification of pathogens [41]. Fruits and vegetables preserved using LAB with antimicrobial properties are perceived as suitable products for the human diet [42].

Dieticians and physicians recommend fermented fruits and vegetables due to the health-promoting properties of these foods. Fermented fruits and vegetables are low-calories foods because they contain considerably lower quantities of sugars compared to their raw counterparts. Fermented vegetables are a source of dietary fibre, which impedes the assimilation of fats and regulates peristalsis in the intestines; they are also a valuable source of vitamin C, B-group vitamins, phenolics and many other nutrients present in the raw material. Lactic acid may also lower gut pH, thereby inhibiting the development of putrefactive bacteria [42].

Many types of fermented fruit and vegetable products exist in the world: sauerkraut, cucumber pickles, and olives in the Western world; Egyptian pickled vegetables in the Middle East; and Indian pickled vegetables, Korean kim-chi, Thai pak-sian-don, Chinese hum-choy, Malaysian pickled vegetables and Malaysian tempoyak. Lactic acid-fermented cereals and tubers (cassava) include Mexican pozol, Ghanaian kenkey, Nigerian gari; boiled rice/raw shrimp/raw fish mixtures such as Philippine balao-balao and burong dalag; lactic-fermented/leavened breads such as sourdough breads in the Western world; Indian idli, dhokla, khaman and Sri Lankan hoppers; Ethiopian enjera, Sudanese kisra and Philippine puto; and Chinese sufu/tofu-ru [9,43].

Commercial distribution of these fermented products lags far behind that of fermented meat and dairy products due to a lack of standardised manufacturing protocols; in addition, their ingredients are subject to limiting and unpredictable weather and geographic conditions [44]. The lactic acid fermentation of vegetables currently has industrial significance only for cucumbers, cabbages and olives [45]. Several other varieties of vegetables cultivated mainly in Southern Italy or, more generally, in the Mediterranean area, such as carrots, French beans, marrows, artichokes, capers and eggplants, may benefit from increased safety, nutritional, sensory and shelf-life properties through standardised industrial lactic acid fermentation [46].

## 5. THE USE OF MICROORGANISMS IN OUR DIET OPENS NEW OPPORTUNITIES

Either as traditional fermented foods or as novel approaches, the rationalised use of microorganisms in our diet could reveal new opportunities. Low dietary quality is an important factor that limits adequate nutrition in many resource-poor settings. Bioavailability is a key aspect of dietary quality with respect to the adequacy of micronutrient intake [47]. Prebiotic food ingredients encourage the growth of probiotic bacteria. The appropriate combination of prebiotics and probiotics manifest in a higher potential for synergistic effects 48]. Probiotic foods are fermented products that contain a sufficient number of a certain live microorganism to favourably modify the intestinal microbiota of the host [49]. Recently developed probiotics tend to be milk-based, although in recent years other substrates have been explored for new probiotic formulations. Amongst these substrates, cereals are becoming one of the most promising alternatives to milk due to their ability to support the growth of probiotic bacteria and their protection against bile resistance [50].

According to Kim et al. [51], cabbage (including the Chinese cabbage), pH-adjusted tomato (pH 7.2), carrot and spinach media give relatively higher fermentability than other vegetables because they have more fermentable saccharides. The tomato (*Lycopersicon esculentum* L.) is one of the most popular and extensively consumed vegetable crops worldwide. The nutritional significance of lycopene, a carotenoid with potent antioxidant activity, has been reported, and accumulating evidence has shown an inverse correlation between the consumption of tomato products rich in lycopene and the risk of several types of cancer and cardiovascular disease [52-54]. Approximately 90 % of the lycopene in dietary sources is found in the linear, all-*trans* conformation, while human tissues mainly contain *cis*-isomers. It has been suggested that the *cis*-isomers of lycopene are better absorbed than the all-*trans* form because they are shorter, have greater solubility in mixed micelles, and have a lower tendency to aggregate [55]. Studies have shown that lycopene levels in plasma increase only after the consumption of red tomato paste and purified lycopene [54]. It has also been revealed that the absorption of lycopene is greater from processed tomatoes than from fresh tomatoes because processing breaks down the tomato cell matrix and makes the lycopene more available [56,57].

The red colour of tomatoes is a result of the degradation of chlorophylls and the increased biosynthesis of carotenoids [58]; thus, a tomato's colour is related to its maturity and post-harvest treatment. Colour is therefore an important attribute indicating the quality of tomato fruit, and it is used in the food industry to predict the colour of finished products. Additionally, the application of instrumental colour measurement to objectively define the colour of tomatoes is an important research topic [59,60]. It was reported that the colour coordinates of a product could relate to its concentration of lycopene and other carotenoids [61,62].

There is an increasing consumer demand for high quality meat products that taste good and are both nutritious and easy to prepare. The diverse nutrient composition of meat makes it an ideal environment for the growth and propagation of meat spoilage microorganisms and common food-borne pathogens. It is therefore essential to apply adequate preservation techniques to maintain its safety and quality [63]. The processes used in meat preservation are principally concerned with inhibiting microbial spoilage, although other methods of preservation seek to minimise additional deteriorative changes in colour and oxidation [64]. The most investigated new preservation technologies for fresh meat involve non-thermal inactivation, such as high hydrostatic pressure (HHP), novel packaging systems, including modified atmosphere packaging (MAP) and active packaging (AP), natural antimicrobial compounds and biopreservation. Storage life is extended and safety is increased by using natural or controlled microflora, including the extensively studied LAB and their antimicrobial products, such as lactic acid and bacteriocins. Bacteriocins are a heterogeneous group of antibacterial proteins that vary in their spectrum of activity, mode of action, molecular weight, genetic origin and biochemical properties [65].

The destruction of the total BLIS (bacteriocin-like inhibitory substances) activity after treatment with proteinase K, trypsin, pepsin and chymotrypsin indicates that antimicrobial substances produced by the tested LAB possess a proteinaceous nature. They might be bacteriocins because protease sensitivity is a key criterion in the classification of antimicrobial substances as BLIS [66]. In our previous studies, BLIS produced by *Lactobacillus sakei* KTU05-6 and *P. pentosaceus* KTU05-9 were designated as sakacin 05-6 and pediocin 05-9 [67]. We proposed that due to their broad inhibition spectrum, the presence of BLIS and organic acids in tested LAB is an indication that these bacteria can be used widely in the food industry as bio-preservatives.

Consumer interest for diverse fermented foods has increased in recent years because of the positive perception of their beneficial impact on health. Hence, there is an evident need to find novel methods and new food preservation agents from natural origins. Biopreservation refers to extending the shelf-life and enhancing the safety of foods using microorganisms or their metabolites [68]. In this aspect, LAB are very good candidates [69].

The food matrices in vegetables offer promising potential as sources and carriers of probiotic strains [70]. Vegetables are fundamental sources of water-soluble vitamins (vitamin C and group B vitamins), provitamin A, phytosterols, dietary fibres, minerals and phytochemicals [71] in the human diet. LAB are a small part (2.0–4.0 $\log_{10}$ CFU $g^{-1}$) of the autochthonous microbiota of raw vegetables [37]. Under favourable conditions of anaerobiosis, water activity, salt concentration and temperature, raw fruits and vegetables may be subject to spontaneous lactic acid fermentation. In some cases, alcoholic fermentation takes place concomitantly [72].

Tomatoes are a rich source of a variety of nutritional compounds, especially key antioxidant components, such as the carotenoid lycopene, vitamin C, and a range of polyphenols. The possible protective characteristics of these antioxidants are of great interest, and consumers have already become aware of their potential importance. A survey of the literature revealed that a great deal of research has been conducted on the biochemical composition of tomatoes and their products [73]. Lycopene, a natural carotenoid found in tomatoes, has been reported to possess various health benefits, such as preventive properties against cardiovascular disease and cancer [74].

# 6. LACTIC ACID FERMENTATION OF TOMATOES: EFFECTS ON *CIS/TRANS* LYCOPENE ISOMERS, Β-CAROTENE CONCENTRATION AND THE FORMATION OF L(+) AND D(-)-LACTIC ACID

The production of L-lactic acid and D-lactic acid isomers during the fermentation of different tomato varieties (var. Ronaldo and var. Cunero) by the bacteriocin-producing LAB *Lactobacillus* and *Pediococcus* spp. have been investigated. The influence of lacto fermentation on the lycopene and β-carotene contents and their relation to the colour characteristics of fermented tomato products were also investigated [75]. Tomato var. Cunero and Ronaldo, the LAB strains were used in this investigation. Tomato var. Cunero and Ronaldo were obtained from the Lithuanian Institute of Horticulture (Babtai, Lithuania) harvested in 2011. Pure cultures of *Lactobaccilus sakei KTU05-6, Pediococcus acidilactici KTU05-7* and *Pediococcus pentosaceus KTU05-8*, characterized as a bacteriocin producing strains [76] are from collection of Kaunas University of Technology (Kaunas, Lithuania) [75].

The LAB strains were propagated in nutrition media (moisture content 72 %), prepared by mixing extruded rice flour (100 g) and tap water. After addition of pure LAB cell suspension (5 g, 10.2 $\log_{10}$colony-forming units (CFU) $g^{-1}$) the mixture was incubated at optimal temperatures (30 °C for *L. sakei*, 32 °C for *P. acidilactici* and 35 °C for *P. pentosaceus*) for 24 h. For comparison purpose control product was prepared using spontaneous fermentation of rice flour without bacterial inoculum at 30 °C for 48 h.

Enumeration of LAB was carried out by plating the diluted samples onto MRS agar at 30 °C for 48 hours. Products obtained after propagation of individual LAB in rice media were used for fermentation of tomato pulp [75].

A rapid and specific Megazyme assay kit for simultaneous determination of L- and D-lactic acid (Megazyme Int., Bray, Ireland) in foods was used as reported by De Lima et al. [77] in this investigation. Extraction of carotenoids and carotenoid analysis by Reverse Phase Liquid Chromatography (RP-HPLC) were used [75] and the colour characteristics of fermented and untreated tomato pulp were evaluated of the surface using CIEL*a*b* [78].

## 6.1. The effect of selected fermentation media on lab viability

As reported in the literature, the behaviour of different LAB depends on substrate composition, where bacteria in different substrates are able to produce different metabolites or increased biomass [79]. For maximum health benefits, it is important to have a significant number of viable LAB present in the probiotic product [80].

Extruded rice flour, a current product of the cereal processing industry, was found to show good fermentability. Counts of viable bacteria cells were measured between 6.62 and 8.50 $\log_{10}$ CFU $g^{-1}$after 48 h of analysed LAB cultivation in selected media (Table 1) [75]. The lowest biomass of bacteria was found in the spontaneously fermented rice media (5.57 $\log_{10}$ CFU $g^{-1}$). According to the obtained results, rice flour is a suitable medium for LAB cultivation to produce a functional food while most likely maintaining the other functional properties of rice. These results are in agreement with Trachoo et al. [81], who showed a biomass increase of lactobacilli over 2.5 $\log_{10}$ CFU $mL^{-1}$during 24 h in a germinated rice broth [75].

*L. sakei, P. acidilactici* and *P. pentosaceus* were found to be capable of sufficient rapid utilisation of tomato pulp for cell synthesis and organic acid production. They reduced the pH to 3.5–3.7 and increased the TTA to as high as 6.4. The viable cell counts reached 6.61 $\log_{10}$ CFU $g^{-1}$ after 48 h of fermentation. In either case, tomato products treated with spontaneous fermentation had pH values that were higher by 7.2% and TTA values that were lower by 17.3 % than products treated with lactofermentation (Table 1) [75].

Acid production depends on the concentration of viable bacteria able to utilise the available carbohydrate sources in the substrate [82]. The viable LAB cells in the fermented tomato products were found to be lower on average by 30% (lactofermentation) or 49.2% (spontaneous fermentation) compared to the rice media (Table 1); however, three LAB counts measured after 48 h of fermentation varied between 4.54 and 6.61 $\log_{10}$ CFU $g^{-1}$. To achieve health benefits, probiotic bacteria must be viable and available at a high concentration, typically approximately 6 $\log_{10}$ CFU $g^{-1}$ of product [80]. According to Sindhu and Khetarpaul [83], probiotic fermentation of indigenous food mixtures containing tomato pulp increases the acidity and improves the digestibility of starch and protein. Our results support the hypothesis that rice media contain the essential nutrients to support the growth of lactobacilli and can be directly used as a fermentation substrate of LAB. The obtained biomass levels are above the minimum required for a probiotic formulation.

**Table 1.The influence of fermentation media on LAB cell counts (log10 CFU g-1), pH and TTA values.**

| Samp-les | Extruded rice | | | Tomato products | | |
|---|---|---|---|---|---|---|
| | LAB count | pH | TTA | LAB count | pH | TTA |
| P.p. | 8.51±0.05d | 3.37±0.01a | 8.2±0.2b | 6.61±0.03c | 3.50±0.01a | 6.4±0.3b |
| P.a. | 6.62±0.03b | 3.40±0.01a | 8.2±0.2b | 4.54±0.04b | 3.71±0.01b | 6.8±0.2c |
| L.s. | 7.75±0.03c | 3.42±0.01a | 8.3±0.2b | 4.83±0.03b | 3.70±0.01b | 7.1±0.2d |
| SF | 5.57±0.02a | 3.73±0.01b | 7.2±0.2a | 2.83±0.02a | 3.92±0.01c | 5.6±0.2a |

[i] - The numbers are means followed by standard deviations (n = 3).

[ii] - Means within a column with different superscript letters are significantly different ($p < 0.05$).

[iii] - Samples: tomato products fermented with: P.p. – *P. pentosaceus,* P.a. – *P. acidilactici,* L. s. – *L. sakei;* SF – spontaneous fermented.

Classic lactic acid vegetable fermentation is a microbial process that involves heterofermentative and homofermentative LAB, generally *Lactobacillus* and *Pediococcus* [82]. At a pH between 3.5 and 3.8, vegetables will be preserved for a long period of time [83]. Tomatoes treated by lactofermentation could be recommended as useful and safe products for human nutrition. Furthermore, fermented tomatoes could serve as a healthy product for vegetarians and consumers who are allergic to dairy products [75].

## 6.2. The production of l- and d-lactic acid during lactofermentation of tomato pulp

Our results showed that all the analysed LAB produced a mixture of L- and D-lactic acid (Figure 1), and the highest amounts of each form were determined in tomato products treated by spontaneous fermentation (7.18±0.03 and 7.67±0.11 mg/100 g, respectively). As reported by Hartman [85] and Li and Cui [86], *Lactobacilli amylophilus, L. bavaricus L. casei, L. maltaromicus,* and *L. salivarius* predominantly yield the L-isomer. Strains such as *L. delbrueckii, L. jensenii,* and *L. acidophilus* yield D-lactic acid or mixtures of both forms. LAB such as *L. pentosus, L. brevis* and *L. lactis* can ferment glucose into lactic acid through homolactic fermentation. The fermentation of rice with two strains of *L. delbrueckii* yielded 3.23 and 5.04 mg/100 g of D-lactic acid [87].

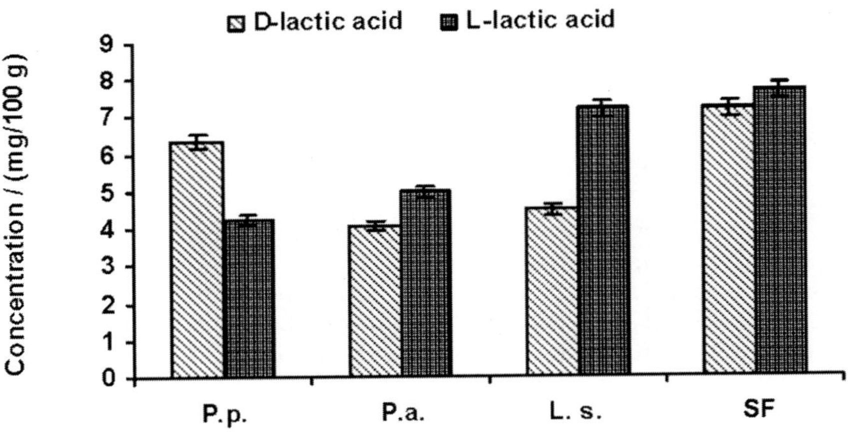


**Figure 1**. Concentrations of L- and D-lactic acid in fermented tomato products. Samples: fermented with LAB: P.p. – *P. pentosaceus*, P.a. – *P. acidilactici*, L.s. – *L. sakei*; SF – spontaneous fermented

The concentration of D-lactic acid in fermented tomato products was measured between 4.05±0.05 and 6.34±0.04 mg/100 g, and the concentration of L-lactic acid ranged from 4.26±0.04 to 7.19±0.08 mg/100 g (Figure 1). The results of our study indicate that compared to spontaneous fermentation, the use of *P. pentosaceus* allowed a reduction in the content of D-lactic acid in tomato products by 11.8% (Figure 1). Fermentation with *P. acidilactici* and *L. sakei* reduced the content of the latter isomer at a higher level (on average by 40.6%).

In summary, *P. pentosaceus* can produce D-rich lactic acid (L/D ratio 0.64), while the other strain, *L. sakei*, produces L-rich lactic acid (L/D ratio 1.61). Fermentation with *P. acidilactici* and spontaneous fermentation gave almost equal amounts of both lactic acid isomers (L/D ratio 1.17 and 1.07, respectively).

By evaluating our knowledge of the potential toxicity of D-lactic acid in terms of nutrition, we can report that tomato products prepared using a pure culture of LAB were found in all cases to be safer than those treated with spontaneous fermentation. The level of D-lactic acid in pure LAB-fermented tomato products was significantly lower ($p<0.05$) than that in those spontaneously fermented (Figure 1). Based on these results, *L. sakei KTU05-6* could be selected as the L-lactic acid bacteria and is recommended for the fermentation of tomatoes [75].

## 6.3. Trans/cis lycopene and β-carotene contents in fermented tomato products

The results from our analysis of lycopene and β-carotene contents in fermented tomato products are presented in Figure 2. The highest concentration of total carotenoids (on average 6.83 mg/100 g) were measured in a var. Cunero sample fermented with *P. pentosaceus* and in a var. Ronaldo sample fermented with *L. sakei*. However, fermentation with the latter bacteria increased the total level of carotenoids by 41.1 and 33.6%, respectively, compared to untreated samples. Compared to untreated tomatoes, fermentation with *P. acidilactici* reduced the concentration of total carotenoids by 3.6% in the samples of var. Cunero and var. Ronaldo (3.96 and 4.61 mg/100 g, respectively), which was accompanied by a reduction in β-carotene content (Figure 2) [75].

On average, the fermented tomato samples of var. Cunero had 24.7 % lower β-carotene and 11.5% higher lycopene content compared to untreated tomatoes. In contrast, the β-carotene concentrations in all the fermented tomato products of var. Ronaldo were generally higher, with an average increase of 69.4% compared to untreated tomatoes (Figure 2) [75].

A 24.8% increase in lycopene content was reached in the var. Ronaldo samples after fermentation with *L. sakei.* Spontaneous fermentation or treatment by *P. pentosaceus* reduced the concentration of lycopene by 11.0 and 4.4%, respectively, compared to the control sample (Figure 2).

According to these results, lactic acid fermentation generally had a positive effect on the lycopene and total carotenoid contents of the fermented tomato products. The β-carotene contents were influenced not only by which LAB was used but also by the variety of tomato. As reported in the literature, compositional variation of lycopene in tomatoes occurs as a consequence of varietal differences, climate conditions, agricultural variables, stage of maturity, harvesting and post-harvest handling and conditions during storage [75]. Other researchers reported lycopene values within the range of 3.1–7.7 mg/100 g for different tomato cultivars [88]. However, Camara et al. [89] reported a lycopene concentration of 6–15 mg/100 g for whole fresh tomato fruit [89], which is higher than the results of this investigation. Lycopene content may be directly affected by the pH of the fruit, as the low pH of red tomatoes accumulates more lycopene [90].

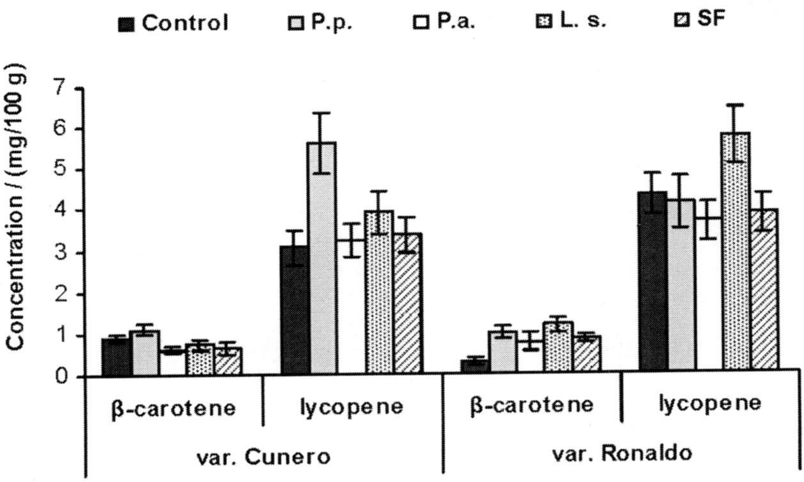


**Figure 2.** Carotenoid contents in untreated and fermented with different LAB tomato products. Samples: Control – untreated tomato pulp; tomato pulp fermented with: P.p. – *P. pentosaceus*; P.a. – *P. acidilactici* MI807; L.s. – *L. sakei*, SF – spontaneous fermented.

Our analysis of all-*trans* and *cis*-lycopene showed that the amounts of both isomers depended significantly on the tomato variety and were slightly affected by the LAB strain used for fermentation (Figure 3). The fermented tomato products of var. Ronaldo had all-*trans*- and *cis*-lycopene contents that were higher on average by 25.9 and 62.6%, respectively, compared to the tomato products of var. Cunero [75].

The control samples of var. Ronaldo had 3.3-fold higher *cis*-lycopene (3.4 mg/kg) compared to var. Cunero (11.3 mg/kg) (Fig. 3). Fermentation by *P. pentosaceus* or *L. sakei* increased the *cis*-lycopene contents on average by

30.6 and 8.5%, respectively in the products of var. Cunero and var. Ronaldo. A lower increase in *cis*-lycopene was noticed during fermentation of the var. Cunero tomatoes with *P. acidilactici* as well as during spontaneous fermentation (an average increase of 9%). Similarly, lactofermentation using *P. acidilactici* and *L. sakei* increased the *cis*-lycopene contents by 5.8% on average in the tomato products [75].

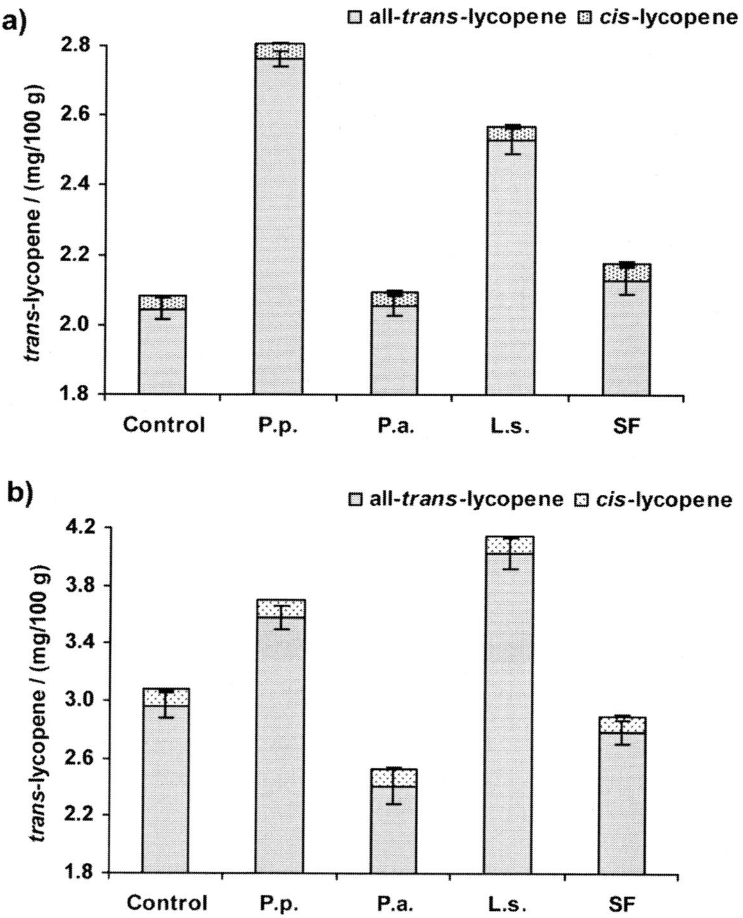


**Figure 3.** Content of all-*trans*- and *cis*-lycopene in fermented tomato of var. Cunero (a) and var. Ronaldo (b) products. Samples: Control – untreated tomato pulp; tomato products fermented with: P.p. – *P. pentosaceus*, P.a. – *P. acidilactici*, L. s. – *L. sakei*; SF – spontaneous fermented.

The fermentation of var. Cunero and var. Ronaldo tomatoes by *P. pentosaceus* and *L. sakei* produced an average of 22.2% more all-*trans*-lycopene compared to the controls (204.6 and 296.0 mg/kg, respectively) (Figure 3) [75].

The *cis/trans* ratio of var. Cunero and var. Ronaldo tomatoes were 1.67 and 3.81, respectively. The highest *cis/trans* ratio was found in the var. Cunero samples fermented by *L. sakei* (2.08), following that of var. Ronaldo samples fermented by *P. acidilactici* (4.90) and spontaneous fermentation (4.09) [75].

It is known from the literature that in human subjects, lycopene from *cis*-isomer-rich tomato sauce is more bioavailable than that from all-*trans*-rich tomato sauce [91]. Because of the positive effect of lactofermentation on the *cis/trans* lycopene ratio, fermented products of the var. Ronaldo tomato, fermented with *P. acidilactici* or *L. sakei*, could be recommended as more biologically accessible products with greater functional value.

## 6.4. Colour characteristics of fermented tomato products

The results from our analysis of the red (a*) and yellow (b*) colour coordinates of fermented tomato products are presented in Table 2. No relation was found in the var. Cunero samples between the yellow colour coordinate (b*) and total carotenoid, lycopene or β-carotene contents ($p > 0.05$) (Table 2). However, the red colour coordinate (a*) slightly correlated ($R^2 = 0.672$) with β-carotene content.

In contrast, a weak relation was noticed between colour coordinate b* of var. Ronaldo and total carotenoid or β-carotene contents ($R^2 = 0.581$ or $R^2 = 0.596$, respectively) (Table 2). In addition, samples of this variety showed a strong relation between colour coordinate b* and lycopene content ($R^2 = 0.825$, $p = 0.03$). No significant relations were observed between a* and β-carotene or lycopene contents ($p > 0.05$) (Table 2) or between total carotenoids and the colour tone ($h^o$) or colour purity (C) values of the var. Cunero and var. Ronaldo samples (Table 3) [75].

The best estimation for β-carotene content was obtained using the b* chromaticity value from the whole fruit measurements or the transformed $a^{*2}$ value from the pure measurements [91]. Neither model, however, could explain more than 55% of the variation in β-carotene levels, suggesting that chromaticity values may not be appropriate for estimating tomato β-carotene content. It has been stated that the inspection of different chromaticity values and regression models suggest that colorimeter readings may not be highly useful for estimating β-carotene content in the tomato fruit [92].

**Table 2. Colour coordinates (a*, b*) of tomato var. Cunero and var. Ronaldo samples and their correlations between total carotenoids, lycopene and β-carotene contents**

| Samples | var. Cunero | | | var. Ronaldo | | |
|---|---|---|---|---|---|---|
| | a* | b* | a*/b* | a* | b* | a*/b* |
| K | 13.97±1.3c | 15.47±0.8a | 0.903 | 13.84±0.9c | 16.71±1.1b | 0.828 |
| P.p. | 14.57±1.1d | 16.46±1.3b | 0.885 | 15.15±0.8e | 18.28±0.9c | 0.829 |
| P.a. | 11.41±0.9a | 17.29±1.3bc | 0.660 | 13.38±1.1b | 19.37±1.3d | 0.691 |
| L. s. | 13.03±1.1b | 15.09±1.3a | 0.864 | 13.85±0.5c | 15.54±0.7a | 0.891 |
| SP | 13.44±1.3b | 19.13±1.4d | 0.703 | 14.26±1.2d | 19.54±1.3d | 0.730 |
| **a; b and a/b correlation with total carotenoid content** | | | | | | |
| R | 0.3905 | 0.04763 | 0.2673 | 0.001697 | 0.5808 | 0.5985 |
| *p* | 0.2597 | 0.7243 | 0.3723 | 0.9476 | 0.1342 | 0.1248 |
| **a; b and a/b correlation with lycopene content** | | | | | | |
| $R^2$ | 0.3186 | 0.02779 | 0.1973 | 0.004398 | 0.8248 | 0.7373 |
| *p* | 0.3215 | 0.7887 | 0.4537 | 0.9156 | 0.0329 | 0.0624 |
| **a; b and a/b correlation with β-carotene content** | | | | | | |
| R | 0.6718 | 0.1955 | 0.6326 | 0.001697 | 0.5808 | 0.5985 |
| *p* | 0.0894 | 0.4560 | 0.1077 | 0.9476 | 0.1342 | 0.1248 |

[i] - The numbers are means followed by standard deviations (n = 3). Means within a column with different superscript letters are significantly different ($p < 0.05$).

[ii] - Samples: control – untreated tomato pulp; tomato products fermented with: P.p. – *P. pentosaceus,* P.a. – *P. acidilactici L.,* s. – *L. sakei;* SF – spontaneous fermented; $R^2$ – correlation coefficient

**Table 3.** Colour tone (h°) and purity (C) of tomato var. Cunero and var. Ronaldo samples and their correlation with total carotenoid contents

| Samples | var. Cunero | | var. Ronaldo | |
|---|---|---|---|---|
| | C | h° | C | h° |
| Control | 22.01±2.3b | 47.03±3.1a | 21.97±2.4b | 46.85±2.4a |
| P.p. | 21.93±1.5b | 48.34±3.2bc | 23.75±1.9c | 50.11±1.9c |
| P.a. | 20.77±1.7a | 56.60±2.7de | 23.44±1.8c | 55.30±2.9d |
| L.s. | 20.04±1.3a | 49.46±2.4c | 20.99±1.6ab | 48.52±1.4b |
| SP | 23.24±2.1c | 55.20±1.7d | 24.35±1.3cd | 53.66±2.3d |
| **Correlation with total caratenoid content** | | | | |
| R | 0.00000565 | 0.2332 | 0.4974 | 0.2043 |
| *p* | 0.9970 | 0.4099 | 0.1834 | 0.4448 |

[i] - The numbers are means followed by standard deviations (n = 3). Means within a column with different superscript letters are significantly different ($p < 0.05$).

[ii] - Samples: control – untreated tomato pulp; tomato products fermented with: P.p. – *P. pentosaceus,* P.a. – *P. acidilactici L.,* s. – *L. sakei;* SF – spontaneous fermented; $R^2$ – correlation coeficient.

The overall results indicate that lycopene content could be measured simply and quite accurately across a wide range of tomato genotypes using chromaticity values taken from fruit puree [91]. In contrast, Liu et al. [93] reported that treating tomatoes with a daily light treatment enhances exocarp lycopene accumulation with minimal effect on the colour. Arias et al. [59] also observed that the b* characteristic was not appropriate for predicting the lycopene content of tomatoes.

According to the obtained results, colour tone (h°) and purity (C) are not suitable indicators of the total carotenoid content in the evaluation of tomato products. We postulate that measuring the yellow coordinate (b*) could be a simple and non-destructive method for predicting lycopene concentration in tomato products [94].

# 7. THE USE OF TOMATO ADDITIVES FERMENTED WITH *PEDIOCOCCUS PENTOSACEUS* KTU05-9 AND *LACTOBACILLUS SAKEI* KTU05-6 TO IMPROVE THE QUALITY OF READY-TO-COOK MINCED MEAT PRODUCTS

The influence of lactic acid fermentation with BLIS-producing lactobacilli (*Pediococcus pentosaceus* KTU05-9, *Lactobacillus sakei* KTU05-6) on the parameters of tomato powder and the impact of fermented tomato products on the acceptability, colour characteristics and carotenoid content of ready-to-cook minced pork meat products (RCMP) have been investigated [95]. In this experiment used tomato powder was obtained from "Obipectin AG" (Bischofszell, Switzerland). The lactic acid bacteria (LAB) P. pentosaceus KTU05-9 and Lactobacillus sakei KTU05-6, previously isolated from spontaneous rye sourdoughs [96] revealed antimicrobial activity against undesirable microorganisms in the food industry by producing organic acids and BLIS [97,98,67] were used for tomato powder fermentation. The LAB were stored at –70 °C and cultured at temperatures of 35 °C (KTU05-9) or 30 °C (KTU05-6) for 48 h in MRS broth (CM0359, Oxoid Ltd, Hampshire, UK) supplemented with 40 mmol $L^{-1}$ fructose and 20 mmol $L^{-1}$ maltose. Solid state fermentation of tomato powder was used [95].

The cell growth results observed at 48 h of fermentation in tomato media are presented in Figure 4. We found the highest amount of LAB in samples treated with *L. sakei* (8.15 $\log_{10}$ CFU $g^{-1}$). The spontaneously fermented samples yielded 6.69 $\log_{10}$ CFU $g^{-1}$ of LAB. The lowest amount of LAB was found in samples fermented with *P. pentosaceus* (4.58 $\log_{10}$ CFU $g^{-1}$).

Different substrates may affect microorganism growth and metabolism [99]. High viable counts are necessary to obtain the desired acid production and pH reduction, which affects the organoleptic properties and shelf-life of the products while preventing contamination. However, the success of fermented products does not rely solely on the ability to provide enough LAB cells; in addition, the consumer must find these organoleptic properties acceptable, which is related in many cases to the organic acid content. We found the lowest pH after 48 h of fermentation in samples fermented with *P. pentosaceus* (pH = 4.1) (Figure 5). Samples fermented with *L. sakei* or through spontaneous fermentation had a pH of 4.16 [95].

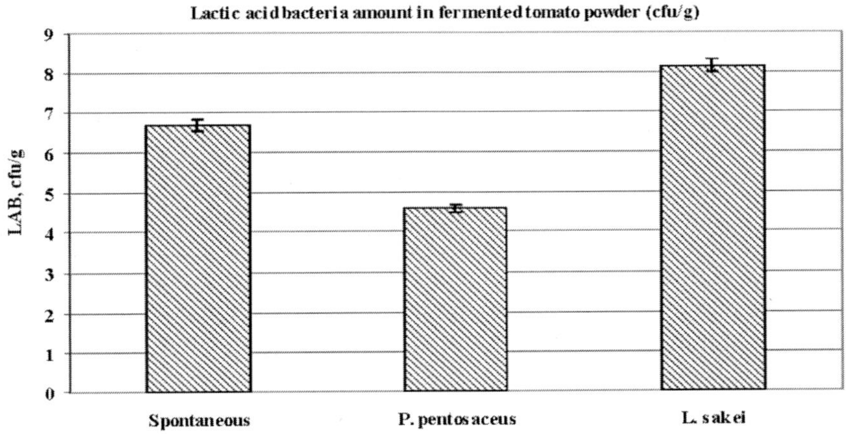


**figure 4.** Lactic acid bacteria (LAB) amount (cfu/g) in fermented tomato powder (Samples: Spontaneous – tomato powder fermented spontaneous; P. pentosaceus - tomato powder fermented with *P. pentosaceus;* L. sakei – tomato powder fermented with *L. sakei*; $p < 0.05$).

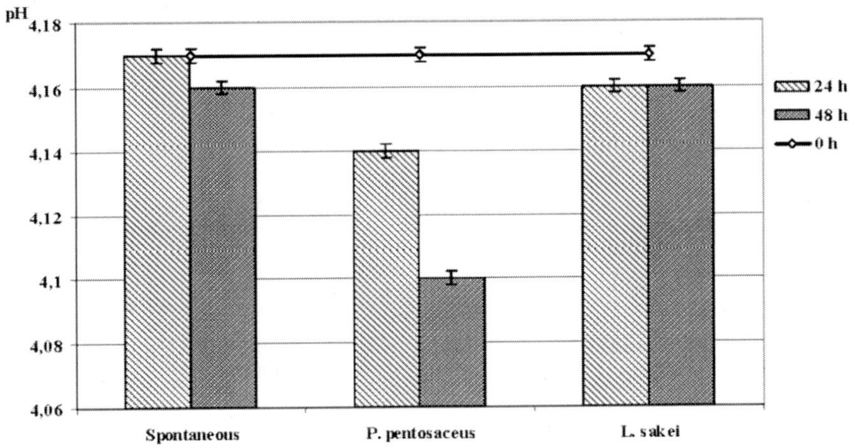


**Figure 5.** pH of fermented tomato powder (Samples: Spontaneous – tomato powder fermented spontaneous; P. pentosaceus - tomato powder fermented with *P. pentosaceus;* L. sakei – tomato powder fermented with *L. sakei*).

## 7.1. Colour parameter relation with carotenoid content in fermented tomato products

By influencing consumer choice and preferences, colour is an important quality attribute in the food and bioprocessing industries. Food colour is governed by the chemical, biochemical, microbial and physical changes that occur during growth, maturation, post-harvest handling and processing. Measuring the colour of food products has been used as an indirect measure of other quality attributes, such as flavour and pigment contents, because it is simple, fast and correlates well with other physicochemical properties [100]. We found that fermentation influenced the colour characteristics and carotenoid content of tomato products (Figure 6) [95]. The highest concentration of carotenoids was found in samples fermented with LAB starters (*P. pentosaceus, L. sakei*). Spontaneous fermentation also increased the content of carotenoids in the tomato samples, but not as effectively (the total carotenoid content in the spontaneously treated samples was 54.78 mg/100 g). A strong and significant relation was found between colour tone (ho) and lycopene content and between colour tone (h°) and total carotenoid content ($R^2 = 0.9045$; $p = 0.0489$ and $R^2 = 0.9035$; $p = 0.0495$, respectively). We found correlations ranging from 0.8922 to 0.5091 between others colour characteristics and β-carotene, lycopene and total carotenoid content, but they were not significant [101].

The beneficial effects of lycopene on health have been reviewed [102-104]. According to our results, fermentation with *L. sakei* and *P. pentosaceus* increases the carotenoid concentration in tomato products by two-fold. We did not study the effects on tomato product fermentation using different LAB starters, and more research is needed to explain the mechanism of increasing carotenoids in fermented tomato products [95].

## 7.2. The influence of fermented tomato additives on the acceptability of ready-to-cook minced meat products (rcmp)

We found significant differences in the acceptability of RCMP with and without 10 or 30 % tomato powder treated with different LAB (*L. sakei* KTU05-06, *P. pentosaceus* KTU05-09) or spontaneous fermentation (Figure 7) [95]. The highest RCMP acceptability was found with 10 % L. sakei fermented tomato powder (an average score of 9.38). Control samples (without additives) were found to be less acceptable (an average score of 5.86) compared to samples with 10 % fermented tomato additives. Ready-to-cook minced pork meat products with 30 % additive were found to be less acceptable than samples with 10 % additive. Compared to samples with 30 % additive, the most acceptable samples were those without fermented tomato products (an average score of 7.72) [95].

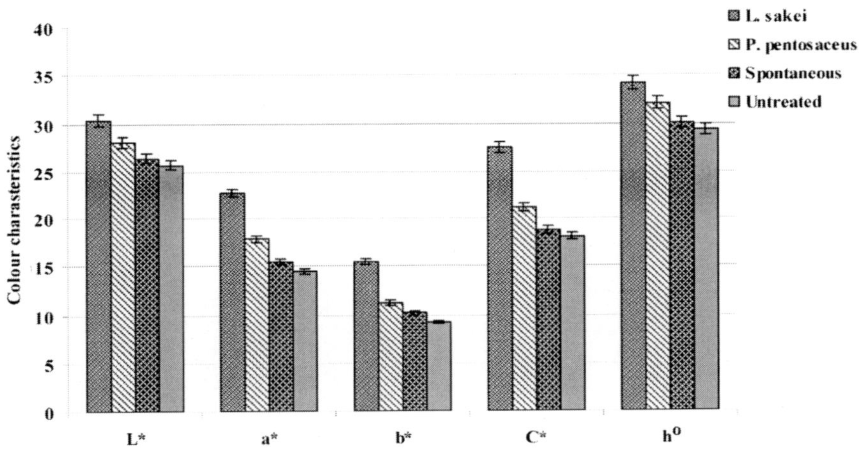


**Figure 6.** Colour characteristics and carotenoids content (mg/100 g) of fermented and untreated tomato products (Samples: Untreated – untreated tomato powder; Spontaneous – tomato powder fermented spontaneous; P. pentosaceus - tomato powder fermented with *P. pentosaceus*; L. sakei – tomato powder fermented with *L. sakei*; $p < 0.05$).

## 7.3. The influence of fermented tomato additives on the colour characteristics of ready-to-cook minced meat products (rcmp), and the influence of carotenoid content on thermal-treated and untreated rcmp

The Joint FAO/WHO Expert Committee on Food Additives (JECFA) previously endorsed the use of lycopene (both natural and synthetic) as a food colour at its eighth, eighteenth, and twenty-first meetings [106-108] but was not able to establish an Acceptable Daily Intake (ADI) due to the limited information available. At its sixty-seventh meeting, JECFA agreed that both synthetic lycopene and lycopene extracted from *Blakeslea trispora* are acceptable as food colours and established a group ADI of 0-0.5 mg/kg bw/day for both preparations [109]. Adding tomato, tomato products or lycopene to meat could lead to products with health benefits. Few studies have been reported regarding the use of tomato products or lycopene in meat products. Candogan [110] reported on the use of tomato paste in beef patties, while Deda, Bloukas, and Fista [111] investigated its use in frankfurters. Calvo et al. [112] reported on the use of lycopene from tomato peel in dry fermented sausages. However, we could not find data on tomato product fermentation with different LAB starters or how fermented tomato products influence RCMP quality parameters [95].

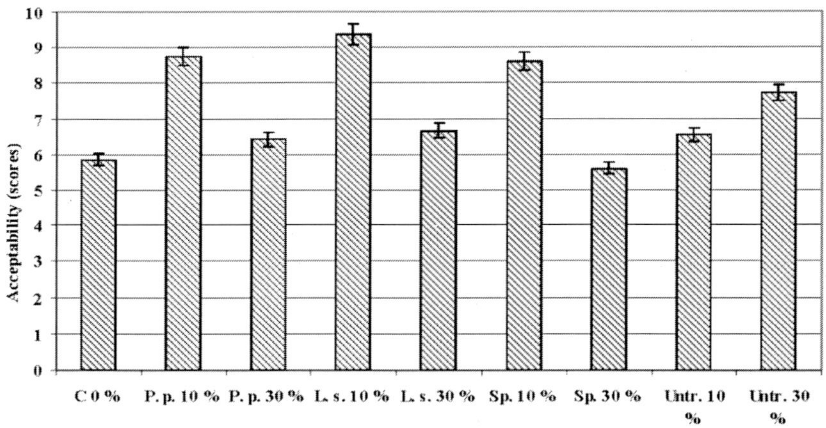


**Figure 7.** Acceptability of ready-to- cook minced meat products (RCMP) (Samples: C0 % - RCMP without tomato products; P.p. 10 % - RCMP with 10 % with P. pentosaceus fermented tomato products; P.p. 30 % - RCMP with 30 % with P. pentosaceus fermented tomato products; L.s. 10 % - RCMP with 10 % with L. sakei fermented tomato products; L.s. 30 % - RCMP with 30 % with L. sakei fermented tomato products; Sp. 10 % - RCMP with 10 % spontaneous fermented tomato products; Sp. 30 % - RCMP with 30 % spontaneous fermented tomato products; Untr. 10 % - RCMP with 10 % untreated tomato powder; Untr. 30 % - RCMP with 30 % untreated tomato powder; $p > 0.05$)

We found that the addition of tomato products significantly affected ($p < 0.05$) all colour parameters (Table 4) of the final product (thermal-treated and untreated). The controls had the highest ($p < 0.05$) lightness and the lowest ($p < 0.05$) redness and yellowness as a consequence of lower hue angle and saturation index. These tendencies were found for both thermal-treated and untreated products [95].

**Table 4. Colour coordinates (a*, b*), L* - lightness, colour tone (ho) and purity (C) of thermal treated (10 min in 100 °C temperature water) and untreated ready- to-cook minced meat products (RCMP)**

| RCMP samples | L* | a* | b* | C | h° |
|---|---|---|---|---|---|
| **Thermal treated** | | | | | |
| P. p. 30 % | 45.64±0.11$^{a}$ | 19.09±0.21$^{a}$ | 28.16±0.69$^{a}$ | 34.02±0.71$^{a}$ | 55.87±0.96$^{a}$ |
| P. p. 10 % | 52.76±0.18 c | 12.8±0.81$^{b}$ | 23.05±0.52$^{b}$ | 29.02±0.32$^{b}$ | 63.83±0.78$^{b}$ |
| L. s. 30 % | 47.96±0.20 c | 19.27±0.93 a | 29.93±0.71$^{a}$ | 35.60±0.23$^{a}$ | 57.22±0.63 $^{a}$ |
| L. s. 10 % | 48.45±0.19 c | 9.31±0.74$^{c}$ | 21.89±0.64$^{b}$ | 23.79±0.18$^{c}$ | 66.96±0.88$^{b}$ |
| Sp. 30 % | 44.54±0.23 a | 12.01±0.63$^{b}$ | 20.33±0.55$^{b}$ | 19.02±0.13$^{c}$ | 54.32±0.49$^{a}$ |
| Sp. 10 % | 43.51±0.11 a | 8.21±0.32$^{b}$ | 17.75±0.39$^{c}$ | 16.57±0.42$^{d}$ | 52.65±0.55$^{a}$ |
| Untr. 30 % | 42.10±0.25 a | 7.56±0.41$^{c}$ | 14.24±0.44$^{c}$ | 14.99±0.54$^{d}$ | 53.66±0.86$^{a}$ |
| Untr. 10 % | 42.14±0.17 b | 4.32±0.37$^{d}$ | 12.32±0.60$^{d}$ | 15.01±0.30$^{d}$ | 52.75±0.97$^{a}$ |
| C 0% | 60.07±0.43 d | 2.41±0.30$^{d}$ | 11.84±0.22$^{d}$ | 12.99±0.21$^{e}$ | 82.71±0.60$^{d}$ |
| **Thermal untreated** | | | | | |
| P. p. 30 % | 46.03±0.25$^{a}$ | 17.27±0.51$^{b}$ | 26.90±0.74$^{a}$ | 31.97±0.65$^{a}$ | 57.30±0.40$^{d}$ |
| P. p. 10 % | 50.64±0.19$^{b}$ | 12.90±0.72$^{a}$ | 28.96±0.83$^{a}$ | 31.70±0.82$^{a}$ | 65.99±0.93$^{c}$ |
| L. s. 30 % | 45.63±0.63$^{a}$ | 18.08±0.61$^{d}$ | 27.60±0.52$^{a}$ | 32.99±0.74$^{b}$ | 56.77±0.68$^{b}$ |

| | | | | | |
|---|---|---|---|---|---|
| L. s. 10 % | 50.9±0.79$^{a}$ | 12.14±0.21$^{a}$ | 28.50±0.48$^{a}$ | 30.98±0.65$^{a}$ | 66.93±0.88$^{c}$ |
| Sp. 30 % | 47.23±0.28$^{a}$ | 15.23±0.34 a | 25.36±0.39$^{b}$ | 29.59±0.54$^{c}$ | 62.39±0.45$^{a}$ |
| Sp. 10 % | 45.21±0.41$^{a}$ | 14.02±0.19 a | 21.45±0.41$^{d}$ | 28.96±0.91$^{c}$ | 61.33±0.46$^{a}$ |
| Untr. 30 % | 46.27±0.52$^{a}$ | 13.25±0.16$^{a}$ | 21.55±0.57$^{d}$ | 28.69±0.58$^{c}$ | 61.45±0.41$^{a}$ |
| Untr. 10 % | 44.25±0.39$^{a}$ | 11.03±0.11$^{c}$ | 19.56±0.59$^{d}$ | 26.98±0.62$^{d}$ | 60.84±0.93$^{a}$ |
| C 0% | 58.38±0.48$^{c}$ | 3.93±0.18$^{e}$ | 15.33±0.78$^{e}$ | 25.63±0.61$^{d}$ | 81.18±0.54$^{e}$ |

[i] - Samples: C 0 % - RCMP without tomato products; P.p. 10 % - RCMP with 10 % with *P. pentosaceus* fermented tomato products; P.p. 30 % - RCMP with 30 % with *P. pentosaceus* fermented tomato products; L.s. 10 % - RCMP with 10 % with *L. sakei* fermented tomato products; *L.s.* 30 % - RCMP with 30 % with *L. sakei* fermented tomato products; Sp. 10 % - RCMP with 10 % spontaneous fermented tomato products; Sp. 30 % - RCMP with 30 % spontaneous fermented tomato products; Untr. 10 % - RCMP with 10 % untreated tomato powder; Untr. 30 % - RCMP with 30 % untreated tomato powder.

[ii] - Means in column with common letter are not different ($p > 0.05$).

High variation in the colour parameters of fermented meat products has been reported [113-115]. These variations could be due to the calibration plate used in the determinations, the composition of the meat products, the size of the meat particles and the ripening time.

Furthermore, the addition of tomato products affects the carotenoid content of RCMP (Table 5) [95]. We found that thermal treatment decreases the carotenoid concentration in RCMP. After thermal treatment, we found 23.71 and 52.03 % less ß-carotene (in samples with 10 % spontaneously treated products and in samples with 30 % L. sakei-fermented tomato products, respectively). Additionally, 10.78 and 50.00 % less lycopene was found in samples with 30 % spontaneously treated products and in samples with 30 % untreated tomato products, respectively, and as a consequence, the highest loss of total carotenoid content was found in samples with 30 % untreated tomato product (49.25 %) [95].

**Table 5.** ß–carotene, lycopene and total carotenoids content in thermal treated and untreated *ready- to-cook minced meat products (RCMP)*

| Samples | ß–carotene | Lycopene | Total carotenoids content |
|---|---|---|---|
| | mg/100 g | | |
| **Thermal untreated** | | | |
| Untr. 10% | 0.19±0.02[a] | 0.48±0.06[a] | 0.67 [a] |
| Untr. 30% | 0.25±0.01[b] | 1.46±0.10[b] | 1.71[b] |
| Sp. 10% | 0.69±0.02[c] | 1.02±0.09[b] | 1.71[b] |
| Sp. 30% | 0.97±0.02[c] | 2.13±0.15[c] | 3.1[c] |
| L. s. 10% | 1.23±0.09[d] | 3.59±0.21[d] | 4.82[c] |
| L. s. 30% | 1.95±0.08[e] | 9.66±0.17 [e] | 11.61[e] |
| P. p. 10% | 1.01±0.07[d] | 3.67±0.21[d] | 4.68[d] |
| P. p. 30% | 1.76±0.03[e] | 10.32±0.11[e] | 12.08[e] |
| **Thermal treated** | | | |
| Untr. 10% | 0.10±0.02[a] | 0.24±0.07[a] | 0.34[a] |
| Untr. 30% | 0.15±0.04[a] | 0.93±0.05[b] | 1.08[b] |
| Sp. 10% | 0.49±0.07[b] | 0.91±0.06[b] | 1.40[b] |
| Sp. 30% | 0.74±0.04[c] | 1.75±0.09[c] | 2.49[c] |

| | | | |
|---|---|---|---|
| L. s. 10% | 0.59±0.06b | 2.15±0.13c | 2.74c |
| L. s. 30% | 1.47±0.08d | 7.32±0.14d | 8.79d |
| P. p. 10% | 0.76±0.06c | 2.62±0.20c | 3.38c |
| P. p. 30% | 1.17±0.10d | 6.31±0.26d | 7.48d |

[i] - Samples: C0 % - RCMP without tomato products; P.p. 10 % - RCMP with 10 % with *P. pentosaceus* fermented tomato products; P.p. 30 % - RCMP with 30 % with *P. pentosaceus* fermented tomato products; L.s. 10 % - RCMP with 10 % with *L. sakei* fermented tomato products; *L.s.* 30 % - RCMP with 30 % with *L. sakei* fermented tomato products; Sp. 10 % - RCMP with 10 % spontaneous fermented tomato products; Sp. 30 % - RCMP with 30 % spontaneous fermented tomato products; Untr. 10 % - RCMP with 10 % untreated tomato powder; Untr. 30 % - RCMP with 30 % untreated tomato powder.

[ii] - Means in column with common letter are not different ($p > 0.05$).

# 8. CONCLUSIONS

Lactic acid fermentation represents the easiest and most suitable way to increase daily consumption of nearly fresh fruits and vegetables. The health and safety of the products can be aided by the development of starter cultures. Progress in the field of antimicrobial LAB strains with multi-functional properties, including the degradation of mycotoxins, can be engineered to significantly improve the quality, safety and acceptability of plant foods.

Tomato processing resulted in several important changes in carotenoid concentration and lycopene isomer profile. Treatment with LAB breaks down the tomato cell matrix and makes carotenoids more available, yielding a higher level of total carotenoids. Moreover, tomatoes subjected to lactic acid fermentation results in high lycopene bioavailability accompanied by increased *cis*-lycopene content. According to our results, *P. pentosaceus* and *L. sakei* may be useful for the preservation of tomatoes. Such products could be recommended as being more biologically accessible with higher functional value.

The results of our tomato product colour analysis offered the possibility of evaluating the level of lycopene using the yellow colour characteristic of tomato products; this method was reproducible and accurate enough to substitute for the chemical extraction determinations and may be a useful tool for the tomato industry.

The direct use of *Pediococcus pentosaceus* KTU05-9 and *Lactobacillus sakei* KTU05-6 for tomato product fermentation increases the carotenoid content in tomato products, which is a beneficial additive that improves the colour, functional value and acceptability of ready-to-cook minced meat products. Ready-to-cook minced meat products that have been enriched with carotenoids, which lend good sensory quality and are produced to contain a high level of lycopene and ß-carotene, can increase the intake of carotenoids in the diet. This is the first time that selected lactobacilli-fermented tomato products have been used as source of lycopene and ß-carotene for food, and more research is needed to explain the mechanism of carotenoid increase in fermented tomato products.

All the tested LAB produced a mixture of L- and D-lactic acid, with the latter isomer at a lower level. Because of the potential toxicity of D-lactic acid in food, we report that the tomato products prepared using pure cultures of tested LAB were found in all cases to be safer than those treated by spontaneous fermentation.

# REFERENCES

1. Klaenhammer T.R., Kullen M.J. Selection and design of probiotics. International Journal of Food Microbiology 1999;50 45-57.
2. The World Health Organization (WHO) (www.who.int) (accessed 3 September 2014).
3. Zia-ur-Rehman Z., Islam M., Shah W.H. Effect of microwave and conventional cooking on insoluble dietary fibre components of vegetables. Food Chemistry 2003;80 237-240.
4. Zhang D.L., Hamauzu Y. Phenolics, ascorbic acid, carotenoids and antioxidant activity of broccoli and their changes during conventional and microwave cooking. Food Chemistry 2004;88 503- 509.
5. Devlieghere F., Vermeiren L., Debevere J. New preservation technologies: possibilities and limitations. International Dairy Journal 2004;14 273-285.
6. Gómez-López V.M., Devlieghere F., Bonduelle V., Debevere J. Intense light pulses decontamination of minimally processed vegetables and their shelf-life. International Journal of Food Microbiology 2005;103 79-89.
7. Elmnasser N., Guillou S., Leroi F., Orange N., Bakhrouf A., Federighi M. Pulsed-light system as a novel food decontamination technology: a review. Canadian Journal of Microbiology 2007;58 813-821.
8. Liu S., Han Y., Zhou Z. Lactic acid bacteria in traditional fermented Chinese foods. Food Research International 2011;44 643-651.
9. Juodeikiene G., Bartkiene E., Viskelis P., Urbonaviciene D., Eidukonyte D., Bobinas C. Advances in Applied Biotechnology. In: Petre M. (ed.) Fermentation Processes Using Bacteriocins Producing Lactic Acid Bacteria for Biopreservation and Improving Functional Properties of Food Products. Rijeka: InTech; 2011. p63-100.
10. Sanchez B., Ruiz L., Gueimonde M., Ruas-Madiedo P., Margolles A. Toward improving technological and functional properties of probiotics in foods. Trends in Food Science and Technology 2012;26 56-63.

11. Clark J.P. Fermentation - new products from an old process. Food Technology 2004;58 75-76.
12. Bourdichon F., Casaregola S., Farrokh C., Frisvad J.C., Gerds M.L., Hammes W.P., Harnett J., Huys G., Laulund S., Ouwehand A., Powell I.B., Prajapati J.B., Seto Y., Ter Schure E., Van Boven A., Vankerckhoven V., Zgoda A., Tuijtelaars S., Hansen E.B. Food fermentations: microorganisms with technological beneficial use. International Journal of Food Microbiology 2012;154 87-97.
13. Caplice E., Fizgerald G. F. Food fermentations: Role of microorganisms in food production and preservation. International Journal of Food Microbiology 1999;50(1-2) 131-149.
14. Ross R. P., Preservation and fermentation: past, present and future. International Journal of Food Microbiology 2002;79 3-16.
15. Gaggia F., Di Gioia D., Baffoni L., Biavati B. The role of protective and probiotic cultures in food and feed and their impact on food safety. Trends in Food Science and Technology 2011;22 58-66.
16. Adams M., Mitchell R. Fermentation and pathogen control: a risk assessment approach. International Journal of Food Microbiology 2002;79 75-83
17. Adams M.R., Nicolaides L. Review of the sensitivity of different foodborne pathogens to fermentation. Food Control 2008;8 227-239.
18. Hammes W.P., Tichaczek P.S. The potential of lactic acid bacteria for the production of safe and wholesome food. Zeitschrift für Lebensmittel-Untersuchung und -Forschung 1994;198 193-201.
19. van Boekel M., Fogliano V., Pellegrini N., Stanton C., Scholz G., Lalljie S., Somoza V., Knorr D., Jasti P.R., Eisenbrand G. A review on the beneficial aspects of food processing. Molecular Nutrition and Food Research 2010;54 1215-1247.
20. Poutanen K., Flander L., Katina K. Sourdough and cereal fermentation in an nutritional perspective. Food Microbiology 2009;7 693-699.
21. Marilley L., Casey M.G. Flavors of cheese products: metabolic pathways, analytical tools and identification of producing strains. International Journal of Food Microbiology 2004;90 139-159.
22. Smit G., Smit B.A., Engels W.J. Flavor formation by lactic acid bacteria and biochemical flavor profiling of cheese products. FEMS Microbiology Reviews 2005;29 591-610.
23. Lacroix N., St Gelais D., Champagne C.P., Fortin J., Vuillemard J.C. Characterization of aromatic properties of old-style cheese starters. Journal of Dairy Science 2010;93 3427-3441.
24. Sicard D., Legras J.L. Bread, beer and wine: yeast domestication in the Saccharomyces sensu stricto complex. Contes rendus biologies 2011;334(3) 229-236.
25. Steinkraus K.H. Introduction to indigenous fermented foods. In: Steinkraus K. H. (ed.) Handbook of indigenous fermented foods. New York: Marcel Dekker; 1996. p1-6.
26. Cleveland J., Thomas J.M., Ingolf F.N., Michael L.C. Bacteriocins: safe, natural antimicrobials for food preservation. International Journal of Food Microbiology 2001;71 1-20.
27. Silva J., Carvalho A. S., Teixeira P., Gibbs P. A. Bacteriocin production by spray-dried lactic acid bacteria. Letters in Applied Microbiology 2002;34(2) 77-81.
28. Zotta T., Parente E., Ricciardi A. Viability staining and detection of metabolic activity of sourdough lactic acid bacteria under stress conditions. World Journal of Microbiology and Biotechnology 2009;25(6) 1119-1124.
29. Corsetti A., Settanni L., Van Sinderen D. Characterization of bacteriocin-like inhibitory substances (BLIS) from sourdough lactic acid bacteria and evaluation of their in vitro and in situ activity. Journal of Applied Microbiology 2004;96(3) 521-534.
30. Rodriguez S.J., Alexopoulos L.G., Epperlein J., Samaga R., Lauffenburger D.A., Klamt S., Sorger P.K. Discrete logic modelling as a means to link protein signalling networks with functional analysis of mammalian signal transduction. Molecular System Biology 2009;5 1-19.
31. Fuller R. Probiotics in man and animals. Journal of Applied Bacteriology 1989;66 365-378.

32. Vaughan A., Eijsink V.G.H., O'Sullivan T.F., O'Hanlon K., van Sinderen D. An analysis of bacteriocins produced by lactic acid bacteria isolated from malted barley. Journal of Applied Microbiology 2001;91 131-138.
33. Altuntas E.G., Cosansu S., Ayhan K. Some growth parameters and antimicrobial activity of a bacteriocin-producing strain Pediococcus acidilactici 13. International Journal of Food Microbiology 2010;141 28-31.
34. Juodeikiene G., Salomskiene J., Basinskiene L., Vidmantiene D., Narbutaite V., Kasnauskyte N. The influence of novel fermented products on wheat bread spoilage and staling. Food Chemistry and Technology 2009;43 36-46.
35. Karovicova J., Kohajdova Z. Lactic acid fermented vegetable juices. Horticultural Science (Prague). 2003;30(4) 152-158.
36. Kalantzopoulos G. Fermented products with probiotic qualities. Anaerobe. 1997;3 185-190.
37. Buckenhüskes H. J. Fermented vegetables. In: Doyle P.D., Beuchat L.R., Montville T.J. (ed.) Food Microbiology: Fundamentals and Frontiers. Washington: ASM Press; 1997. p595-609.
38. Bartkiene E., Juodeikiene G., Vidmantiene D., Viskelis P., Urbonaviciene D. Nutritional and quality aspects of wheat sourdough bread using L. luteusand L. angustifolius flours fermented by Pedioccocus acidilactici. International Journal of Food Science and Technology 2011;46 1724-1733.
39. Karocicova J., Drdak M., Greif G., Hybenova E. The choice of strains of Lactobacillusspecies for the lactic acid fermentation of vegetable juices. European Food Research and Technology. 1999;210 53-56.
40. Buckenhuskes H.J. Selection criteria for lactic acid bacteria to be used as starter cultures for various food commodities. FEMS Microbiological Reviews 1993;12 253-272.
41. Maki M. Lactic acid bacteria in vegetables fermentation. In: Salminen S., Von Wright A., Ouwehand A. (ed.) Lactic acid bacteria microbiological and functional aspects. New York: Marcel Dekker; 2004. p419-430.
42. Jazwiak K.P., Rozmierska J., Chablowska B., Stecka K.M., Skapska S., Kliszcz M., Rzeszowiak E.S. Starter cultures for lactic acid fermentation of sweet pepper, pattypan squash and tomatoes. Polish Journal of Food and Nutrition Science. 2013; 63(2) 95-102.
43. Peres C. M., Peres C., Hernandez-Mendoza A., Malcata F. X. Review on fermented plant materials as carriers and sources of potentially probiotic lactic acid bacteria – With an emphasis on table olives. Trends in Food Science and Technology 2012;26 31-42.
44. Cetin B. Production of probiotic mixed pickles (Tursu) and microbiological properties. African Journal of Biotechnology 2011;10 14926-14931.
45. Montet D., Loisea, G., Kakhia-Rozis N. Microbial technology of fermented vegetables. In: Ray R.C., Ward O.P. (ed.) Microbial Biotechnology in Horticulture. New Hampshire: Science Publishers; 2006; p309-343.
46. Di Cagno R., Surico R.F., Paradiso A., De Angelis M., Salmon J.-C., Buchin S., De Gara L., Gobbetti M. Effect of autochthonous lactic acid bacteria starters on health-promoting and sensory properties of tomato juices. International Journal of Food Microbiology 2008;128 473-483.
47. Hotz C., Gibson R.S. Traditional food-processing and preparation pactices to enhance the bioavailability of micronutrients in plant-based diets. Journal of Nutrition 2007;137 1097-1100.
48. Ranadheera R.D.C.S., Baines S.K., Adams M.C. Importance of food in probiotic efficacy. Food Research International 2010;43 1-7.
49. Yoon K.Y., Woodams E.E., Hang Y.D. Probiotication of tomato juice by lactic acid bacteria. Journal of Microbiology 2004;42 315-318.
50. Parrado J., Miramontes E., Jover M., Gutierrez J.F., Collantes de T.L., Bautista J. Preparation of a rice bran enzymatic extract with potential use as functional food. Food Chemistry 2006;98 742-748.
51. Kim H.Y., Min J.H., Lee J.H., Ji G.E. Growth of lactic acid bacteria and Bifidobacteria in natural media using vegetables, seaweeds, grains and potatoes. Food Science Biotechnology, 2000;9: 322-324.

52. Rao A.V., Agarwal S. Role of lycopene as antioxidant carotenoid in the prevention of chronic diseases: A review. Nutritional Research 1999;99 305-323.
53. Giovannucci E. Tomato products, lycopene and prostate cancer: a review of the epidemiological literature. The Journal of Nutrition 2005;135 2030-2315.
54. Talvas J., Caris-Veyrat C., Guy L., Rambeau M., Bernard L., Minet-Quinard R., Lobaccaro J.-M.A., Vasson M.-P., George S., Mazur A., Rock E. Differential effects of lycopene consumed in tomato paste and lycopene in the form of a purified extract on target genes of cancer prostatic cells. The American Journal of Clinical Nutrition 2010;91 1716-1724.
55. Van Breemen R.B., Sharifi R., Viana M., Pajkovic N., Zhu D., Yuan L., Yang Y., Bowen P.E., Stacewicz-Sapuntzakis M. Antioxidant effects of lycopene in African American men with prostate cancer or benign prostate hyperplasia: a randomized, controlled trial. Cancer Prevention Research (Phila.) 2011;4 711-418.
56. Gardner N.J., Savard T., Obermeier P., Caldwell G., Champagne C.P. Selection and characterization of mixed starter cultures for lactic acid fermentation of carrot, cabbage, beet and onion vegetable mixtures. International Journal of Food Microbiology 2001;64 261-275.
57. Richelle M., Bortlik K., Liardet S., Hager C., Lambelet P., Baur M., Applegate L.A., Offord E.A. A food-based formulation provides lycopene with the same bioavailability to humans as that from tomato paste. Journal of Nutrition 2002;132 404-408.
58. Harris W.M., Spurr A.R. Chromoplasts of tomato fruits. II. The red tomato, American Journal of Botanic 1969;56 380-389.
59. Arias R., Lee T.C., Logendra L., Janes H. Correlation of lycopene measured by HPLC with the L*, a*, b* color readings of a hydroponic tomato and the relationship of maturity with color and lycopene content. Journal of Agriculture and Food Chemistry 2000;48 1697-1702.
60. Raffo A., Leonardi C., Fogliano V., Ambrosino P., Salucci M., Gennaro L. Nutritional value of cherry tomatoes (Lycopersicon esculentum Cv. Naomi F1) harvested at different ripening stages. Journal of Agriculture and Food Chemistry 2002;50 6550-6556.
61. Heredia A., Peinado I., Barrera C., Grau A.A. Influence of process variables on colour changes, carotenoids retention and cellular tissue alteration of cherry tomato during osmotic dehydration. Journal of Food Composition and Analysis 2009;22 285-294.
62. Melendez-Martinez A.J., Escudero-Gilete M.L., Vicario I.M., Heredia F.J. Study of the influence of carotenoid structure and individual carotenoids in the qualitative and quantitative attributes of orange juice colour. Food Research International 2010;43 1289-1296.
63. Aymerich T., Picouet P. A., Monfort J. M. Decontamination technologies for meat products, Meat Science 2008;7 114-129.
64. Zhou G. H., Xu X. L., Liu Y. Preservation technologies for fresh meat – A review. Meat Science 2010;86 119-128.
65. Stiles M. E., Hastings J. W. Bacteriocin production by lactic acid bacteria: potential for use in meat preservation Trends in Food Science and Technology 1991;2 247-251.
66. Klaenhammer T. R. Genetics of bacteriocins produced by lactic acid bacteria. FEMS Microbiology Reviews 1993;12 39-86.
67. Cizeikiene D., Juodeikiene G., Paskevicius A., Bartkiene E. Antimicrobial activity of lactic acid bacteria against pathogenic and spoilage microorganism isolated from food and their control in wheat bread. Food Control 2013;31 539-545.
68. Ross R. P., Morgan S., Hill C. Preservation and fermentation: past, present and future. International Journal of Food Microbiology 2002;79 3-16.
69. El-Ghaish S., Ahmadova A., Hadji-Sfaxi I., El Mecherfi K.E., Bazukyan I., Choiseta Y., Rabesona H., Sitohy M., Popov Y.G., Kuliev A.A., Mozzig F., Chobert J.-M., Haertle T. Potential use of lactic acid bacteria for reduction of allergenicity and for longer conservation of fermented foods. Trends Food Sciences Technology 2011(22) 509-516.
70. Peres C.M., Peres C., Hernández-Mendoza A., Malcata F.X. Show more Review on fermented plant materials as carriers and sources of potentially probiotic lactic acid bacteria – With an emphasis on table olives. Trends in Food Science and Technology 2012;26 31-42.

71. Gebbers J. O. Atherosclerosis, cholesterol, nutrition, and statins – a critical review. German Medical Science 2007;5 1-11.
72. Di Cagno R., Coda R., De Angelis M., Gobbetti M. Exploitation of vegetables and fruits through lactic acid fermentation. Food Microbiology 2013;33 1-10.
73. Capanoglu E., Beekwilder J., Boyacioglu D., De Vos R.C., Hall R.D. The Effect of Industrial Food Processing on Potentially Health Beneficial Tomato Antioxidants. Critical Reviews in Food Science and Nutrition 2010;50 919-931.
74. Palozza P., Parrone N., Catalano A., Simone R. Tomato Lycopene and Inflammatory Cascade: Basic Interactions and Clinical Implications. Current Medicinal Chemistry 2010;17 2547-2563.
75. Bartkiene E., Vidmantiene D., Juodeikiene G., Viskelis P., Urbonaviciene D. Lactic acid fermentation of tomato: effects on cis/trans lycopene isomer and β-carotene concentration and formation of L(+) and D(-)-lactic acid. Food Technology and Biotechnology 2013;51(4) 471-478.
76. Digaitiene A., Hansen A.S., Juodeikiene G., Eidukonyte D., Josephsen J. Lactic acid bacteria isolated from rye sourdouhs produce bacteriocin-like inhibitory substances active against Bacillus subtilis and fungy. Journal of Applied Microbiology. 2012;112 732-742.
77. De Lima C.J.B., Coelho L.F., Blanco K.C., Contiero J. Response surface optimization of D(-)-lactic acid production by Lactobacillus SMI8 using corn steep liquor and yeast autolysate as an alternative nitrogen source. African Journal of Biotechnology 2009;8 5842-5846.
78. McGuire R.G. Reporting of objective color measurements. Horticulture Science 1992;27 1254-1255.
79. Kedia G., Vázquez J.A., Pandiella S.S. Fermentability of whole oat flour, PeriTec flour and bran by Lactobacillus plantarum. Journal of Food Enginiering 2008;89 246-249.
80. Shah N.P. Functional Foods from probiotics and prebiotics. Food Technology 2001;55 46-53.
81. Trachoo N., Boudreaux C., Moongngarm A., Samappito S., Gaensakoo R. Effect of germinated rough rice media on growth of selected probiotic bacteria. Pakistan Journal of Biological Science 2006;9 2657-2661.
82. Steinkraus K.H. Classification of fermented foods: worldwide review of household fermentation techniques. Food Control 1997;8 311-317.
83. Sindhu S.C., Khetarpaul N. Probiotic fermentation of indigenous food mixture: effect on antinutrients and digestibility of starch and protein. Journal of Food Composition and Analysis 2001;14 601-609.
84. Gartner C., Stahl W., Sies H. Lycopene is more bioavailable from tomato paste than from fresh tomatoes American Journal of Clinical Nutrition 1997;66 116-22.
85. Hartmann M.H. Biopolymers from Renewable Resources. In: Kaplan D.L. (ed.) Berlin: Springer. 1998; p367-411.
86. Li Y.B., Cui F. Sustainable Biotechnology: Sources of Renewable Energy. In: Singh O., Harvey S. (ed.) Dordrecht: Springer. 2010; p211-228.
87. Lee C.W. Production of D-lactic acid by bacterial fermentation of rice. Fiber and Polymers 2007;8 571-578.
88. Nguyen M.L., Schwartz S.J. Lycopene: chemical and biological properties. Food Technology 1999;53(2) 38-45.
89. Camara M., Matallana M.C., Sánchez-Marta M.C., Liko Ayne R., Labra E. Lycopene and hydroxymethylfurfural (HMF) evaluation in tomato products. Acta Horticulturale 2003;613 365-371.
90. Gould W.A. Quality evaluation of processed tomato juice. Journal of Agricultural and Food Chemistry 1978;26 1006-1011.
91. Unlu N.Z., Bohn T, Francis D.M., Nagaraja H.N., Clinton S.K., Schwartz S.J. Lycopene from heat-induced cis-isomer-rich tomato sauce is more bioavailable than from all-trans-rich tomato sauce in human subjects. British Journal of Nutrition 2007;98 140-146.
92. Hyman R.J., Gaus J., Foolad R.M. A rapid and accurate method for estimating tomato lycopene content by measuring chromaticity values of fruit puree. Journal of the American Society for Horticultural Science 2004;129(5) 717-723.

93. Liu L.H., Zabaras D., Bennett L.E., Aguas P., Woonton B.W. Effects of UV-C, red light and sun light on the carotenoid content and physical qualities of tomatoes during post-harvest storage. Food Chemistry 2009;115 495-500.
94. Fernandez-Ruiz V., Torrecilla J.S., Camara M., Mata M.C., Shoemaker C. Radial basis network analysis of color parameters to estimate lycopene content on tomato fruits. Talanta 2010;83 (2010) 9-
95. In press: Bartkiene E., Juodeikiene G., Vidmantiene D., Viskelis P., Urbonaviciene D. The use of fermented with Pediococcus pentosaceus KTU05-9 and Lactobacillus sakei KTU05-6 tomato additives for ready-to- cook minced meat products quality improving. 2015.
96. Parrado J., Miramontes E., Jover M., Gutierrez J.F., Collantes de Teran L., Bautista J. Preparation of a rice bran enzymatic extract with potential use as functional food. Food Chemistry 2006;98 742-748.
97. Yun J.S., Wee Y.J., Kim J.N., Ryu H.W. Fermentative production of dl-lactic acid from amylase-treated rice and wheat brans hydrolyzate by a novel lactic acid bacterium Lactobacillus spp., Biotechnol. Lett. 2004;26 1613-1616.
98. Digaitiene A., Hansen A.S., Juodeikiene G., Eidukonyte D., Josephsen J. Lactic acid bacteria isolated from rye sourdoughs produce bacteriocin-like inhibitory substances active against Bacillus subtilis and fungi. Journal of Applied Microbiology 2012;112 732-742.
99. Rathore S., Rathore S., Salmeron I., Pandiella S.S. Production of potentially probiotic beverages using single and mixed cereal substrates fermented with lactic acid bacteria cultures. Food Microbiology 2012;30 239-244.
100. Pathare P. B., Opara U. L., Al-Julanda Al-Said F. Colour Measurement and Analysis in Fresh and Processed Foods: A Review. Food and Bioprocess Technology 2013;6 36-60.
101. Urbonaviciene D., Viskelis P., Viskelis J., Jankauskiene J., Bobinas C. Lycopene and β-carotene in non-blanched and blanched tomatoes. Journal of Food, Agriculture and Environment 2012;10 (2): 142-146.
102. Choski P. M., Joshi V. Y. A review on lycopene – Extraction, purification, stability and applications. International Journal of Food Properties 2007;10 289-298.
103. Kavanaugh C. J., Trumbo P. R., Ellwood K. C. The US food and drug administration's evidence-based review for qualified health claims: Tomatoes, lycopene and cancer. Journal of the National Cancer Institute 2007;99 1074-1085.
104. Khachik F. Carvalho L., Bernstein P.S., Muir G.J., Zhao D.Y., Katz N.B. Chemistry, distribution and metabolism of tomato carotenoids and their impact in human health. Experimental Biological and Medicine 2002;227 845-851.
105. Agarwal S., Rao A.V. Tomatoes, lycopene and human health and chronic disease. CMAJ 2000;163(6) 739-44.
106. FAO/WHO, 1965. Specification for the identity and purity of food additives and their toxicological evaluation: food colours and some antimicrobials and antioxidants (Eighth report on the Joint FAO/WHO Expert Committee on Food Additives). Geneva, World Health Organization (WHO Technical Report Series, No. 309).
107. FAO/WHO, 1975. Evaluation of certain food additives (Eighteenth report of the Joint FAO/WHO Expert Committee on Food Additives). Geneva, World Health Organization (WHO Technical Report Series, No. 557).
108. FAO/WHO, 1978. Evaluation of certain food additives (Twenty-first report of the Joint FAO/WHO Expert Committee on Food Additives). Geneva, World Health Organization (WHO Technical Report Series, No. 617).
109. FAO/WHO, 2006. Evaluation of certain food additives (Sixty-seventh report of the Joint FAO/WHO Expert Committee on Food Additives). Geneva, World Health Organization (WHO Technical Report Series, No. 940).
110. Candogan K. The effect of tomato paste on some quality characteristics of beef patties during refrigerated storage. European Food Research and Technology 2002;215 305-309.
111. Deda M. S., Bloukas J. G., Fista G. A. Effect of tomato paste and nitrite level on processing and quality characteristics of frankfurters. Meat Science 2007;76 501-508.
112. Calvo M. M., Garcia M. L., Selgas M. D. M. Dry fermented sausages enriched with lycopene from tomato peel. Meat Science 2007;80(2) 167-172.

113. Gimeno O., Astiasaran I., Bello J., Calcium ascorbate as a potential partial substitute for NaCl in dry fermented sausage: Effect on colour, texture and hygienic quality at different concentration. Meat Science 2001;57 23-29.
114. Jakobsen M.,. Bertelsen G. Colour in food. Modelling colour stability. London: Woodhead Publishing Limited. 2002. p233–247.
115. Perez-Alvarez J. A. Physicochemical characteristics of spanishtype dry-cured sausage. Food Research International 1999;32 599-607.

# CHAPTER 4

# New traits in crops produced by genome editing techniques based on deletions

**C. C. M. van de Wiel[1] J. G. Schaart[1] L. A. P. Lotz[1] M. J. M. Smulders[1]**

*1 Wageningen University and Research, Wageningen, The Netherlands*

## ABSTRACT

One of the most promising New Plant Breeding Techniques is genome editing (also called gene editing) with the help of a programmable site-directed nuclease (SDN). In this review, we focus on SDN-1, which is the generation of small deletions or insertions (indels) at a precisely defined location in the genome with zinc finger nucleases (ZFN), TALENs, or CRISPR-Cas9. The programmable nuclease is used to induce a double-strand break in the DNA, while the repair is left to the plant cell itself, and mistakes are introduced, while the cell is repairing the double-strand break using the relatively error-prone NHEJ pathway. From a biological point of view, it could be considered as a form of targeted mutagenesis. We first discuss improvements and new technical variants for SDN-1, in particular employing CRISPR-Cas, and subsequently explore the effectiveness of targeted deletions that eliminate the function of a gene, as an approach to generate novel traits useful for improving agricultural sustainability, including disease resistances. We compare them with examples of deletions that resulted in novel functionality as known from crop domestication and classical mutation breeding (both using radiation and chemical mutagens).

Finally, we touch upon regulatory and access and benefit sharing issues regarding the plants produced.

**Keywords:** Nuclease SDN SSN Precision breeding Sustainable agriculture

# 1. INTRODUCTION

New plant breeding techniques (NPBT) encompass a set of diverse techniques and concepts that all aim to improve efficiency and/or precision of plant breeding. Most make use of transgenic plant lines at some point in the breeding process, but the final product generally contains only small mutations and in specific cased no modifications at all, and is often indistinguishable from the conventional breeding products (Lusser et al. 2012; Schaart et al. 2016). One of the most promising among these, genome editing (also called gene editing or gene targeting) with the help of a programmable nuclease (SDN: Site-directed nuclease, or SSN: sequence-specific nuclease), recently has led to a deluge of creative applications with the introduction of the CRISPR-Cas9 system. While the thorough review of NPBT by Lusser et al. (2012) not yet mentioned CRISPR-Cas, 4 years later a search using CRISPR and plants as keywords provided 246 publications in Web of Science (on 31-10-2016). Similarly to zinc finger nucleases (ZFN) and TALENs, CRISPR-Cas is able to make a double-strand break (DSB) at a precisely specified location in the genome, but it is much more versatile and easy to use, because the specificity of the target sequence is achieved by a separate guide RNA (gRNA) that can be easily designed and readily synthesised rather than by the protein structure itself (ZFN, TALEN). The use in plants has recently been reviewed by Luo et al. (2016), Paul and Qi (2016), Hilscher et al. (2016), and Rani et al. (2016).

At an early stage of the development of genome editing, three types of uses based on the DSB repair mechanism have been distinguished for regulatory purposes (Lusser et al. 2011). In SDN-1, the DSB is repaired by the non-homologous end joining (NHEJ) DNA repair machinery of the cell, during which small mistakes may be introduced, mostly consisting of small indels. Plants with such mistakes can be identified by screening. Indels in genes frequently lead to gene knockout as a result of reading frame shifts causing premature translational stops. In SDN-2, an oligonucleotide is provided to assist in the repair of the DSB that is identical to the sequence in which the DSB is made, except that a desired mutation is included. Some plants may use this oligo in an alternative cellular repair mechanism, homology-directed repair (HDR). This may lead to a higher frequency of plants with the desired change, e.g., a single amino-acid substitution. In SDN-3, the DSB is repaired in the same way, but the repair template is a longer sequence that may include a complete gene or a promoter. This will lead to the insertion of native or foreign sequences at a precisely specified location in the genome. SDN-3 is thus similar to genetic transformation and it generally results in transgenic plants, but they are produced with improved precision as the inserted sequence is targeted to a specific site in the genome, or exchanged with existing sequences at that site, e.g., in promoter swaps. Thus, SDN-3 holds much promise for diverse applications, including stacking desirable (transgenic) traits, so that they are passed on to progeny as a single block during breeding, which is desirable for efficient introgression into diverse elite materials (Kumar et al. 2016). HDR is still a relatively difficult to control mechanism. An alternative technique is called ODM (oligo-directed mutagenesis). It may produce results similar to SDN-2 but does not employ a nuclease to make a DSB. Combinations of both, improving efficiency, are also explored (Sauer et al. 2016).

Although SDN-1 is precise only in the targeted site and not in the resulting type of mutation, it appears presently to be the most frequently implemented, and is, therefore, the subject of this review. We first discuss improvements and new variants of SDN-1, in particular employing CRISPR-Cas, and subsequently explore the possibilities of targeted deletions that eliminate the function or part of the function of a gene, as an approach to generate novel traits useful for improvement of agricultural sustainability. We compare them with examples of deletions that resulted in novel functionality as known from crop domestication and classical mutation breeding. Finally, we touch upon regulatory and IP issues around the plants produced, as well as access and benefit sharing.

# 2. IMPROVEMENTS AND NEW VARIANTS OF THE TECHNOLOGY

## 2.1 Off-target effects

CRISPR-Cas genome editing has been reported to be accompanied by off-target effects, i.e., mutations arising from repairing DSBs induced elsewhere in the genome than the targeted sequence. Several research groups have made changes to the Cas protein and/or gRNA design to enhance the sequence target specificity (Kleinstiver et al. 2016; Slaymaker et al. 2016; Doench et al. 2016). In addition, alternative endonucleases (e.g. Cpf1) could increase specificity (e.g., Ran et al. 2015; Kim et al. 2016a).

Schiml et al. (2014) turned Cas9 into a nickase, so that two adjacent gRNAs were required to make a DSB with sticky ends. This improved the specificity, alone and in combination with using truncated gRNAs (Fu et al. 2014). Tolerance for mismatches in the gRNA (actually off-target effects) may also be used to specifically target related genes from gene families (Endo et al. 2015).

Various methods for genome-wide detection of off-target mutations have been developed, including BLESS (Crosetto et al. 2013), GUIDE-seq (Tsai et al. 2015), Digenome-seq (Kim et al. 2015, 2016b), and END-seq (Canela et al. 2016). For the few plant species assessed so far, little off-target effects appear to occur using CRISPR-Cas, but this would need to be studied in more species (Peterson et al. 2016). For example, off-target effects could be limited using an optimal molar ratio of Cas9 to gRNAs by Woo et al. (2015), while Peterson et al. (2016) did not detect any in a multiplex approach by re-sequencing in *Arabidopsis*, including in computationally predicted sites.

Off-target mutations are considered a problem in, e.g., applications in humans, but they are less likely to pose a problem in plant breeding, as usually multiple plants are produced that are subsequently screened and selected. In many cases, there are multiple generations between the genome editing step and the variety produced, in which off-target mutations are selected against and in seed-propagated crops, segregate out. In addition, the frequency of off-target mutations made across the genome will be much lower than in classical mutagenesis. The type and frequency of mutations in classical mutagenesis depend on the method (chemical or ionizing radiation) used, type and concentration of chemical or type of radiation, and the duration of the treatment, which are generally adjusted case by case (Suprasanna et al. 2015). Polyploid crops, in general, tolerate a higher frequency than diploid crops (Uauy et al. 2009; Shu et al. 2011; Oladosu et al. 2016). A frequency of one mutation induced every 78 kb (what Tsuda et al. 2015 used in soybean) would in soybean or tomato mean that classical mutagenesis may introduce for every desired mutation more than 10,000 other mutations.

## 2.2 Delivery

Commonly plants are transformed using *Agrobacterium* or biolistic systems to introduce the Cas9- and gRNAs-encoding DNA stably into the plant genome. This may work best in tissue culture systems, as exemplified by the high success rate in rice (Paul and Qi 2016). When using floral dip, germline editing was shown to be improved in *Arabidopsis* by driving the expression of Cas9 with promoters specific for egg cells (Wang et al. 2015) or for actively dividing tissues, such as meristems and embryo sacs (Hyun

et al. 2015; Yan et al. 2015). Alternatively, viral vectors for gRNA (Ali et al. 2015; Yin et al. 2015) may improve efficiency.

Clasen et al. (2016) used a temporary expression system, i.e., a TALEN transcribed from plasmids introduced into potato protoplasts. Regenerated plants with mutations were checked for the absence of plasmid sequence insertions in their genome. Woo et al. (2015) transfected protoplasts of *Arabidopsis thaliana*, tobacco, lettuce, and rice with ribonucleoprotein complexes containing Cas9 nuclease protein and appropriate gRNAs and obtained genome-edited regenerated plants. They called their approach "DNA free genome editing". It has also been used in the unicellular green alga *Chlamydomonas reinhardtii* (Baek et al. 2016) and it was also shown to work in *Petunia hybrida* protoplasts (Subburaj et al. 2016). The approach requires protocols for protoplasting and plant regeneration from protoplasts, which are not available for many crops, or, if existing, not for all varieties within a crop. The desire to use genome editing in research and in plant breeding urges for a surge of activity into transformation, protoplasting, and regeneration protocols for crops and model species that up to now have remained recalcitrant, such as wheat (Altpeter et al. 2016).

## 2.3 Multiplexing

CRISPR-Cas9 is particularly suitable to multiplexing because of the versatility of the gRNAs, but also TALENs have been multiplexed: four *Nicotiana benthamiana* genes involved in glycosylation were knocked out simultaneously using TALENs for the benefit of biopharmaceutical production of glycoproteins devoid of plant-specific residues (Li et al. 2016). Peterson et al. (2016) targeted 14 loci simultaneously in *A. thaliana* with CRISPR-Cas using stacked gRNA expression arrays (Peterson et al. 2016). Steinert et al. (2015) simultaneously used two alternative modified Cas proteins, and each was directed by its own gRNA.

## 2.4 Larger deletions

Using two gRNAs targeted to sites at some distance from each other on the same chromosome, larger deletions become possible based on annealing the distant sites through NHEJ. In this way, Zhou et al. (2014) managed to delete a cluster of up to 10 loci in the terpenoid synthesis pathway in rice. Deleting clusters of related genes is useful, e.g., for the removal of alpha- or gamma-gliadins, which are organized in gene repeats on different chromosomes, while trying to produce wheat that is safe for people with celiac disease (Jouanin et al., in preparation).

### 2.4.1 Generation of new traits for crop improvement through deletions

'Classical' genetic modification was developed to introduce genes into plants to obtain a gain of function, although it soon was also used to silence endogenous genes, e.g., using RNAi. The reason for this is that the loss of a gene or gene function may also result in a new plant phenotype that is useful for man. During crop domestication, many traits have been selected for that are inherited as recessive, so mostly loss-of-function (Lu et al. 2006; Hancock 2012; Martínez-Ainsworth and Tenaillon 2016). They often comprised knockouts of genes. Examples include loss of seed shattering, loss of seed dormancy, reduction of shoots improving the harvest index, reduction of chemical and physical defences, and loss of photoperiodicity and/or the vernalisation requirement (Nakamichi 2015). Similar mutations have been selected for independently in different crops (e.g., Cheng et al. 2016).

Since the 1930s, (knockout) mutagenesis has been used to generate new traits, boosted by increased knowledge on the effects of various sources of radiation (which mostly induce deletions) and chemical mutagens (for example, EMS induces C > T mutations). This had led to the development of more than 3000 crop varieties, and it is still popular, for instance, for generating new flower colours in ornamentals (it is a standard procedure in Chrysanthemum). Interestingly, mutation breeding also produces, at useable frequencies, gain of function traits, such as new disease resistances (Oladosu et al. 2015). Resistance to biotic and abiotic stresses is an important goal of breeding new varieties, as one of the elements to feed the world in a sustainable way.

The SDN-1 systems make it also easier to deal with the recessive nature of the mutations, which in classical mutagenesis generally means that mutations do not show a phenotype in the mutants themselves, only in their progeny. SDN-1 can mutate all alleles of the targeted locus simultaneously, which is a particularly significant improvement in mutating polyploids. Examples are deletions in the three homoeologues (six alleles) of *MLO* for powdery mildew resistance in hexaploid wheat using TALENs by Wang et al. (2014) and the four copies of GBSS for altering starch composition in tetraploid potato by Andersson et al. (2016).

## 2.5 Disease resistance

Increasing pathogen resistance in crops is an important way to improve agricultural sustainability. Resistance may be generated by deleting plant genes that are used ('hijacked') by pathogens and on which the pathogens depend for growth and development (Pavan et al. 2010). These are called S genes (for Susceptibility, in contrast to dominant R genes which confer

Resistance through recognition of the pathogen). The classical example is *mlo*, a recessive mutant which has been effective for mildew resistance in barley for over 30 years (reviewed by Acevedo-Garcia et al. 2014).

Knocking out S genes generated disease resistance in multiple crop species, but this may be accompanied by poor plant performance. Precisely targeted deletions using SDN-1 can make it possible to generate specific pathogen resistance without fitness costs for the plant. Using TALENs, Li et al. (2012) were able to identify and delete a small part of the rice SWEET14 promoter that was targeted by an effector protein secreted by the causal agent of Bacterial Leaf Blight, *Xanthomonas oryzae* pv. oryzae (Xoo), apparently without affecting adequate expression of the gene by the plant itself. Thus, a rice line was generated that was resistant to Xoo strains that use this effector. Later on, also CRISPR-Cas9 was implemented for mutating SWEET-type S genes (Zhou et al. 2014). S genes have been identified in a range of species, including *Arabidopsis*, rice, soybean, and potato and tomato (Sun et al. 2016a, 2016b; Zheng et al. 2016).

Resistance to viruses can also be generated following an S gene approach. Disrupting the functionality of eIF4E (eukaryotic translation initiation factor 4E) effected broad potyvirus resistance in cucumber (Chandrasekaran et al. 2016). Pyott et al. (2016) likewise achieved potyvirus resistance in *Arabidopsis*.

## 2.6 Product quality

In many crops, including legumes and cereals, phytate in seeds interferes with phosphorous uptake. Decreasing phytate content in feed will increase net uptake of P by livestock which may reduce losses of P to the environment. Liang et al. (2014) used both TALEN and CRISPR-Cas9 successfully in maize to obtain mutations in *IPK*, which is involved in phytate production. Reducing phytate comes with trade-offs for plant performance that may be addressed more precisely through genome editing. Recently, Yamaji et al. (2017) reported an alternative approach of reducing seed P content in rice using retrotransposon-insertional knockout mutants of *SPDT*, a P distribution transporter controlling the allocation of P to the grains. In these mutant plants, yield and seed performance were apparently not affected. An alternative TALEN-based approach was explored by Wendt et al. (2013): introducing small indels into the promoter of the most important barley grain phytase gene, *HvPAPhy_a*, which could be used to test for the possibilities of changing gene expression in the grain. Mutations were achieved in the targeted promoter site, but no further testing of plants was reported.

Clasen et al. (2016) used gene editing to knock out *Vnvl* to reduce acrylamide (a potential carcinogen) levels of potato products after heating.

A similar result could be obtained by silencing asparagine synthetase genes, in this case using RNAi (Zhu et al. 2016). Some other examples of traits improving product quality targeted by SDN-1 are high amylopectin ("waxy") maize (by Pioneer, https://www.aphis.usda.gov/biotechnology/downloads/reg_loi/15-352-01_air_response_signed.pdf) and potato (Andersson et al. 2016) based on knockout of *GBSS*, and high oleic acid oilseed crops soybean (Haun et al. 2014) and camelina (Jiang et al. 2016) based on knockouts of *FAD2* and/or *FAD3*. Both traits had been addressed earlier using classical mutagenesis or RNAi. Rice fragrancy was improved by knockout of *BADH2* (Shan et al. 2015). For bio-based crops, reducing lignin contents has been studied using a transgenic silencing method up to the level of field trials in poplar for improving biofuel production (Van Acker et al. 2014). Zhou et al. (2015) reported successful SDN-1 in two poplar 4CL genes, the knockout of one of which led to lower lignin levels in stem wood.

## 2.7 Allergens

Removing allergens through genome editing would benefit specific groups of consumers. Dubois et al. (2015) showed that silencing Mal d 1 reduced the allergenicity of apple, which, in most patients, is a mild allergy resulting from cross-reactivity of the birch pollen Bet v 1 allergens to PR-10 proteins in Rosaceous fruits, such as apple, cherry, and strawberry. PR-10 proteins are encoded by a large gene family. Apple has 31 Mal d 1 genes of which 20 may be expressed in the fruit (Pagliarani et al. 2013) and may be targeted by genome editing. Peanut allergy is a life-threatening food allergy. Dodo et al. (2008) managed to reduce the allergenicity of the immunodominant Ara h 2 protein in peanuts using RNAi. For hypoallergenic peanuts to be safe for consumption by many patients, all genes coding for allergens would need to be silenced or removed, and genome editing offers the tools to efficiently do this.

Sensitivity of individuals with coeliac disease to cereal gluten is particularly difficult to address as gluten comprises of large gene families, with several epitopes in mainly alpha-, gamma-, and omega-gliadins that can be recognised by human T cells (Van Herpen et al. 2006; Van den Broeck et al. 2009; Salentijn et al. 2012) and that all are targets for controlled deletion or modification (Smulders et al. 2015; Jouannin et al. in prep.). Barro et al. (2016) succeeded in removing highly immunogenic gliadin proteins from wheat using RNAi. Clinical trials using breads from these silenced wheat plants were planned in Spain. Silencing will in most cases not completely remove the gliadin, and a complete knockout approach by genome editing may be more effective, and possibly more easily accepted by consumers as no foreign DNA is introduced (Laursen 2016).

## 2.8 Hybrid varieties

Pivotal to successful production of high-quality hybrid varieties is the availability of reliably male-sterile maternal lines. Pioneer is using a male sterility system in maize involving mutants of the male fertility gene *Ms45*. Knockout of the *Ms45* gene using CRISPR-Cas9 was recently reported by Svitashev et al. (2015).

## 2.9 Yield

Interestingly, a complex trait par excellence, yield, was also shown to be amenable to an SDN-1 approach. Li et al. (2017) used CRISPR-Cas9 in rice to mutate the regulatory genes *Gna1, DEP1*, and *GS3* and obtained plants with increased grain numbers, dense erect panicles plus semi-dwarf phenotype, and larger grains, respectively.

# 3. REGULATORY AND IP ISSUES

Currently, there is worldwide a discussion whether genome-edited plants should fall under existing regulatory systems that have been designed for transgenic (GM or GE) plants (for an overview, see Sprink et al. 2016). In the US, no regulatory oversight was deemed necessary by USDA-APHIS (e.g., waxy maize, high oleic acid soybean mentioned above under 'Product quality' and *sweet14*-based blight resistance in rice under 'Disease resistance', see for further examples Table 1 of Hilscher et al. 2016). In Europe, there is no definitive legal analysis yet. The EFSA GMO unit (2015) considered SDN-1 a form of mutagenesis; further analysis may be needed upon technological advancement. Among EU Member States, the German BVL (Federal Office of Consumer Protection and Food Safety) had a similar judgement, including for CRISPR, as gRNAs were not seen as recombinant DNA (i.e., no novel combination of genetic material) and the Swedish Board of Agriculture (SBA) saw CRISPR-Cas as equivalent to mutagenesis provided that no "foreign" DNA was left in plants. The German BfN (Federal Agency for Nature Conservation) held a different view, as were NGOs and organic farming organizations (see IFOAM 2015 statement): they consider the process as most relevant and the process involves molecules not occurring naturally (Sprink et al. 2016). Mutagenesis was exempted from regulation in EU Directive 2001/18/EC as having a history of safe use (see for detailed comparison with SDN-1, the off-targets section above).

The EU GM regulation is perceived as prohibitive to most applications and to small companies because of costs and uncertainties around timing and outcomes of the application procedure. When extensive GM regulation would be applicable to SDN-1 techniques, this may lead to a paradoxical result. Parties aiming at commercializing the plant products could produce a gene-edited plant with deletions to establish whether it generates the desired functionality, and then would "re-produce" a similar plant using classical mutagenesis. Companies would be able to do that also because the quickly increasing efficiency of next-generation DNA sequencing for mutation screening in recent times has stimulated them to (re)develop mutated populations in their crops (Van de Wiel et al. 2016). However, this would not be an efficient and optimal use of resources in breeding, and would give up on advantages hardly feasible by classical mutagenesis, such as the possibility of inducing recessive mutations in all alleles of a targeted gene in polyploid crops. There is a need for regulatory systems duly addressing biosafety that are balanced in terms of workability and costs, so that they do not unduly limit the benefits of producing innovative plant products with potential advantages to agricultural sustainability.

The possibilities offered by genome editing also have ramifications to IP, access, and benefit issues, even when it concerns deletions and loss-of-function mutations. It may also raise questions in the context of the use of genetic resources: can copying a mutation already existing in a genebank accession into elite material be considered use in the sense of the '*Nagoya Protocol on Access to Genetic Resources and the Fair and Equitable Sharing of Benefits Arising from their Utilization to the Convention on Biological Diversity*'? These ramifications may attract a lot of attention in the coming years when the possibilities of SDN-1 are realized in the form of improved crop varieties (or even varieties from underutilized or novel crops). These discussions are important for the realisation of the possibilities that SDN-1 using genome editing offers for increasing the sustainability of agriculture.

## ACKNOWLEDGEMENTS

Writing of this review was supported by the Ministry of Economic Affairs of The Netherlands (BO-20-003-006). The funders had no role in study design, data collection and analysis, decision to publish, or preparation of the manuscript.

# REFERENCES


1. Acevedo-García J, Kusch S, Panstruga R (2014) Magical mystery tour: MLO proteins in plant immunity and beyond. New Phytol 204:273–281
2. Ali Z, Abul-faraj A, Li LX, Ghosh N, Piatek M, Mahjoub A, Aouida M, Piatek A, Baltes NJ, Voytas DF, Dinesh-Kumar S, Mahfouz MM (2015) Efficient virus-mediated genome editing in plants using the CRISPR/Cas9 system. Mol Plant 8:1288–1291
3. Altpeter F, Springer NM, Bartley LE, Blechl AE, Brutnell TP, Citovsky V, Conrad LJ, Gelvin SB, Jackson DP, Kausch AP, Lemaux PG, Medford JI, Orozco-Cardenas ML, Tricoli DM, Van Eck J, Voytas DF, Walbot V, Wang K, Zhang ZYJ, Stewart CN (2016) Advancing crop transformation in the era of genome editing. Plant Cell 28:1510–1520. doi:10. 1105/tpc.16.00196
4. Andersson M, Turesson H, Nicolia A, Fält AS, Samuelsson M, Hofvander P (2016) Efficient targeted multiallelic mutagenesis in tetraploid potato (*Solanum tuberosum)* by transient CRISPRCas9 expression in protoplasts. Plant Cell Rep. doi:10.1007/s00299-016-2062-3
5. Baek K, Kim DH, Jeong J, Sim SJ, Melis A, Kim JS, Jin E, Bae S (2016) DNA-free two-gene knockout in Chlamydomonas reinhardtii via CRISPR-Cas9 ribonucleoproteins. Sci Rep 6:30620
6. Barro F, Iehisa JCM, Giménez MJ, García-Molina MD, Ozuna CV, Comino I, Sousa C, Gil-Humanes J (2016) Targeting of prolamins by RNAi in bread wheat: effectiveness of seven silencingfragment combinations for obtaining lines devoid of celiac disease epitopes from highly immunogenic gliadins. Plant Biotechnol J 14:986–996
7. Canela A, Sridharan S, Sciascia N, Tubbs A, Meltzer P, Sleckman BP, Nussenzweig A (2016) DNA breaks and end resection measured genome-wide by end sequencing. Mol Cell 63:898–911
8. Chandrasekaran J, Brumin M, Wolf D, Leibman D, Klap C, Pearlsman M, Sherman A, Arazi T, Gal-On A (2016) Development of broad virus resistance in non-transgenic cucumber using CRISPR/Cas9 technology. Mol Plant Pathol 17:1140–1153
9. Cheng F, Sun R, Hou X, Zheng H, Zhang F, Zhang Y et al (2016) Subgenome parallel selection is associated with morphotype diversification and convergent crop domestication in Brassica rapa and Brassica oleracea. Nature Genet 48:1218–1224. doi:10.1038/ng.3634
10. Clasen BM, Stoddard TJ, Luo S, Demorest ZL, Li J, Cedrone F, Tibebu R, Davison S, Ray EE, Daulhac A, Coffman A, Yabandith A, Retterath A, Haun W, Baltes NJ, Mathis L, Voytas DF, Zhang F (2016) Improving cold storage and processing traits in potato through targeted gene knockout. Plant Biotechnol J 14:169–176
11. Crosetto N, A Mitra, MJ Silva, M Bienko, N Dojer, Q Wang, E Karaca, R Chiarle, M Skrzypczak, K Ginalski, P Pasero, M Rowicka, I Dikic (2013) Nucleotide-resolution DNA double-strand break mapping by next-generation sequencing. Nature Methods 10:361–365. doi:10.1038/nmeth.2408
12. Dodo HW, Konan KN, Chen FC, Egnin M, Viquez OM (2008) Alleviating peanut allergy using genetic engineering: the silencing of the immunodominant allergen Ara h 2 leads to its significant reduction and a decrease in peanut allergenicity. Plant Biotechnol J 6:135–145
13. Doench JG, Fusi N, Sullender M, Hegde M, Vaimberg EW, Donovan KF, Smith I, Tothova Z, Wilen C, Orchard R, Virgin HW, Listgarten J, Root DE (2016) Optimized sgRNA design to maximize activity and minimize off-target effects of CRISPR-Cas9. Nat Biotechnol 34:184–191
14. Dubois AEJ, Pagliarani G, Brouwer RM, Kollen BJ, Dragsted LO, Eriksen FD, Callesen O, Gilissen LJWJ, Krens FA, Visser RGF, Smulders MJM, Vlieg-Boerstra BJ, Flokstra-de Blok BJ, van de Weg WE (2015) First successful reduction of clinical allergenicity of food by genetic modification: Mal d 1 silenced apples cause fewer allergy symptoms than the wild type cultivar. Allergy 70:1406–1412. doi:10.1111/all.12684

15. EFSA GMO UNIT (2015) EFSA response to DG SANTE request to provide technical assistance on issues related to the legal analysis of new plant breeding techniques. Mandate Number: M-2015-0183. P 4. http://registerofquestions.efsa.europa.eu/ roq Frontend/ListOfQuestionsNoLogin?1
16. Endo M, Mikami M, Toki S (2015) Multigene knockout utilizing off-target mutations of the CRISPR/Cas9 system in rice. Plant Cell Physiol 56:41–47
17. Fu YF, Sander JD, Reyon D, Cascio VM, Joung JK (2014) Improving CRISPR-Cas nuclease specificity using truncated guide RNAs. Nat Biotechnol 32:279–284
18. Hancock JF (2012) Plant evolution and the origin of crop species. CABI, Wallingford Haun W, Coffman A, Clasen BM, Demorest ZL, Lowy A, Ray E, Retterath A, Stoddard T, Juillerat A, Cedrone F, Mathis L, Voytas DF, Zhang F (2014) Improved soybean oil quality by targeted mutagenesis of the fatty acid desaturase 2 gene family. Plant Biotechnol J 12:934–940
19. Hilscher J, Bürstmayr H, Stoger E (2016) Targeted modification of plant genomes for precision crop breeding. Biotechnol J. doi:10.1002/biot.201600173
20. IFOAM EU Group (2015) IFOAM EU position paper on New Plant Breeding Techniques. IFOAM EU Group, p 5. http://www.ifoam-eu.org/sites/default/files/ifoameu_policy_npbts_position_final_20151210.pdf
21. Jiang WZ, Henry IM, Lynagh PG, Comai L, Cahoon EB, Weeks DP (2016) Significant enhancement of fatty acid composition in seeds of the allohexaploid, Camelina sativa, using CRISPR/Cas9 gene editing. Plant Biotechnol J. doi:10.1111/pbi.12663
22. Kim D, Bae S, Park J, Kim E, Kim S, Yu HR, Hwang J, Kim JI, Kim JS (2015) Digenome-seq: genome-wide profiling of CRISPR-Cas9 off-target effects in human cells. Nat Methods
23. 12:237–243
24. Kim D, Kim J, Hur JK, Been KW, Yoon S-H, Kim J-S (2016a) Genome-wide analysis reveals specificities of Cpf1 endonucleases in human cells. Nat Biotechnol 34:863–868. doi:10.1038/nbt.3609
25. Kim D, Kim S, Kim S, Park J, Kim J-S (2016b) Genome-wide target specificities of CRISPR-Cas9 nucleases revealed by multiplex Digenome-sEq. Genome Res 26:406–415. doi:10.1101/gr.199588.115
26. Kleinstiver BP, Pattanayak V, Prew MS, Tsai SQ, Nguyen NT, Zheng Z, Joung JK (2016) High-fidelity CRISPR-Cas9 nucleases with no detectable genome-wide off-target effects. Nature 529:490–495
27. Kumar S, Barone P, Smith M (2016) Gene targeting and transgene stacking using intra genomic homologous recombination in plants. Plant Methods 12:11
28. Laursen L (2016) Will Europe toast GM wheat for gluten sufferers? Nat Biotechnol 34:369–371. doi:10.1038/nbt.3533
29. Li T, Liu B, Spalding MH, Weeks DP, Yang B (2012) High-efficiency TALEN-based gene editing produces disease-resistant rice. Nat Biotechnol 30:390–392
30. Li J, Stoddard TJ, Demorest ZL, Lavoie PO, Luo S, Clasen BM, Cedrone F, Ray EE, Coffman AP, Daulhac A, Yabandith A, Retterath AJ, Mathis L, Voytas DF, D'Aoust MA, Zhang F (2016) Multiplexed, targeted gene editing in *Nicotiana benthamiana* for glyco-engineering and monoclonal antibody production. Plant Biotechnol J 14:533–542
31. Li J, Sun Y, Du J, Zhao Y, Xia L (2017) Generation of targeted point mutations in rice by a modified CRISPR/Cas9 system. Mol Plant. doi:10.1016/j.molp.2016.12.001
32. Liang Z, Zhang K, Chen K, Gao C (2014) Targeted mutagenesis in *Zea mays* using TALENs and the CRISPR/Cas system. J Genet Genom 41:63–68
33. Lu J, Tang T, Tang H, Huang JZ, Shi SH, Wu CI (2006) The accumulation of deleterious mutations in rice genomes: a hypothesis on the cost of domestication. Trends Genet 22:126–131
34. Luo M, Gilbert B, Ayliffe M (2016) Applications of CRISPR/Cas9 technology for targeted mutagenesis, gene replacement and stacking of genes in higher plants. Plant Cell Rep 35:1439–1450
35. Lusser M, Parisi C, Plan D, Rodríguez-Cerezo E (2011) New plant breeding techniques. State-of-the-art and prospects for commercial development. European Commission, JRC-IPTS, p 218. doi:10.2791/54761

36. Lusser M, Parisi C, Plan D, Rodríguez-Cerezo E (2012) FEATURE— Deployment of new biotechnologies in plant breeding. Nat Biotechnol 30:231–239
37. Martinez-Ainsworth NE, Tenaillon MI (2016) Superheroes and masterminds of plant domestication. (Special issue: trajectories of genetics, 150 years after Mendel.). CR Biol 339:268–273
38. Nakamichi N (2015) Adaptation to the local environment by modifications of the photoperiod response in crops. Plant Cell Physiol 56:594–604. doi:10.1093/pcp/pcu181
39. Oladosu Y, Rafii MY, Abdullah N, Hussin G, Ramli A, Rahim HA, Miah G, Usman M (2016) Principle and application of plant mutagenesis in crop improvement: a review. Biotechnol Biotechnol Equip 30:1–16. doi:10.1080/13102818.2015.1087333
40. Pagliarani G, Paris R, Arens P, Tartarini S, Ricci G, Smulders MJM, van de Weg WE (2013) A qRT-PCR assay for the expression of all Mal d 1 isoallergen genes. BMC Plant Biol 13: 51. DOI:10.1186/1471-2229-13-51
41. Paul JW III, Qi Y (2016) CRISPR/Cas9 for plant genome editing: accomplishments, problems and prospects. Plant Cell Rep 35:1417–1427
42. Pavan S, Jacobsen E, Visser RGF, Bai Y (2010) Loss of susceptibility as a novel breeding strategy for durable and broad-spectrum resistance. Mol Breed 25:1–12
43. Peterson BA, Haak DC, Nishimura MT, Teixeira PJPL, James SR, Dangl JL et al (2016) Genome-Wide Assessment of efficiency and specificity in CRISPR/Cas9 mediated multiple site targeting in *Arabidopsis*. PLoS One 11:e0162169
44. Pyott DE, Sheehan E, Molnar A (2016) Engineering of CRISPR/ Cas9-mediated potyvirus resistance in transgene-free *Arabidopsis* plants. Mol Plant Pathol 17:1276–1288
45. Ran FA, Cong L, Yan WX, Scott DA, Gootenberg JS, Kriz AJ, Zetsche B, Shalem O, Wu XB, Makarova KS, Koonin EV, Sharp PA, Zhang F (2015) *In vivo* genome editing using *Staphylococcus aureus* Cas9. Nature 520:186–191
46. Rani R, Yadav P, Barbadikar KM, Baliyan N, Malhotra EV, Singh BK, Kumar A, Singh D (2016) CRISPR/Cas9: a promising way to exploit genetic variation in plants. Biotechnol Lett 38:1991–2006
47. Salentijn EMJ, Mitea DC, Goryunova SV, Van der Meer IM, Padioleau I, Gilissen LJWJ, Koning F, Smulders MJM (2012) Celiac disease T cell epitopes from gamma-gliadins: immunoreactivity depends on the genome of origin, transcript frequency, and flanking protein variation. BMC Genom 13:277. doi:10.1186/1471-2164-13-277
48. Sauer NJ, Mozoruk J, Miller RB, Warburg ZJ, Walker KA, Beetham PR, Schöpke CR, Gocal GFW (2016) Oligonucleotide-directed mutagenesis for precision gene editing. Plant Biotechnology J 14:496–502
49. Schaart JG, Van de Wiel CCM, Lotz LAP, Smulders MJM (2016) Opportunities for products of new plant breeding techniques. Trends Plant Sci 21:438–449
50. Schiml S, Fauser F, Puchta H (2014) The CRISPR/Cas system can be used as nuclease for *in planta* gene targeting and as paired nickases for directed mutagenesis in Arabidopsis resulting in heritable progeny. Plant J 80:1139–1150
51. Shan Q, Zhang Y, Chen K, Zhang K, Gao C (2015) Creation of fragrant rice by targeted knockout of the *OsBADH2* gene using TALEN technology. Plant Biotechnol J 13:791–800
52. Shu QY, BP Forster, H Nakagawa (eds) (2011) Plant Mutation breeding and biotechnology. FAO, Rome. ISBN 978-92-5-105000-0
53. Slaymaker IM, Gao L, Zetsche B, Scott DA, Yan WX, Zhang F (2016) Rationally engineered Cas9 nucleases with improved specificity. Science 351:84–88
54. Smulders MJM, Jouanin A, Schaart J, Visser RGF, Cockram J, Leigh F, Wallington E, Boyd LA, van den Broeck HC, van der Meer IM, Gilissen LJWJ (2015) Development of wheat varieties with reduced contents of celiac-immunogenic epitopes through conventional and GM strategies. Proceedings of the 28th meeting of the Working Group on Prolamin Analysis and Toxicity (Peter Koehler, ed.), 25–27 September 2014, Nantes, France, pp 47–56. http://www.wgpat.com/proceeding_28th.html; http://www.wgpat.com/proceeding_28th.pdf
55. Sprink T, Eriksson D, Schiemann J, Hartung F (2016) Regulatory hurdles for genome editing: process- vs. product-based approaches in different regulatory contexts. Plant Cell Rep 35:1493–1506

56. Steinert J, Schiml S, Fauser F, Puchta H (2015) Highly efficient heritable plant genome engineering using Cas9 orthologues from *Streptococcus thermophilus* and *Staphylococcus aureus*. Plant J 84:1295–1305
57. Subburaj S, Chung SJ, Lee C, Ryu S-M, Kim DH, Kim J-S, Bae S, Lee G-J (2016) Site-directed mutagenesis in *Petunia hybrid* protoplast system using direct delivery of purified recombinant Cas9 ribonucleoproteins. Plant Cell Rep 35:1535–1544. doi:10. 1007/s00299-016-1937-7
58. Sun K, Wolters AMA, Loonen AEHM, Huibers RP, Van de Vlugt R, Goverse A, Jacobsen E, Visser RGF, Bai Y (2016a) Downregulation of *Arabidopsis DND1* orthologs in potato and tomato leads to broad-spectrum resistance to late blight and powdery mildew. Transgenic Res 25:123–138
59. Sun K, Wolters AMA, Vossen JH, Rouwet ME, Loonen AEHM, Jacobsen E, Visser RGF, Bai Y (2016b) Silencing of six susceptibility genes results in potato late blight resistance. Transgenic Res 25:731–742
60. Suprasanna P, SJ Mirajkar, SG Bhagwat (2015) Induced mutations and crop improvement. Bahadur INB et al. (eds.), Plant biology and biotechnology: Volume I: plant diversity, 593 Organization, Function and Improvement, doi:10.1007/978-81-322-2286-6$423, pp 593–617
61. Svitashev S, Young JK, Schwartz C, Gao HR, Falco SC, Cigan AM (2015) Targeted mutagenesis, precise gene editing, and site-specific gene insertion in maize using Cas9 and guide RNA. Plant Physiol 169:931–945
62. Tsai SQ, Zheng Z, Nguyen NT, Liebers M, Topkar VV, Thapar V, et al. (2015) GUIDE-seq enables genome-wide profiling of off-target cleavage by CRISPR-Cas nucleases. Nat Biotechnol 33:187–197. doi:10.1038/nbt.3117
63. Tsuda M, Kaga A, Anai T, Shimizu T, Sayama T, Takagi K, Machita K, Watanabe S, Nishimura M, Yamada N, Mori S, Sasaki H, Kanamori H, Katayose Y, Ishimoto M (2015) Construction of a high-density mutant library in soybean and development of a mutant retrieval method using amplicon sequencing. BMC Genom 16:1014. doi:10.1186/s12864-015-2079-y
64. Uauy C, Paraiso F, Colasuonno P, Tran RK, Tsai H, Berardi S, Comai L, Dubcovsky J (2009) A modified TILLING approach to detect induced mutations in tetraploid and hexaploid wheat. BMC Plant Biol 9:115. doi:10.1186/1471-2229-9-115
65. Van Acker R, Leple JC, Aerts D, Storme V, Goeminne G, Ivens B, Legee F, Lapierre C, Piens K, Van Montagu MCE, Santoro N, Foster CE, Ralph J, Soetaert W, Pilate G, Boerjan W (2014) Improved saccharification and ethanol yield from field-grown transgenic poplar deficient in cinnamoyl-CoA reductase. Proc Natl Acad Sci USA 111:845–850
66. Van de Wiel CCM, Smulders MJM, Visser RGF, Schaart JG (2016) New developments in green biotechnology—an inventory for RIVM. Report, Wageningen UR Plant Breeding, Wageningen, p 51. doi:10.18174/385481
67. Van den Broeck HC, van Herpen TWJM, Schuit C, Salentijn EMJ, Dekking L, Bosch D, Hamer RJ, Smulders MJM, Gilissen LJWJ, van der Meer IM (2009) Removing celiac disease-related gluten proteins from bread wheat while retaining technological properties: a study with Chinese Spring deletion lines. BMC Plant Biol 9:41. doi:10.1186/1471-2229-9-41
68. Van Herpen TWJM, SV Goryunova, J van der Schoot, M Mitreva, E Salentijn, O Vorst, MF Schenk, PA van Veelen, F Koning, LJM van Soest, B Vosman, D Bosch, RJ Hamer, LJWJ Gilissen, MJM Smulders (2006) Alpha-gliadin genes from the A, B, and D genomes of wheat contain different sets of celiac disease epitopes. BMC Genom 7:1. doi:10. 1186/1471-2164-7-
69. Wang Y, Cheng X, Shan Q, Zhang Y, Liu J, Gao C, Qiu J-L (2014) Simultaneous editing of three homoeoalleles in hexaploid bread wheat confers heritable resistance to powdery mildew. Nat Biotechnol 32:947–951
70. Wang Z, Xing H, Dong L, Zhang H, Han C, Wang X, Chen Q (2015) Egg cell-specific promoter-controlled CRISPR/Cas9 efficiently generates homozygous mutants for multiple target genes in *Arabidopsis* in a single generation. Genome Biol 16:144

71. Wendt T, Holm PB, Starker CG, Christian M, Voytas DF, Brinch- Pedersen H, Holme IB (2013) TAL effector nucleases induce mutations at a pre-selected location in the genome of primary barley transformants. Plant Mol Biol 83:279–285
72. Woo J, Kim J, Kwon S, Corvalán C, Cho S, Kim H, Kim S, Kim S, Choe S, Kim J (2015) DNA-free genome editing in plants with preassembled CRISPR-Cas9 ribonucleoproteins. Nat Biotechnol 33:1162–1164
73. Yamaji N, Takemoto Y, Miyaji T, Mitani-Ueno N, Yoshida KT, JF Ma (2017) Reducing phosphorus accumulation in rice grains with an impaired transporter in the node. Nature 541:92–95. doi:10.1038/nature20610
74. Yan LH, Wei SW, Wu YR, Hu RL, Li HJ, Yang WC, Xie Q (2015) High-efficiency genome editing in *Arabidopsis* using *YAO* promoter- driven CRISPR/Cas9 system. Mol Plant 8:1820–1823
75. Zheng Z, Appiano M, Pavan S, Bracuto V, Ricciardi L, Visser RGF, Wolters AM, Bai Y (2016) Genome-wide study of the tomato SlMLO gene family and its functional characterization in response to the powdery mildew fungus *Oidium neolycopersici*. Front Plant Sci 7:380. doi:10.3389/fpls.2016.00380
76. Zhou HB, Liu B, Weeks DP, Spalding MH, Yang B (2014) Large chromosomal deletions and heritable small genetic changes induced by CRISPR/Cas9 in rice. Nucl Acids Res 42:10903–10914
77. Zhou XH, Jacobs TB, Xue LJ, Harding SA, Tsai CJ (2015) Exploiting SNPs for biallelic CRISPR mutations in the outcrossing woody perennial *Populus* reveals 4-coumarate: CoA ligase specificity and redundancy. New Phytol 208:298–301
78. Zhu X, Gong H, He Q, Zeng Z, Busse JS, Jin W, Bethke PC, Jiang J (2016) Silencing of vacuolar invertase and asparagine synthetase genes and its impact on acrylamide formation of fried potato products. Plant Biotechnol J 14:709–718. doi:10.1111/pbi.12421

# CHAPTER 5

# Biotechnology of Agricultural Wastes Recycling Through Controlled Cultivation of Mushrooms

**Marian Petre and Alexandru Teodorescu**

*Department of Natural Sciences, Faculty of Sciences, University of Pitesti, Romania*

## 1. INTRODUCTION

The agricultural wastes recycling with applications in agro-food industry is one of the biological challenging and technically demanding research in the biotechnology domain known to humankind so far. Annually, the accumulation of huge amounts of vineyard and winery wastes causes serious environmental damages nearby winemaking factories. Many of these ligno-cellulose wastes cause serious environmental pollution effects, if they are allowed to accumulate in the vineyards or much worse to be burned on the soil. At the same time, the cereal by-products coming from the cereal processing and bakery industry are produced in significant quantities all over the world (Moser, 1994; Verstraete & Top, 1992).

To solve the environmental troubles raised by the accumulation of these organic wastes, the most efficient way is to recycle them through biological means (Smith, 1998). As a result of other recent studies, the cultivation of edible and medicinal mushrooms was applied using both the solid state cultivation and controlled submerged fermentation of different natural by-products of agro-food industry that provided a fast growth as well as high biomass productivity of the investigated strains (Petre& Teodorescu, 2009 ; Stamets, 2000).

These plant wastes can be used as the main ingredients to prepare the organic composts for edible and medicinal mushrooms growing in order to get organic food and biological active compounds from the nutritive fungal biomass resulted after solid state cultivation or submerged fermentation of such natural materials (Petre & Petre, 2008; Petre et al., 2010).

Taking into consideration this biological advantage there were tested some variants of biotechnology for agricultural wastes recycling through the controlled cultivation of edible and medicinal mushrooms *Ganoderma lucidum* (Curt.:Fr.) P. Karst (folk name: Reishi or Ling-zhi), *Lentinus edodes* (Berkeley) Pegler (folk name: Shiitake) and *Pleurotus ostreatus* (Jacquin ex Fries) Kummer (folk name: Oyster Mushroom) on organic composts made of cereal grain by-products as well as winery and vineyard wastes ( Petre & Teodorescu, 2010 ).

# 2. THE SOLID STATE CULTIVATION OF MUSHROOMS ON WINERY AND VINEYARD WASTES

The main aim of this work was focused on screening the optimal biotechnology of edible and medicinal mushrooms growing through the solid-state cultivation by recycling different kind of agricultural by-products and wastes coming from vineyard farms and winemaking industry (Petre *et al.*, 2011).

Taking into consideration that most of the edible and medicinal mushrooms species requires a specific micro-environment including complex nutrients, the influence of all physical and chemical factors upon fungal pellets production and mushroom fruit bodies formation has been studied by testing new biotechnological procedures (Oei, 2003).

To establish the laboratory biotechnology of recycling the winery and vineyard wastes by using them as a growing source for edible mushrooms, two mushroom species of Basidiomycetes group, namely *L. edodes* (Berkeley) Pegler and *P. ostreatus* (Jacquin ex Fries) Kummer were used as pure mushroom cultures isolated from the natural environment and being preserved in the local collection of the University of Pitesti. The stock cultures were maintained on malt-extract agar (MEA) slants (20% malt extract, 2% yeast extract, 20% agar-agar). Slants were incubated at 25 C for 120-168 h and stored at 4 C.

The pure mushroom cultures were expanded by growing in 250-ml flasks containing 100 ml of liquid malt-extract medium at 23 C on rotary shaker incubators at 110 rev min $^{-1}$ for 72-120 h. After expanding, the pure mushroom cultures were inoculated into 100 ml of 3-5% (v/v) malt-yeast extract liquid medium, previously poured in 250 ml rotary shake flasks and then were maintained at 23-25 C (Petre & Teodorescu, 2010 ).

The experiments of inoculum preparation were set up under the following conditions: constant temperature, 25 C; agitation speed, 90-120 rev min$^{-1}$; initial pH, 5.5–6.5. All the seed mushroom cultures were incubated for 120–168 h.

After that, the seed cultures of these mushroom species were inoculated in liquid culture media (20% malt extract, 10% wheat bran, 3% yeast extract, 1% peptone) at pH 6.5 previously distributed into rotary shake flasks of 1,000 ml. During the incubation time, all the spawn cultures were maintained in special culture rooms, designed for optimal incubation at 25 C. Three variants of culture compost were prepared from marc of grapes and vineyard cuttings in the following ratios: 1:1, 1:2, 1:4 (w/w).

The winery and vineyard wastes were mechanically pre-treated by using an electric grinding device to breakdown the lignin and cellulose structures in order to make them more susceptible to the enzyme actions. All the culture compost variants made of winery and vineyard wastes were transferred into 1,000 ml glass jars and disinfected by steam sterilization at 120 C for 60 min. When the jars filled with composts were chilled they were inoculated with the liquid spawn already prepared ( Petre *et al.*, 2010).

Each culture compost variant for mushroom growing was inoculated using such liquid spawn having the age of 72–220 h and the volume size ranging between 3–9% (v/w). During the period of time of 18–20 d after this inoculation, the mushroom cultures had developed a significant mycelia biomass on the culture substrates (Carlile & Watkinson, 1996).

According to the registered results of the performed experiments the optimal laboratory-scale biotechnology for edible mushroom cultivation on composts made of marc of grapes and vineyard cuttings was established (Fig. 1).

The effects induced by the composts composition, nitrogen and mineral sources as well as the inoculum amount upon the mycelia growing during the incubation period were investigated. There were made three variants of composts which were tested by comparing them with the control sample made of poplar sawdust ( Petre & Teodorescu, 2010 ).

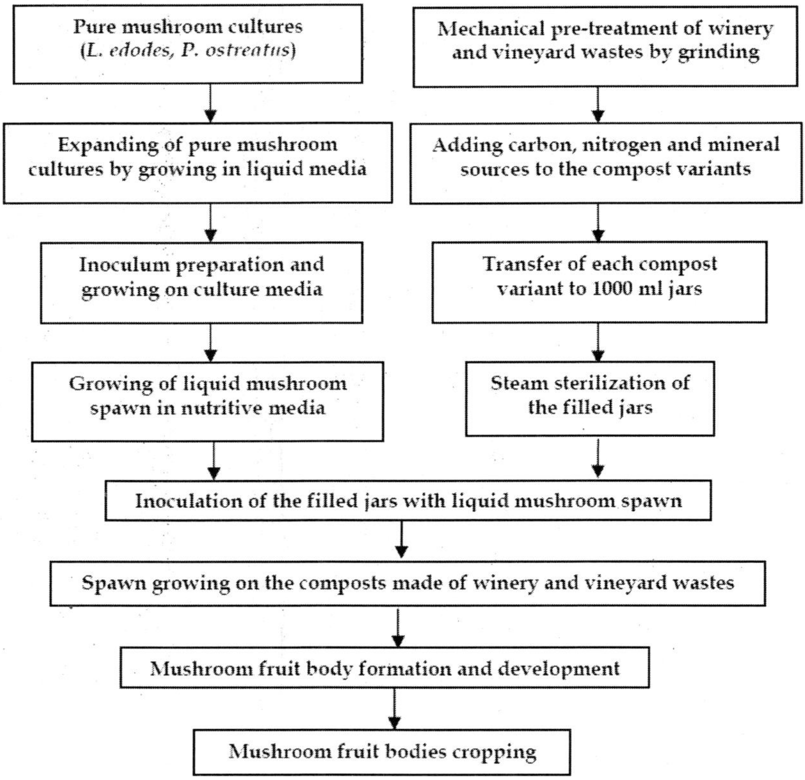


**Figure 1.** Scheme of laboratory-scale biotechnology for edible mushroom production by recycling winery and vineyard wastes.

The first variant of compost composition was prepared from vineyard cuttings, the second one from a mixture between marc of grapes and vineyard cuttings in equal proportions and the third one was made only from marc of grapes as full compost variant. The experiments were carried out for 288 h at 25 C with the initial pH 6.5 and the incubation period lasted for 168-288 h (Petre *et al.*, 2007).

## 2.1. Results and discussion

As it can be noticed in figure 2, the registered results show that from all tested compost variants the most suitable substrate for mycelia growing was that one prepared from marc of grapes, because it showed the highest influence upon the mycelia growing and fresh mushroom production (32–35 g%). All registered data represent the means of triple determinations.

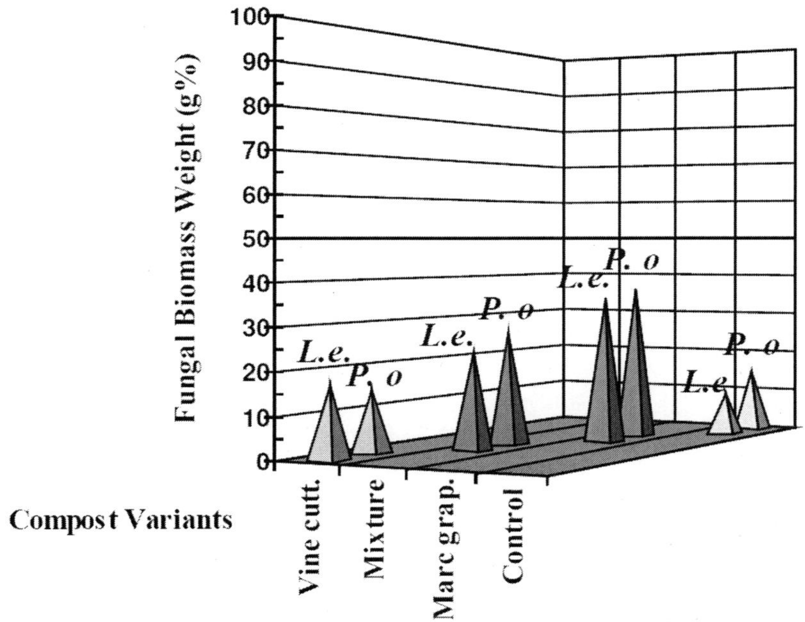


**Figure 2.** Comparative effects of composts composition upon mycelia growing of *P. ostreatus* (*P.o.*) and *L. edodes* (*L.e.*).

This compost variant was followed by the mixture prepared from marc of grapes and vineyard cuttings in equal amounts (20-23 g %) and, finally, by the variant made of only vineyard cuttings (12-15 g %). From the tested nitrogen sources, barley bran was the most efficient upon the mycelia growing and fruit mushroom producing at 35-40 g % fresh fungal biomass weight, being closely followed by rice bran at 25–30 g %. Wheat bran is also a well known nitrogen source for fungal biomass synthesis but its efficiency in these experiments was relatively lower than the ones induced by the barley and rice bran added as natural organic nitrogen sources (Stamets, 2000). All registered data are the means of triple determinations. The effects of nitrogen sources were registered as they are presented in figure 3. Among the tested mineral sources, the natural calcium carbonate ($CaCO_3$) from marine shells yielded the best mycelia growing as well as fungal biomass production at 28-32 g% and, for this reason, it was registered as the most appropriate mineral source, being followed by the natural gypsum ($CaSO_4$ 2 $H_2O$) at 20-23 g %, as it is shown in figure 4.

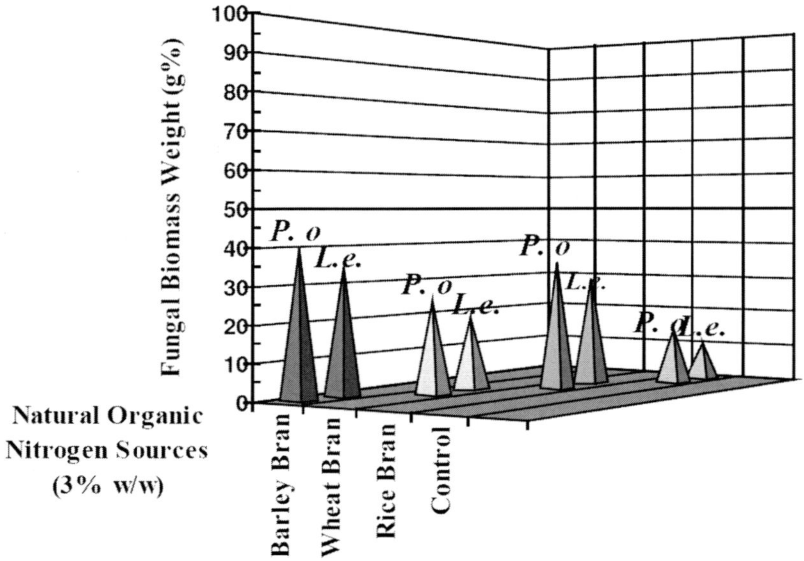


**Figure 3.** Comparative effects of organic nitrogen sources upon mycelia growing of *P. ostreatus* (*P.o.*) and *L. edodes* (*L.e.*).

The mineral sources like hepta-hydrate magnesium sulfate ($MgSO_4$ 7 $H_2O$) showed a quite moderate influence upon the fungal biomass growing as other researchers have already reported so far. All data are the means of triple determinations (Stamets, 2000; Chahal, 1994).

The whole period of mushroom growing from the inoculation to the fruit body formation lasted between 25-30 d in case of *P. ostreatus* cultivating and 50-60 d for *L. edodes*, depending on each fungal species used in experiments (Chahal, 1994). However, during the whole period of fruit body formation, the culture parameters were set up and maintained at the following levels, depending on each mushroom species: air temperature, 15–17°C; the air flow volume, 5–6$m^3$/h; air flow speed, 0.2–0.3 m/s; the relative moisture content, 80–85%, light intensity, 500–1,000 luces for 8–10 h/d. The final fruit body production of these mushroom species used in experiments was registered between 1.5 kg for *L.edodes* and 2.8 kg for *P. ostreatus*, relative to 10 kg of composts made of vineyard and winery wastes, comparing with 0.7-1.2 kg on 10 kg of poplar sawdust used as control samples.

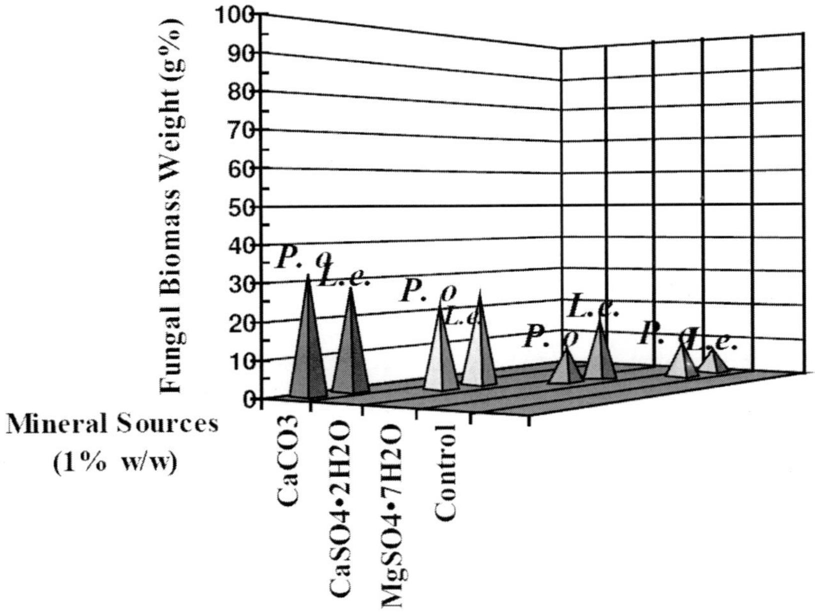


**Figure 4.** Comparative effects of mineral sources upon mycelia growing of *P. ostreatus* (*P. o.*) and *L. edodes* (*L.e.*).

## 3. THE CONTROLLED SUBMERGED CULTIVATION OF MUSHROOMS ON WINERY WASTES

The submerged cultivation of mushroom mycelium is a promising biotechnological procedure which can be used for synthesis of pharmaceutical substances with anticancer, antiviral and immuno-stimulatory effects from the nutritive mushroom biomass (Wasser & Weis, 1994). As result of other recent studies, the continuous cultivation of edible and medicinal mushrooms was applied by using the submerged fermentation of different natural by-products of agro-food industry (Bae, et al., 2000; Jones, 1995; Moo-Joung, 1993). The biotechnology of controlled cultivation of medicinal mushrooms was established and tested in different variants of culture media that were made of different sorts of bran and broken seeds resulted from the industrial food processing of wheat, barley and rye seeds. This biotechnology can influence the faster growth as well as higher biomass productivity of *G. lucidum* and *L. edodes* mushroom species ( Petre *et al.*, 2010 ).

The main stages of biotechnology to get high nutritive fungal biomass by controlled submerged fermentation were the followings:

1) Preparation of culture media and pouring them into the cultivation vessel of the bioreactor.

2) Steam sterilization of bioreactor vessel at 121 C and 1.1 atm. for 20 min.

3) Inoculation of sterilized culture media with mycelium from pure cultures of selected strains inside the bioreactor vessel for submerged cultivation, using the sterile air hood with laminar flow.

4) Running the submerged cultivation cycles under controlled conditions: temperature 23 ± 2 C, speed 70 rpm and continuous aeration at 1.1 atm.

5) Collecting, cleaning and filtering the fungal pellets obtained by the submerged fermentation of substrates made of by-products resulted from cereal grains processing.

Two mushroom species belonging to Basidiomycetes Class, namely *G. lucidum* (Curt.:Fr.) P. Karst and *L. edodes* (Berkeley) Pegler were used as pure cultures in experiments. The stock cultures were maintained on malt-extract agar (MEA) slants. Slants were incubated at 25 C for 5-7 d and then stored at 4 C. The fungal cultures were grown in 250-ml flasks containing 100 ml of MEYE (malt extract 20%, yeast extract 2%) medium at 23 C on rotary shaker incubators at 110 rev $min^{-1}$ for 5-7 d. The fungal cultures were prepared by aseptically inoculating 100 ml in three variants of culture media by using 3-5% (v/v) of the seed culture and then cultivated at 23-25 C in 250 ml rotary shake flasks. The biotechnological experiments were conducted under the following conditions: temperature, 25 C; agitation speed, 120-180 rev $min^{-1}$; initial pH, 4.5–5.5. After 10–12 d of incubation the fungal cultures were ready to be inoculated aseptically into the glass vessel of 20 l laboratory-scale bioreactor, that was designed to be used for controlled submerged cultivation of edible and medicinal mushrooms on substrata made of wastes resulted from the industrial processing of cereal grains (Fig. 5).

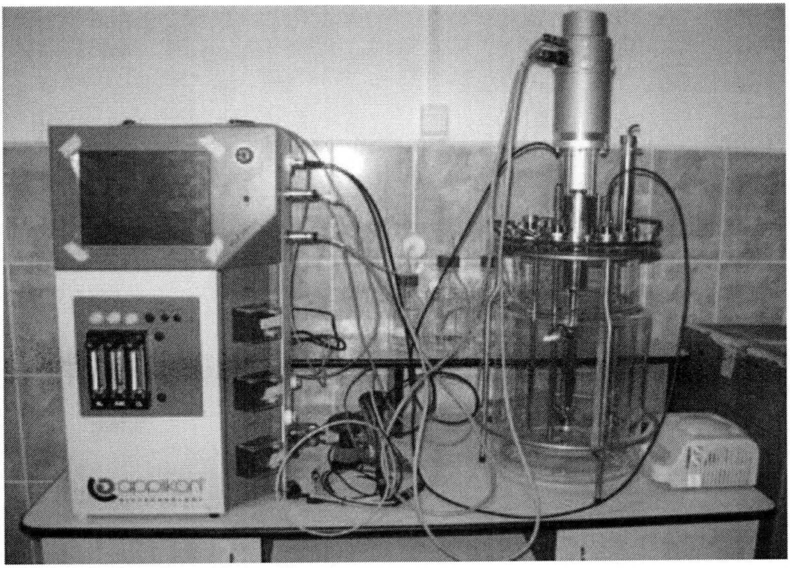

**Figure 5.** General view of the Laboratory scale bioreactor (15 L).

**Figure 6.** Mycelial biomass of *G. lucidum* collected after submerged fermentation.

**Figure 7.** Mycelial biomass in the shape of fungal pellets of *L. edodes,* collected after submerged fermentation.

After a period of submerged fermentation lasting up to 120 h, small mushroom pellets developed inside the nutritive broth (Fig. 6, 7).

The fermentation process was carried out by inoculating the growing medium volume (10,000 ml) with mycelia inside the culture vessel of the laboratory-scale bioreactor. The whole process of growing lasts for a single cycle between 5-7 days in case of *L. edodes* and between 3 to 5 days for *G. lucidum*. The strains of these fungal species were characterized by morphological and cultural stability, proven by their ability to maintain the phenotypic and taxonomic identities. The experiments were carried out in three repetitions. Observations on morphological and physiological characters of these two tested species of fungi were made after each culture cycle, highlighting the following aspects:

- sphere-shaped structure of fungal pellets, sometimes elongated, irregular, with various sizes (from 2 to 5 mm in diameter), reddish-brown colour – *G. lucidum* culture (Fig. 8).
- elliptically-shaped structures of fungal pellets, with irregular diameters of 4 up to 7 mm showing mycelia congestion, which developed specific hyphae of *L. edodes* (Fig. 9).

**Figure 8.** Stereomicroscopic view of *G. lucidum* pellets after controlled submerged fermentation.

**Figure 9.** Stereomicroscopic view of *L. edodes* pellets after controlled submerged fermentation.

Samples for analysis were collected at the end of the fermentation process, when pellets formed specific shapes and characteristic sizes. The fungal biomass was washed repeatedly with double distilled water in a sieve with 2 mm diameter eye, to remove the remained bran in each culture medium.

## 3.1. Results and discussion

Biochemical analyses of fungal biomass samples obtained by submerged cultivation of mushrooms were carried out separately for the solid fraction and liquid medium remained after the separation of fungal biomass by filtering. The percentage distribution of solid substrate and liquid fraction in the samples of fungal biomass are shown in table 1.

**Table 1.** Percentage distribution of solid substrate and liquid fraction in the preliminary samples of fungal biomass.

| Mushroom species | Total volume of separated liquid per sample (ml) | Total biomass weight per sample (g) | Water content after separation (%) |
|---|---|---|---|
| *L. edodes* | 83 | 5.81 | 83.35 |
| *L. edodes* | 105 | 7.83 | 82.50 |
| *L. edodes* | 95 | 7.75 | 82.15 |
| *L. edodes* | 80 | 5.70 | 79.55 |
| *G. lucidum* | 75 | 7.95 | 83.70 |
| *G. lucidum* | 115 | 6.70 | 82.95 |
| *G. lucidum* | 97 | 5.45 | 80.75 |
| *G. lucidum* | 110 | 6.30 | 77.70 |

In each experimental variant the amount of fresh biomass mycelia was determined. The percentage amount of dry biomass was determined by dehydration at 70 C, up to constant weight. Total protein content was determined by biuret method, whose principle is similar to the Lowry method, this method being recommended for the protein content ranging from 0.5 to 20 mg/100 mg sample. In addition, this method required only one sample incubation period (20 min) and by using them was eliminated the interference with various chemical agents (ammonium salts, for example).

The principle method is based on reaction that takes place between copper salts and compounds with two or more peptides in the composition in alkali, which results in a red-purple complex, whose absorbance is read in a spectrophotometer in the visible domain (λ - 550 nm). The registered results are presented as the amounts of fresh and dry biomass as well as protein contents for each fungal species and variants of culture media (Tables 2, 3).

**Table 2.** Fresh and dry biomass and protein content of *L. edodes* after submerged fermentation.

| Culture variants | Fresh biomass (g) | Dry biomass (%) | Total protein (g % d.w.) |
|---|---|---|---|
| I | 20.30 | 5.23 | 0.55 |
| II | 23.95 | 6.10 | 0.53 |
| III | 22.27 | 4.79 | 0.73 |
| IV | 20.10 | 4.21 | 0.49 |
| Control | 4.7 | 0.5 | 0.2 |

**Table 3.** Fresh and dry biomass and protein content of *G. lucidum* after submerged fermentation.

| Culture variants | Fresh biomass (g) | Dry biomass (%) | Total protein (g % d.w.) |
|---|---|---|---|
| I | 25.94 | 9.03 | 0.67 |
| II | 22.45 | 10.70 | 0.55 |
| III | 23.47 | 9.95 | 0.73 |
| IV | 21.97 | 9.15 | 0.51 |
| Control | 5.9 | 0.7 | 0.3 |

According to the registered data, using wheat bran strains the growth of *G. lucidum* biomass was favoured, while the barley bran led to the increased growth of *L. edodes* mycelium and *G. lucidum* as well. In contrast, dry matter content was significantly higher when using barley bran for both species used. Protein accumulation was more intense in case of using barley bran compared with those of wheat and rye, at both species of mushrooms.

The sugar content of dried mushroom pellets collected at the end of experiments was determined by using Dubois method (Wasser & Weis, 1994). The mushroom extracts were prepared by immersion of dried pellets inside a solution of NaOH pH 9, in the ratio 1:5.

All dispersed solutions containing the dried pellets were maintained 24 h at a precise temperature of 25$^0$C, in full darkness, with continuous homogenization to avoid the oxidation reactions. After removal of solid residues by filtration, the samples were analyzed by the previous mention method. The nitrogen content of mushroom pellets was analyzed by Kjeldahl method (Table 4).

**Table 4.** The sugar, total nitrogen and total protein contents of dried mushroom pellets.

| Mushroom species | Culture variant | Sugar content (mg/ml) | Kjeldahl nitrogen (%) | Total protein (g % d.w.) |
|---|---|---|---|---|
| *L. edodes* | I | 5.15 | 6.30 | 0.55 |
| *L. edodes* | II | 4.93 | 5.35 | 0.53 |
| *L. edodes* | III | 4.50 | 5.70 | 0.73 |
| *L. edodes* | IV | 4.35 | 5.75 | 0.49 |
| | Control | 0.55 | 0.30 | 0.2 |
| *G. lucidum* | I | 4.95 | 5.95 | 0.67 |
| *G. lucidum* | II | 5.05 | 6.15 | 0.55 |
| *G. lucidum* | III | 5.55 | 6.53 | 0.73 |
| *G. lucidum* | IV | 4.70 | 5.05 | 0.51 |
| | Control | 0.45 | 0.35 | 0.3 |

Comparing all registered data resulted from triple determinations, it can be noticed that the biochemical correlation between dry weight of mushroom pellets and their sugar and nitrogen contents is kept at a balanced ratio for each tested mushrooms (Stamets, 2000).

Among all mushroom samples that were tested in biotechnological experiments *G. lucidum* G-3 showed the best values of their composition in sugars, total nitrogen and total protein contents. In this stage, 70-80% of the former fungal pellets were separated by collecting them from the culture vessel of the bioreactor and separating from the broth by slow vacuum filtration. On the base of these results, the optimal values of physical and chemical factors which influence the mushroom biomass synthesis were taken into consideration in order to established the following schematic flow of the biotechnology for mushroom biomass producing by submerged fermentation, as it is shown in figure 10.

The main advantages of the submerged fermentation of winery wastes under the metabolic activity of selected mushrooms, by comparison with the solid state cultivation are the followings:

- the shortening of the biological cycle and cellular development in average from 8-10 weeks to at mostly one week per cellular culture cycle;
- the ensuring of the optimal control of physical and chemical parameters which are essential for producing important amounts of mushroom pellets in a very short time;
- 20–30% reduction of energy and work expenses as well as the volume of the volume of raw materials materials which are manipulated during each culture cycle;
- 15-20% increasing of fungal biomass amount per medium volume unit for each mushrooms species;
- the whole removing of any pollutant sources during the biotechnological flux;
- the culture media for mushroom growing are integrally natural without using of artificial additives as it is used in classical cultivating procedures;
- the mushroom pellets produced by applying this biotechnology for ecological treatment of agricultural wastes was 100% made by natural means and will be used for food supplements production with therapeutic properties which will contribute to the increasing of health level of human consumers having nutritional metabolic deficiencies.
- the biochemical correlation between the dry weight of mushroom pellets and their sugar and nitrogen contents is kept at a balanced ratio for each tested mushroom species.

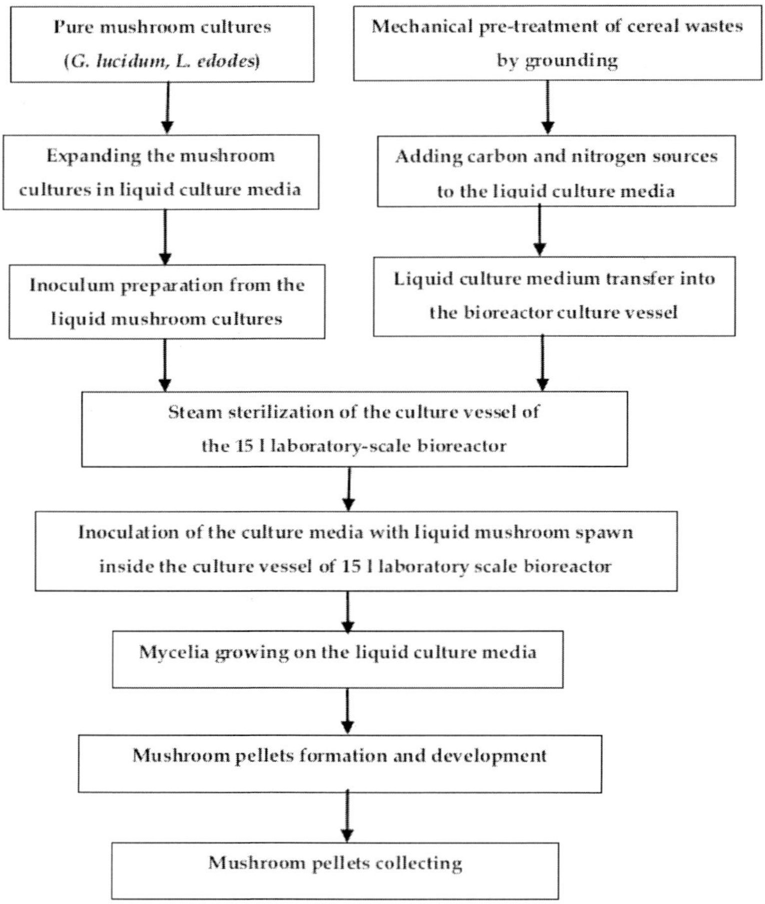


**Figure 10.** Schematic flow of the biotechnology for mushroom biomass producing by submerged fermentation.

# 4. THE CONTROLLED CULTIVATION OF MUSHROOMS IN MODULAR ROBOTIC SYSTEM

The agricultural works as well as industrial activities related to plant crops and their processing have generally been matched by a huge formation of wide range of lignocellulose wastes. All these vegetal wastes cause serious environmental troubles if they accumulate in the agro-ecosystems or much worse to be burned on the soil. For the human–operational farms, all processes are made by human personnel exclusively, starting from filling of cultivation beds with compost, up to fruit-bodies harvesting (Reed *et al.,* 2001).

In this respect, a strong tendency for increasing the number of researches in the field of mushroom's automated cultivation, harvesting and processing technologies as well as for continuously development of new robotic equipments can be noticed (Reed *et al.,* 2001).

The solid state cultivation of edible and medicinal mushrooms *Lentinula edodes* and *Pleurotus ostreatus* could be performed by using a modular robotic system that provides the following fully automatic operations: sterilization of composts, inoculation in aseptic chamber by controlled injection device containing liquid mycelia as inoculum, incubation as well as mushroom fruit bodies formation in special growing chambers with controlled atmosphere and the picking up of edible and medicinal mushroom fruit bodies ( Petre *et al.,* 2009 ).

The biotechnology concerning the controlled cultivation of edible mushrooms in continuous flow depends on the strictly maintaining of biotic as well as physical and chemical factors that could influence the bioprocess evolution. The proceeding of edible mushroom cultivation consists in a continuous biotechological flow, having a chain of succesive stages that are working in the non-sterile zone and mostly in the sterile zone of the modular robotic system. In this way, there is provided the technological security both from the structural and functional points of view in order to produce organic foods in highest security and food quality. The functional biotechnological model of the modular robotic system was designed for controlled cultivation and integrated processing of edible mushrooms to get ecological food in highest safety conditions ( Petre *et al.,* 2009 ).

The modular robotic system designed for edible mushroom cultivation provides the automatic sterilization of composts, the automatic inoculation inside the aseptic room by a special device of controlled injection of liquid mycelia, the incubation and fruit bodies formation in special chambers under controlled atmosphere as well as the automatic harvesting of mushroom fruit bodies (Petre *et al.,* 2011).

This system includes three major zones, respectively, the non-sterile zone, the sterile zone and the fruit-body processing zone (Fig. 11).

Thus, during the first stage of the biotechnological flow, in the non-sterile zone of the cultivation system, a natural and nutritive compost is prepared from sawdust or shavings of deciduous woody species in the ratio of 30-40 parts per weight (p.p.w.), marc of grapes chemically untreated, in 20-30 p.p.w., brans of organic cereal seeds (wheat, barley, oat, rye, rice), in 10-20 p.p.w., yeasts, in 3-5 p.p.w., and powder of marine shells, in 1-3 p.p.w., for pH adjustment, which then, it is hidrated with demineralized water, in 20-30 p.p.w. In the next stage, such prepared compost is decanting in polyethylene thermoserilizable bags, which have round orifices of 0,3-0,5 mm in diameter, uniform distribuited between them, at 10-15 cm distance, each one of them having a working volume of 10-20 kg (Petre *et al.,* 2011).

Beforehand, special devices for uniform distribution of mycelia as liquid inoculum are mounted inside of these bags. Then, these bags are fitted out with supporting devices on the transfer and transport systems and special devices for coupling to the automatic inoculation subdivision by controlled injection of liquid mycelia (Fig. 11).

Each one of these zones is linked with next one by an interfacing zone. In this way, the non-sterile zone is linked with the sterile zone through the first interfacing zone and this one is connected with the fruit body processing zone by the second interfacing area, as it is shown in figure 11.

Inside the non-sterile zone, the bags filled with composts are placed on the supporting devices, mounted on the transfer pallets, which are inserted in the first part of the sterile zone, respectively, in the module of the automatic sterilization with microwave at 120-125°C, and the pallets with bags are automatically chilled in the zone of controlled cooling of sterilized composts up to the room temperature. These pallets with sterilized bags are automatically transferred into the aseptic room to make the inoculation with liquid mycelia by using a robotic device of controlled injection. Further on, the pallets with the inoculated bags either are evacuated from the sterile zone or they are automatically transferred to the incubation and fruit body formation rooms. In these rooms of incubation and fruit body formation, both the optimal temperature of mycelia growing and the relative air humidity are provided as well as a constant steril air flow introduced under pressure by using an automatic device and an adecquate lighting level (Petre *et al.,* 2011; Petre *et al.*, 2009).

In this way, the bags are maintained from 15 up to 30 days, during this time a mycelial net being formed from the hypha anastomosis having a compact structure and a white-yelowish color, that covers the whole surface of compost and from which the mushroom fruit bodies will emerge and develop soon as specific morphological structures of the origin species. These mushroom fruit bodies were grown and maturated in almost 3-10 days, depending on the cultivated mushroom species, at constant temperature of 18-21°C, air relative humidity 90-95% and controlled aeration at 3-5 air volume exchanges per hour and the suitable lighting at 2.000-3.000 luxes per hour, for 12 h daily. For the fruit bodies picking-up, the pallets are automatically discharged by the same robotic system and transferred to the automatic harvesting zone, where another robotic system automatically collects all the mushroom fruit bodies by a special designed device to be conditioned and packaged aseptically (Fig. 11). The modular robotic system designed for edible mushroom cultivation provides the automatic sterilization of composts, the automatic inoculation inside the aseptic room by a special device of controlled injection of liquid mycelia, the incubation and fruit bodies formation in special chambers under controlled atmosphere as well as the automatic picking-up of mushroom fruit bodies (Reed *et al.*, 2001).

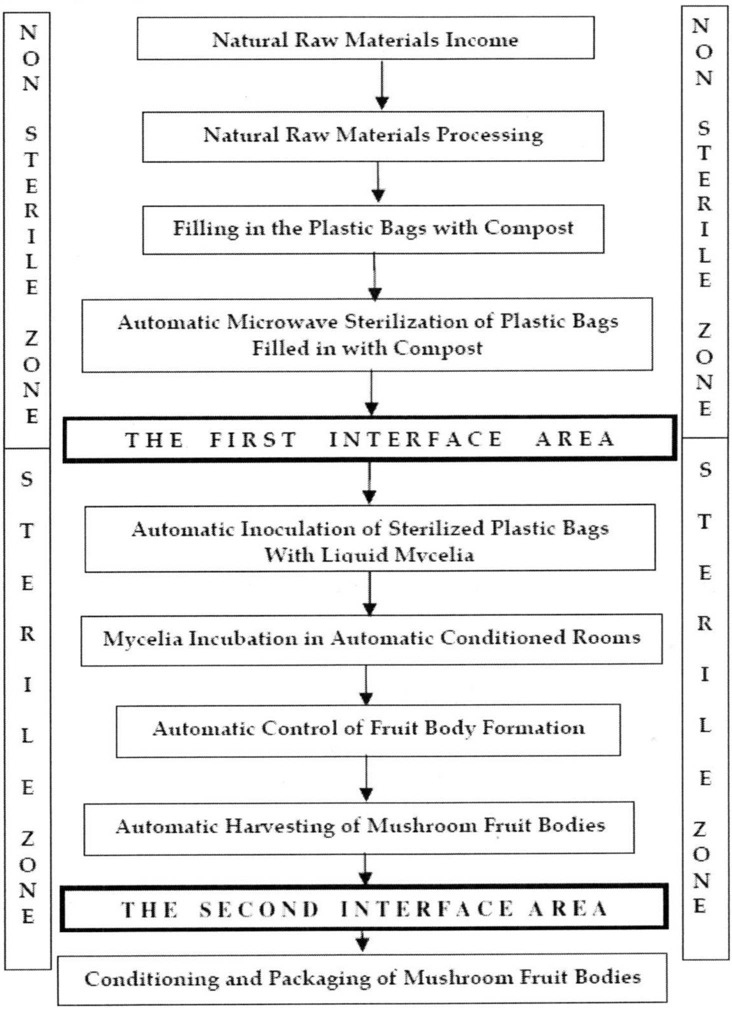


**Figure 11.** Schematic flow of the modular robotic system for controlled cultivation of edible mushrooms.

Both interfacing zones were designed to keep the sterile zone at the highest level of food safety against the microbial contamination. Using this robotic biotechnological model of mushroom cultivation, the economical efficiency can be significantly increased comparing to the actual conventional technologies, by shorting the total time of mushroom cultivation cycles in average with 5-10 days, depending on the mushroom strains that were grown and providing high quality mushroom fruit bodies produced in complete safety cultivation system ( Petre *et al.,* 2009 ).

## 4.1. Results and discussion

To increase the specific processes of cellulose biodegradation of winery and vineyard wastes and finally induce their bioconversion into protein of fungal biomass, there were performed experiments to cultivate the mushroom species of *P. ostreatus* and *L. edodes* on the following variants of culture substrata (see Table 5).

**Table 5.** The composition of five compost variants used in mushroom culture.

| Variants of culture substrata | Composition |
|---|---|
| S1 | Winery wastes |
| S2 | Mixture of winery wastes and rye bran 2.5% |
| S3 | Mixture of winery wastes and rise bran 2% |
| S4 | Mixture of vine cuttings and wheat bran 1% |
| S5 | Mixture of vine cuttings and barley bran 1.5% |
| Control | Pure cellulose |

The fungal cultures were grown by inoculating 100 ml of culture medium with 3-5% (v/v) of the seed culture and then cultivated at 23-25 C in 250 ml rotary shake flasks. The experiments were conducted under the following conditions: temperature, 25 C; agitation speed, 120-180 rev min $^{-1}$; initial pH, 4.5–5.5. After 10–12 d of incubation the fungal cultures were inoculated aseptically into glass vessels containing sterilized liquid culture media in order to produce the spawn necessary for the inoculation of 10 kg plastic bags filled with compost made of winery and vineyard wastes (Petre *et al.,* 2011; Petre *et al.,* 2009).

These compost variants were mixed with other natural ingredients in order to improve the enzymatic activity of mushroom mycelia and convert the cellulose content of winery and vineyard wastes into protein biomass. Until this stage, all the technological operations were handmade. In the next production phases, all the operations were designed to be carried out automatically by using a robotic modular system, which makes feasible the safety culture of edible mushrooms in continuous flow using as composts the winery and vineyard wastes.

The modular robotic system designed for edible mushrooms cultivation provides the automatic sterilization of composts, the automatic inoculation inside the aseptic room by a special device of controlled injection of liquid mycelia, the incubation and fruit bodies formation in special chambers under controlled atmosphere and the automatic picking-up of mushroom fruit bodies. In this way, the whole bags filled with compost have to be sterilized at 90-100$^{0}$C, by introducing them in a microwave sterilizer. In the next stage, all the sterilized bags must be inoculated with liquid mycelia, which have to be pumped through an aseptic injection device (Fig. 12).

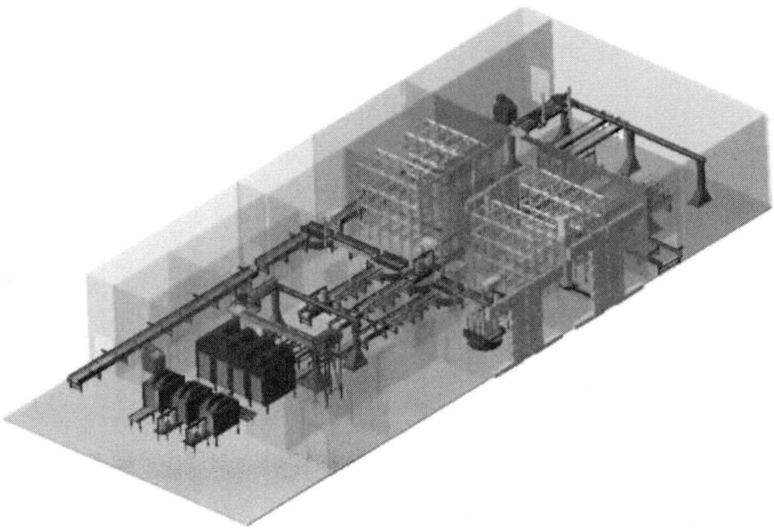

**Figure 12.** General overview of the modular robotic system for controlled cultivating of mushrooms.

Then, all the inoculated bags have to be transferred inside the growing chambers for incubation. After a time period of 10-15 d from the sterilized plastic bags filled with compost, the first buttons of the mushroom fruit bodies emerged.

For a period of 20-30 d there were harvested between 1.5 – 3.5 kg of mushroom fruit bodies per 10 kg compost bag. The specific rates of cellulose biodegradation were determined using the direct method of biomass weighing the results being expressed as percentage of dry weight (d.w.) before and after their cultivation. The registered data are presented in Table 6.

**Table 6.** The rate of cellulose degradation of culture substrata during the growing cycles of *L. edodes and P. ostreatus.*

| Variants of culture substrata | Before cultivation (g% d.w.) | | After cultivation (g% d.w.) | |
|---|---|---|---|---|
| | L. edodes | P. ostreatus | L. edodes | P. ostreatus |
| S1 | 2.6-2.7 | 2.7-2.9 | 0.5 | 0.9 |
| S2 | 2.3-2.5 | 2.5-2.8 | 0.4 | 0.7 |
| S3 | 2.3-2.5 | 2.3-2.5 | 0.5 | 0.4 |
| S4 | 2.5 -2.7 | 2.5 -2.7 | 0.7 | 0.8 |
| S5 | 2.7-2.9 | 2.5-2.7 | 0.5 | 0.7 |
| Control | 3.0 | 3.0 | 1.4 | 1.5 |

The registered data revealed that by applying this biotechnology, the winery and vineyard wastes can be recycled as useful raw materials for mushroom compost preparation in order to get significant production of mushrooms.

In this respect, the final fruit body production during the cultivation of these two mushroom species was registered as being between 20–28 kg relative to 100 kg of composts made of winery wastes.

Significant bioconversion increasing of the winery and vineyard wastes by using the modular robotic system of continuous controlled cultivation of edible mushrooms can be achieved by:

a) using pure strains of the mushroom species *P. ostreatus* and *L. edodes* whose biomass has got nutritive and functional properties proved by the research results of some achieved projects or others that are running now;

b) excluding any potential contamination sources for the edible mushrooms by using total sterilization or filtration equipments in each production module, by controlling all raw and auxiliary materials, water and air;

c) keeping the high precision and accuracy of the inoculation operations, incubation and fruit body formation of edible mushrooms which induce constant biomass composition of either fungal mycelia or mushroom fruit bodies;

d) avoiding all errors in the sterile zone of production flow as well as the potential risk of edible mushroom contamination by the human operators.

# 5. CONCLUSIONS

According to the previous mentioned results, the following conclusions can be drawn

1) Most suitable organic compost for mycelia growing was prepared from marc of grapes, showing the highest influence upon the mycelia growing and fresh mushroom production of 32–35 g%.

2) From the tested nitrogen sources, barley bran was the most efficient upon the mycelia growing and fruit mushroom producing at 35-40 g%, being closely followed by rice bran at 25–30 g% both in case of *P. ostreatus* and *L. edodes*, all data being reported as fresh biomass.

3) Among the tested mineral sources, the natural calcium carbonate ($CaCO_3$) yielded the best mycelia growing as well as fungal biomass production at 28-32 g%; for this reason it was registered as the most appropriate mineral source being followed by natural gypsum ($CaSO_4$ 2 $H_2O$) at 20-23 g%.

4) The originality and novelty of this biotechnology of winery and vineyard wastes recycling was confirmed by the Patents no 121717/2008 and 121718/2008 issued by the Romanian Office of Patents and Trade Marks

5) The mushroom pellets produced by applying the controlled cultivation of mushrooms as biotechnology for ecological treatment of winery wastes was 100% made by natural means and will be used for food supplements production with therapeutic properties which will

contribute to the increasing of health level of human consumers with nutritional metabolic deficiencies.

6) The biochemical correlation between the dry weight of mushroom pellets and their sugar and nitrogen contents was kept at a balanced ratio for each tested mushroom species. Among all mushroom samples that were tested in biotechnological experiments *G. lucidum* G-3 had shown the best values of its composition in sugars, total nitrogen and total protein content.

7) The originality and novelty of these biotechnological procedures to recycle the cereal wastes in order to get high nutritive biomass of mushroom pellets were confirmed through the Patents no 121677/2008, 121678/2008 and 121679/2008 issued by the Romanian Office of Patents and Trade Marks

8) By applying the biotechnology of controlled cultivation of edible mushrooms in modular robotic system, the final fruit body productions of both mushroom species *P. ostreatus* as well as *L. edodes* were registered as being between 20–28 kg relative to 100 kg of composts made of winery wastes.

9) The continuous controlled cultivation of edible mushrooms by using the modular robotic system can be achieved by:

   a) using pure strains of the mushroom species *P. ostreatus* and *L. edodes* whose biomass has got nutritive and functional properties proved by the research results of some achieved projects or others that are running now;

   b) excluding any potential contamination sources for the edible mushrooms by using total sterilization or filtration equipments in each production module, by controlling all raw and auxiliary materials, water and air

   c) keeping the high precision and accuracy of the inoculation operations, incubation and fruit body formation of edible mushrooms which induce constant biomass composition of either fungal mycelia or mushroom fruit bodies

   d) avoiding all errors in the sterile zone of production flow as well as the potential risk of edible mushroom contamination by the human operators.

10) The originality and novelty of this biotechnology of controlled cultivation of edible mushrooms in modular robotic system were confirmed by the Patent no 123132/20010, issued by the Romanian Office of Patents and Trade Marks.

## ACKNOWLEDGEMENTS

All these works were supported by the Romanian Ministry of Education and Research through the Projects no. 51-002/2007 and 52143/2008 in the frame-work of the 4th Programme of Research and Development – „Partnership in priority domains“

## REFERENCES

1. Bae, J.T.; Sinha, J.; Park, J.P.; Song, C.H. & Yun, J.W. (2000). Optimization of submerged culture conditions for exo-biopolymer production by Paecilomyces japonica. Journal of Microbiology and Biotechnology, Vol. 10, pp. 482-487, ISSN: 1017-7825
2. Carlile, M.J. & Watkinson, S.C. (1996). Fungi and biotechnology. In: The Fungi, M.J. Carlile, S.C. Watkinson (Eds.), 253-264, Academic Press, ISBN: 0-12-159960-4, London, England
3. Chahal, D.S. (1994). Biological disposal of lignocellulosic wastes and alleviation of their toxic effluents. In: Biological Degradation and Bioremediation of Toxic Chemicals, G.R. Chaudry (Ed.), 347-356, Chapman & Hall, ISBN: 978-0-412-62290-8, London, England
4. Jones, K. (1995). Shiitake – The Healing Mushroom. Healing Arts Press, Rochester, ISBN: 0- 89281-499-3, Vermont, USA Moo-Young, M. (1993). Fermentation of cellulose materials to mycoprotein foods, Biotechnology Advances, Vol. 11, No. 3, pp. 469-482, ISSN: 0734-9750
5. Moser, A. (1994). Sustainable biotechnology development: from high-tech to eco-tech. Acta Biotechnologica, Vol. 12, No. 2, pp. 10-15, ISSN: 0138-4988
6. Oei, P. (2003). Mushroom Cultivation. 3rd Edition, Backhuys Publishers, ISBN: 90-5782-137- 0, Leiden, The Netherlands
7. Petre, M.; Teodorescu, A.; Bejan, C.; Giosanu, D. & Andronescu, A.(2011). Enhanced Cultivation of Edible and Medicinal Mushrooms on Organic Wastes from Wine Making Industry. In: Proceedings of the International Conference „Environmental Engineering and Sustainable Development“, pp. 234-239, ISBN: 978-606-613-002-8, Alba Iulia, Romania, May 26-28, 2011
8. Petre, M. & Teodorescu, A. (2010). Handbook of submerged cultivation of eatable and medicinal mushrooms. CD Press, ISBN: 978-606-528-087-8, Bucharest, Romania
9. Petre, M.; Teodorescu, A.; Tuluca, E.; Bejan, C. & Andronescu, A. (2010). Biotechnology of Mushroom Pellets Producing by Controlled Submerged Fermentation. Romanian Biotechnological Letters, Vol 12, No. 2, pp. 50-56, ISSN: 1224-5984
10. Petre, M. & Teodorescu, A. (2009): Ecological Biotechnology for Agro-Food Wastes Valorization. In: Biotechnology of Environmental Protection, M. Petre (Ed.), vol. 2, 2nd Edition, 143-150, CD Press, ISBN: 978-606-528-040-3; 978-606-528-042-7, Bucharest, Romania
11. Petre, M.; Teodorescu, A.; Nicolescu, A.; Dobre, M. & Giosanu , D. (2009). Food biotechnology for edible mushrooms producing by using a modular robotic system. In: Proceedings of 2nd International Symposium „New Researches in Biotechnology",
12. SimpBTH 2009, Biotechnology Series F–Suppl., pp. 261-269, ISSN:1224-7774, Bucharest, Romania, November 18-19, 2009
13. Petre, M. & Petre, V. (2008). Environmental Biotechnology to Produce Edible Mushrooms by
14. Recycling the Winery and Vineyard Wastes. Journal of Environmental Protection and Ecology, Vol. 9, No.1, pp. 88-95, ISSN: 1311-5065

15. Petre, M.; Bejan, C.; Visoiu, E.; Tita, I. & Olteanu, A. (2007). Mycotechnology for optimal recycling of winery and vine wastes. International Journal of Medicinal Mushrooms, Vol. 9, No. 3, pp. 241-243, ISSN: 1521-9437
16. Reed, J.N.; Miles, S.J; Butler, J.; Baldwin, M. & Noble, R. (2001). Automation and Emerging Technologies for Automatic Mushroom Harvester Development. Journal of Agricultural Engineering Research, Vol 33, pp. 55-60, ISSN: 1095-9246
17. Smith, J. (1998). Biotechnology. 3rd Edition. Cambridge University Press, ISBN: 0-521-44911- 1, London, EnglandStamets, P. (2000). Growing Gourmet and Medicinal Mushrooms. Ten Speed Press, ISBN: 1- 58008-175-4, Berkeley, Toronto, Canada
18. Verstraete, W. & Top, E. (1992). Holistic Environmental Biotechnology. Cambridge University Press, ISBN: 0-521-42078-4, London, England
19. Wasser, S.P. & Weis, A.L. (1994). Therapeutic effects of substances occurring in higher Basidiomycetes mushrooms: a modern perspective. Critical Reviews in Immunology, Vol. 19, pp. 65-96, ISSN: 1040-840

# CHAPTER 6

# Perception of climate extreme trends over three Ethiopian eco-environments: Comparison with records and analysis of determinants

**Aklilu Mekasha[1*], Chillot Yirga[1], Kindie Tesfaye[4], Lisanework Nigatu[2], Alan J. Duncan[3]**

[1]*Ethiopian Institute of Agricultural Research (EIAR) P. O. Box 2003, Addis Ababa, Ethiopia.*

[2]*Haramaya University, P. O. Box 138, Dire Dawa, Ethiopia.*

[3]*International Livestock Research Institute (ILRI) P. O. Box 5689, Addis Ababa, Ethiopia.*

[4]*International Maize and Wheat Improvement Center (CIMMYT), ILRI Campus, P. O. Box 5689, Addis Ababa, Ethiopia.*

## ABSTRACT

Understanding household perceptions of climate change and determinants of such perceptions are important for planning community/household based climate change mitigation and adaptation strategies. In this study, herding/farming households' perceptions were studied and compared with recorded trends of extreme rainfall and temperature indicators from nearby weather stations across three eco-environments (pastoral, agro-pastoral and mixed crop-livestock highland system) in Ethiopia. Factors influencing household perceptions were assessed using a multinomial logit model. Results indicated that the majority of households (52.5-98.8%) across the three eco-environments perceived increasing numbers of extreme warm days and warm nights and decreasing numbers of extreme cool days and cool nights. In most cases, the household perceptions agreed with the recorded extreme temperature trends. Household perceptions of the studied extreme events were significantly affected by literacy, eco-environment, contact with the agricultural extension service, and presence of relief aid. We conclude that policy programs that enhance the literacy level of household and strengthen eco-environment-based extension services may increase the level of awareness and understanding of climate change by households which could help them to better adapt to climate change.

**Keywords:** Determinant, Ethiopia, household, perception, rainfall, season, temperature.

# 1. INTRODUCTION

Climate change as a reality has been increasingly recognized with the advent of a growing number of scientific studies (Henry, 2000; Thornton et al., 2006; Trenberth et al., 2007). In many cases, analysis of weather monitoring station data is a primary source of evidence (Trenberth et al., 2007). In Sub-Saharan Africa, however, scientific studies based on station data lag behind other parts of the world mainly because of low station density, lack of data continuity and heterogeneity in the quality of records (Seleshi and Zanke, 2004). Similarly weather information from monitoring stations seldom reaches herders and farmers who rather rely on age old traditional knowledge and perceptions accumulated through long historical exposure to the different facets of local climate (Nyong et al., 2007; Ishaya and Abaje, 2008).

Numerous studies have investigated how local knowledge and perceptions (Hansen et al., 2004; Viscusi and Zekhauser, 2006; Maddison, 2007; Semenza et al, 2008;Gbetibouo,2009**;** Deressa et al., 2011; Ofuoku, 2011;Piya et al., 2012; Silvestri et al., 2012) or social awareness (Saroar and Routray, 2010; Sarkar and Padaria, 2010; Acquah, 2011; Mandleni and Anim, 2011; Akerlof et al., 2013) are related to weather monitoring stations' records from climate change perspectives (Maddison, 2007; Benedicta et al., 2010). Among others, studies by Maddison (2007) in eleven African countries showed that significant number of African farmers' perceptions of increased temperature and decreased rainfall are somewhat equivocal with records from weather monitoring stations. Farmers' perceptions that climate is changing were found to be consistent among neighborhoods (Maddison, 2007; Bryan et al., 2013) and overwhelming in some cases (Benedicta et al., 2010; Enujeke, and Ofuoku, 2012) while trends derived from weather station data were found to show a much less clear picture of climate change (New et al., 2006).

The joint CCl/CLIVAR/JCOMM Expert Team (ET) on Climate Change Detection and Indices (ETCCDI) (http://cccma.seos.uvic.ca/ETCCDI) have developed and recommended 27 precipitation and temperature extreme indices for detecting climate change. Yet there are no agreed standard indices available for using human perceptions in climate change studies. Available literature to date reports perceptions based on general descriptions of temperature and rainfall trends (Hassan and Nhemachena, 2008; Maddison, 2007; Deressa et al., 2011). Such generalizations, however, are ambiguous to understand as they are not validated against weather station data. Secondly, people seem more able to detect and remember extreme weather events (Bento et al., 2013) than more gradual changes in averages (Hulme et al., 2009). Thirdly, because of media exposure, peoplemay falsely attribute occasional but normal events such as yield reductions, changes in vegetation phenologies and/or droughts to climate change where in reality they represent extremes of a time series whose mean is stable (Byrd et al., 2001; Weber, 2010). However, these short comings of perception studies could be partially dealt with by matching perceptions with analysis of standard meteorological extreme indices using records from nearby weather observatory stations.

Climate change perceptions can also be shaped by psychometric, cultural, demographic, social and institutional factors (Vedwan, 2006; Dhaka et al., 2010; Acquah, 2011; Hasan and Akhter, 2011; Silvestri et al., 2012). For example the way in which the climate change issue is addressed in mass media, education and extension determines households' awareness. Farming experience is also highly related to experience with local climatic normal, extreme events and general environ-mental responses. For instance, the Borana and Guji pastoralists of southern Ethiopia have an ecologically sound range management culture. Their seasonal movements, grazing and watering resources are managed by traditional rules and regulations that have evolved in response to local climate (Coppock, 1994; Abebe, 2000; Desta, 2000; Angassa and Oba 2007; Abate et al., 2010). Similarly the mixed crop-livestock farming highlanders who have adopted a sedentary life based on crop farming have also developed cropping and farming cultures in response to their local climate (Gebru, 2001; Vedwan, 2006; Lamma and Devkota, 2009). The agro-pastoral households have also developed efficient herding and farming cultures based on opportunistic use of experience from both pastoralism and farming.

Despite this rich traditional knowledge, there are few rigorous studies that relate household perceptions to weather station data in Sub-Saharan Africa. A prior understanding of these households perceptions on climate extreme trends and their relationship with weather monitoring station data and other factors at play could help to create community sensitization on climate change; such understanding could also enable policy makers and development planners to design and popularize community based climate resilient adaptation strategies. Thus this paper presents results of relating household perceptions on selected rainfall and temperature extremes indicators and comparing them with values computed from nearby daily weather station data over three seasons and major eco-environments of Ethiopia. It also assesses major environmental, social and/or institutional factors that influenced household perceptions taking the case of pastoralists in Liben, agro-pastoralists in Mieso and mixed crop-livestock highland farmers in Tiyo districts of Oromia National Regional State in Ethiopia.

# 2. MATERIALS AND METHODS

## 2.1 Analysis of extreme rainfall and temperature trends

### 2.1.1 Data source, station selection and quality control

Long term (1967-2008) daily rainfall and temperature data from stations located at Negele-Borana in Liben district in the pastoral eco-environment, Mieso district in the agro-pastoral eco-environment, and Asela and Kulumsa in Tiyo district in the mixed crop-livestock highland eco-environments of the country were sourced from the National Meteorological Agency of Ethiopia. Prior to analysis, the data of all these stations were plotted against time in days of the year format and subjected to visual examination for quality control. Special codes for missing values were removed. Typing errors, duplicated years and outliers defined as values above or below the mean plus or minus 4 times the standard deviation (Tank et al., 2009) were treated on a case by case using information from the day before and after the event and also by reference to nearby stations.

### 2.1.2 Defining the extreme parameters and trend analysis

The quality controlled data were subjected to the RClimDex package developed to run under the open source R software (R Development Core Team, 2012) to compute number of extreme cool days (TX10p), number of extreme cool nights (TN10p), number of extreme warm days (TX90p) and number of extreme warm nights (TN90p), seasonal total rainfall (PRCPTOT) and simple daily intensity index (SDDI) for the major rainy, small rainy and dry seasons as defined in Table 1 after ETCCDI's (http://cccma.seos.uvic.ca/ETCCDI) definition. The period from 1971 to 2000 was used as the base period in the analysis. A linear trend was then fitted (on mean values of the two station in case of Tiyo district) using Kendal's *tau* and the slope of the line was computed using the Sen's slope estimator in order to determine the rate of change in extreme events. The statistical significance of the slopes was tested at 5% probability level.

## 2.2 Household perception survey

### 2.2.1 Site and household selection

A total of 217 households were selected through a mix of purposive and stratified random sampling. First, the three districts, one from each of the three major eco-environments of the country were selected purposively so as to represent the pastoral, agro-pastoral and mixed crop-livestock highland eco-environments. Once the districts are selected, the lowest administrative units (Kebeles) in each of these districts were stratified into three strata using subjective expert judgment based on relative proportion of land allocated to crops and livestock, and agro-ecological settings of the Kebeles. From each stratum one Kebele was randomly selected and household censuses conducted to collect names and ages of the household heads. Household heads above 50 years old (supposed to have rich knowledge of local environment) were identified and later sampled randomly for selection of households. The number of households surveyed was proportional to the number of households in the district; a total of 81, 44 and 92 households in the pastoral, agro-pastoral and mixed crop-livestock highland systems were randomly selected for interview.

### 2.2.2 Household interview

A structured questionnaire was prepared, pretested and administered with 217 selected household heads whose age range from 51 to 82 with mean of 63 years old. Data collected included household level data such as literacy level of the household head, land holding, livestock ownership, social and /or institutional responsibility of the household head, distance from market, access to extension services and relief aid (Table 2). The interview also included information on perceptions of the respondent about trends of rainfall amount, daily rainfall intensity, and frequency of extreme cool days, cool nights, warm days and warm nights for the major rainy, small rainy and dry seasons (Table 3).

**Table 1.** Extreme precipitation and temperature indices and their definition

| S/N | Index | Definition of the index | Unit |
|---|---|---|---|
| 1 | PRCPTOT | Seasonal total rainfall: determined by summing daily precipitation events in a season with daily rainfall ≥1 mm. | mm/day |
| 2 | SDII | Simple daily intensity index: determined as season's seasonal total rainfall on wet days (precipitation ≥1 mm) divided by number of rainy days with rainfall ≥1 mm in a season. | mm/day |
| 3 | TN10p | Cool nights: Percentage of days in a season when the daily minimum temperature is less than 10$^{th}$ percentile of base period (1971-2000). | Days |
| 4 | TX10p | Cool days: Percentage of days in a season when the daily maximum temperature is less than 10$^{th}$ percentile of base period (1971-2000). | Days |
| 5 | TN90p | Warm nights: Percentage of days in a season when the daily minimum temperature is greater than 90$^{th}$ percentile of base period (1971-2000). | Days |
| 6 | TX90p | Warm days: Percentage of days in a season when the daily maximum temperature is greater than 90$^{th}$ percentile of base period (1971-2000). | Days |

**Table 2.** Description of the independent variables.

| Variable | Description | Value |
|---|---|---|
| **Household characteristics** | | |
| Literacy | Household head literacy status. | 1= literate, 0= illiterate |
| Land holding | Household head's land holding. | hectare |
| Livestock ownership | Household head's livestock holding | Tropical livestock unit (TLU) |
| Distance from market | Household head's residential house distance from nearby market place | km |
| **Institutional factors** | | |
| Social/institutional responsibility | Household head's social and / or institutional responsibility such as serving as priest, 'kebele administration, etc. | 1= yes, 0=no |
| Access to ext. service | Household head's access to extension service including material benefits on livestock and crop over the last decades in his farm life. | 1= yes, 0=no |
| Access to relief aid | Household head food relief aid and or safetynet assistance received over the last decades in his farm life | 1= yes, 0=no |
| **Environmental factors** | | |
| **Eco-environment** | | |
| Pastoral | Household head living in pastoral eco-environment. | 1 yes, 0= otherwise |
| Agro-pastoral | Household head living in agro-pastoral eco-environment. | 1 yes, 0= otherwise |
| Mixed crop-livestock highland | Household head living in mixed crop-livestock highland eco-environment | 1 yes, 0= otherwise |
| **Season** | | |
| Major rain season | Household head's response for the major rain season | 1 yes, 0= otherwise |
| Small rain season | Household head's response for the small rain season | 1 yes, 0= otherwise |
| Dry season | Household head's response for the dry season | 1 yes, 0= otherwise |

**Table 3.** The dependent variables and definition used for the study.

| S/N | Variable | Definition of the variable | Unit |
|---|---|---|---|
| 1 | TR | Seasonal total rainfall: Perceived amount of rainfall in a season | polycotomous: 1 = increasing; 2, decreasing; 3 = no change |
| 2 | IR | Intensity of rainfall: Perceived strength of rainfall during raining time in a season | Polycotomous: 1= increasing; 2, decreasing; 3 = no change |
| 3 | CD | Frequency of cool days: Number of days with extreme coolness of the day time hours in a season | Polycotomous: 1= increasing; 2, decreasing; 3 = no change |
| 4 | CN | Frequency of cool nights: Number of days with extreme coolness of the night time hours in a season | Polycotomous: 1 = increasing; 2, decreasing; 3 = no change |
| 5 | HD | Frequency of warm days: Number of days with extreme warmness of the day time hours in a season | Polycotomous: 1 = increasing; 2, decreasing 3 = no change |
| 6 | HN | Frequency of warm night: Number of days with extreme warmness of the night time hours in a season | Polycotomous: 1 = increasing; 2, decreasing; 3 = no change |

### 2.2.3 Perception analysis: Theoretical framework

According to the Oxford dictionary perception is defined as the way in which something is regarded, understood, or interpreted. In the climate literature models two perception theories emerge prominently (Tansey and O'riordan, 1999). These are the psychometric and cultural theories. The psychometric theory is all about individualism and perception is treated as an exclusive property of individuals (Tansey and O'riordan, 1999). The cultural theory on the other hand is about people. It focuses on what is shared by people who form their outlook through their interaction in the social world (Tansey and O'riordan, 1999). According to Tansey and O'riordan (1999) culture is a shared interpretative framework for such groups or the common way that a group of persons make sense of the world. They share common sets of plans, laws, rules, regulations, customs, belief, norms and rituals to which individuals abide. According to the psychometric theory a human being uses close observation to assess his local climate and makes day to day decisions about farming, travel, clothing and others to match his local weather conditions. Thus, from experience *per se* human beings can assess significant changes in local climate.

Those who oppose this theory, however, say that "the deviations in the long term mean termed climate change is difficult to recognize unless one uses certain statistical analysis and that climate change is a constructed issue" (Storch, 2011). According to this group there are different classes of constructions (Stehr and Storch, 1995; Pasquaré and Oppizzi, 2012). One is through objective analysis of observations and interpretation by theories and the other is what is maintained and transformed by the public media (Stehr and Storch, 1995; Pasquaré and Oppizzi, 2012). Thus, according to this theory farmers' and herders' perceptions of changes in the extremes of rainfall and temperature indices might be derived from external sources such as extension services or mass media. According to Weber (2010), Frank et al. (2011) and Akerlof et al. (2013) knowledge of local climate is derived from personal experiences, local sources of knowledge and external sources of techno-scientific information. This indicates that herding/farming household perceptions of changes in those indices could be a result of personal experiences and influences from external agents. Based on these theories we used empirical models to identify what might be responsible for the observed household perceptions of rainfall and temperature extremes from sets of given environmental, social/ institutional and personal variables as given subsequently.

### 2.2.4 The empirical model and model specification

A multinomial logit (MNL) model commonly used for climate change adaption studies (Deressa et al., 2009, Hassan and Nhemachena, 2008; Bryan et al., 2013) and adoption decision studies involving multiple choices was used to identify the determinants of household perceptions of rainfall and temperature extreme indicators. The model was estimated based on households' responses using three choices namely increases (j=1), decreases (j=2) and no change (j=3) in specified climate extreme variables. Each household head was asked to give as single perception choice (j=1...J) to each rainfall and temperature extreme indicator denoted as y=1...Y. Thus for each extreme indicator taking one perception choice with sets of conditioning factors and household characteristics, the MNL model takes the following form:

$$P(y = j \setminus x) = \frac{\exp(x\beta j)}{1 + \sum_{h=1}^{J} \exp(x\beta h), j = 1,\ldots J}$$

The MNL model requires assumption of independence of irrelevant alternatives (IIA) to hold which states that the probability of choosing a certain perception alternative by a given household needs to be independent from the probability of choosing another perception alternative (that is, Pj/Pk is independent of the remaining probabilities).

The parameter estimates of the MNL model offer only the direction of the effect of the independent variables on the dependent (response) variable, and estimates do not signify either the actual magnitude of change nor probabilities. Hence differentiating the equation above with respect to the explanatory variables provides marginal effects of the explanatory variables given as:

$$\frac{\partial Pj}{\partial x_k} = Pj(\beta_{jk} - \sum_{J=1}^{J-1} P_j \beta_{jk})$$

Where X is a vector of perception characteristics (specified in Table 3), β is a set of estimated parameters and J is a number of choices.

The marginal effects or marginal probabilities are functions of the probability itself and measure the expected change in probability of a particular choice being made with respect to a unit change in an independent or explanatory variable (Greene, 2000).

### 2.2.5 Dependent variables

A total of six dependent variables (Table 3) which included household perceptions of trends in seasonal total rainfall, intensity of rainfall, frequency of extreme cool days, frequency of extreme cool nights, frequency of extreme warm days and frequency of extreme warm nights were identified. Six independent MNL models were run to regress these dependent variables on sets of environmental and social and/or institutional factors hypothesized to affect household perceptions.

### 2.2.6 Independent variables

Nine independent variables namely, household head characteristics (literacy level, land holding and livestock ownership), access to institutional services (social and /or institutional responsibility, distance from market, access to extension and relief aid services), environmental factors (eco-environment and season) were used in the model to assess the effect of changes in the above response variables on household perceptions (Table 2).

## 2.3 Hypotheses to be tested

### 2.3.1 Household characteristics

According to Hidalgo and Pisano (2010) environmental perceptions in relation to climate change are related to knowledge of the respondents. In line with this studies have shown that education increases climate change perception and awareness (Maddison, 2007; Deressa et al., 2009; Acquah 2011; Hasan and Akhter, 2011; Enujeke and Ofuoku, 2012; Tesso et al., 2012). Hence literate household heads might have more access to information and also analysis of environmental factors than their illiterate or uneducated counterparts. Literate household heads are expected to perceive changes in rainfall and temperature extreme indicators more than the illiterate household heads.

Land and livestock holdings represent wealth and farm activities. The size of these holdings influences farmers planning and execution of activities which could be affected by climate (Deressa et al., 2009). Studies have shown that livestock ownership is positively related to climate change perception and adaptation decisions (Deressa et al., 2011; Silvestri et al., 2012; Mandleni and Anim, 2011). We hypothesize that farming and herding households with large land and livestock holding sizes may know climate change better than those with few livestock for several reasons. Firstly they have more social interactions and communications with neighbors for the management of their land and livestock. Secondly, they need to adjust their farming and herding practices and operational calendars in response to the changing local level climate extremes. Thirdly, they may be wealthier through sale of their farm produce and livestock and this wealth can improve access to communication media.

### 2.3.2 Household access to institutional services

Household heads with social and /or institutional responsibilities such as serving the community as religious leaders, arbitration, and involvement in various capacities within the 'Kebele' administration gives opportunity for communication with various people who might include those informed about climate change. Hence, we hypothesis that household heads with social and /or institutional responsibility could perceive rainfall and temperature extremes changes better than others with fewer responsibility.

Distance from input markets has been found to positively affect household perceptions of climate change and adaptation to the changes (Mandleni and Anim, 2011). In contrast, studies by Tesso et al. (2012) showed that distance from market was negatively related to farmers' perceptions of climate change. We hypothesize that households close to market outlets may have more awareness about climate change than distant ones. Those household heads close to market centers have a tendency to frequent market areas where they meet many people and share ideas about changes in rainfall and temperature extremes.

Extension services are an important source of information on climate and climate related issues (Deressa et al., 2009, 2011; Hassan and Nhemachena, 2008). Access to information on rainfall and temperature has been found to positively relate to climate change awareness and hence adaptation measures (Deressa et al., 2009; Dhaka et al., 2010; Mandleni and Anim, 2011; Enujeke and Ofuoku, 2012; Tesso et al., 2012). In contrast, studies by Silvestri et al. (2012) showed that climate change perceptions of agro-pastoralists are negatively affected by livestock extension field visits. We hypothesize that herders and farmers close to extension services have better interaction with extension agents and hence have information on rainfall and temperature extremes. This is because extension agents themselves could teach them about climate change.

Studies have shown that food or other relief aid positively affects climate change perceptions of agro-pastoral households (Silvestri et al., 2012). We hypothesize that relief aid recipient households have a better perception of rainfall and temperature extremes than the non-recipient households. This is because on the one hand relief aid recipients understand that increased frequencies of climatic extremes might expose them to repeated crop failures and livestock mortalities which lead them to be relief aid recipients. On the other hand donors and other informed relief aid workers could tell them of changes in rainfall and temperature extreme trend.

### 2.3.3 Environmental factors

Ethiopia is topographically very diverse but three major eco-environments with distinct climate and land use patterns have been specified. These diverse eco-environmental settings could affect perceptions of change in patterns of rainfall and temperature extremes among households living in each eco-environment. Previous studies in the Ethiopian mixed crop-livestock highlands have shown that farmers living in different eco-environments perceive climate change and hence take certain adaptation measures (Deressa et al., 2011) appropriate to their conditions. However studies by Tesso et al. (2012) did not find a significant effect of eco-environment on farmers' perceptions of climate change. We hypothesize that households living in different eco-environment may perceive changes in frequencies of occurrence of rainfall and temperature extremes from experience. Pastoral and agro-pastoral households have developed system behaviours focusing on mobility in response to the warm dry climate of pastoral areas and have knowledge on the occurrence of climatic extremes to which they have responded with certain management strategies. This is less important among the more stable and sedentary mixed crop-livestock highlanders.

Rainfall and temperature patterns differ among seasons over the different regions of the country. The rainfall of small rainy season plays vital role for land preparation and planting of long season crops such as sorghum and maize while the major rain season is important for planting short duration crops over most parts of the country. Studies by McSweeney et al. (2010) indicated significant increase in frequency of hot days and hot nights and conversely significant declining trends in frequencies of cold days and nights for the different seasons of Ethiopia. Seasonal differences in duration and magnitude of occurrence of rainfall and temperature variables affect choices of adaptation options to climate change (Hassan and Nhemachena, 2008). This difference could be one factor to affect household perceptions of changes in patterns of rainfall and temperature extremes. We hypothesize that because of their agricultural significance households perceive patterns of change in rainfall and temperature extremes better for the major and small rainy seasons than the dry season in the country.

### 2.3.4 Estimation of empirical model parameters

For all response variables, the multinomial logit model was estimated by taking the no change perception response as a baseline category against which other alternatives were compared. A multicollinearity test was conducted by employing an Ordinary Least Square (OLS) model using a Variance Inflation Factor (VIF) (Myers, 1990; O'brien, 2007). The VIF was found to be less than 10 for all variables indicating multicollinearity was not a serious problem in all the cases. We then ran Hausman's test to check the validity of the models for Independence of the Irrelevant Alternatives (IIA) assumption. The Hausman's test, however, failed to reject the null hypothesis of independence of the rainfall and temperature change percep-

**Table 4.** Model fitting diagnostic characteristics of the multinomial logit model.

| Diagnostics | TR | IR | CD | CN | HD | HN |
|---|---|---|---|---|---|---|
| Base category | No change | No change | No change | No change | No change | No change |
| Number of observation | 651 | 651 | 651 | 651 | 651 | 651 |
| Wald $X^2$ | 375068.67 | 248.16 | 61251.05 | 107282.83 | 1116.16 | 574.34 |
| $P>X^2$ | 0.0000 | 0.0000 | 0.0000 | 0.0000 | 0.0000 | 0.0000 |
| Pseudo $R^2$ | 0.3918 | 0.326 | 0.2555 | 0.2869 | 0.3236 | 0.2651 |
| Log pseudo likelihood | -375.6573 | -424.9048 | -349.1884 | -427.9930 | -268.6830 | -416.2042 |

TR = Seasonal total rainfall; IR = intensity of rainfall; CD = number of extreme cool days; CN = number of extreme cool night; HD = number of extreme warm days, HN = number of extreme warm night.

tion options, indicating the multinomial logit model specifications are adequate to model rainfall and temperature extreme perception of the study household. As evidenced from Table 4 the likelihood ratio test of the models were found to be highly significant (P<0.0000) indicating strong explanatory power of the models. For each of the six multinomial logit models we ran, the marginal effect and interpretations of model output are based on marginal effects of the multinomial logit. For all models the output for the increased and decreased perceptions are interpreted by comparing with the no change perception category.

# 3. RESULTS

## 3.1 Household perception of extreme weather events vis-à-vis observed records

### 3.1.1 Pastoral eco-environment

In the pastoral eco-environment, the majority of households perceived an increasing number of extreme warm days (WD) and warm nights (WN), and conversely a decreasing number of extreme cool days (CD) and cool nights (CN) across all seasons. The perceptions were in line with recorded significant increasing trends in WD and WN across seasons. The decreasing trends from observed data were significant only for the major and small rainy seasons but were broadly in line with the decreasing trends in CD and CN perceived by the household. The majority of pastoral households (61.0-94.0%) also perceived decreasing trends in seasonal total rainfall (TR) and daily rainfall intensity (IR) across all seasons. Perceptions on IR were in line with observed data which also showed decreasing trends in all seasons. On the other hand, the observed data for TR showed a significant decline only in the major rainy season (Table 5).

### 3.1.2 Agro-pastoral eco-environment

Similar to the pastoral eco-environment, in the agro-pastoral eco-environment, the majority of households perceived increasing numbers of extreme warm days (WD) and warm nights (WN), and conversely decreasing numbers of extreme cool days (CD) and cool nights (CN) across all seasons. However, the station data for CD indicated significant decreases only in the major rainy and dry seasons while the observed decrease was significant only in the major rainy season. Contrary to household perceptions, the recorded trends in WD revealed significant decreasing trends in the small rainy season. The majority of agro-pastoral households also perceived decreasing seasonal total rainfall (TR) and daily rainfall intensity (IR) across all seasons. However, the recorded trends showed no significant change in these parameters across all seasons (Table 6).

### 3.1.3 Mixed crop-livestock highland eco-environment

In the mixed crop-livestock highland eco-environment, most of the households perceived increasing numbers of extreme warm days (WD) and warm nights (WN), and conversely decreasing numbers of extreme cool days (CD) and cool nights (CN). The perceptions for the major rainy and dry seasons agree with the recorded station data for WN which also showed increasing trends. For the small rainy season, however, station data indicated that only WD showed significant increasing trends. On the other hand, the recorded station data did not show significant changes in CD and CN except for the small rainy season where a decrease was seen in the small rainy season. With regard to rainfall extremes (TR and IR), households were equally divided between perceiving these to be increasing and decreasing in the major rainy seasons plus also the small rainy season in case of TR. Similarly for dry season, the respondents were equally divided between perceiving decreasing trends and no change (Table 7).

### 3.1.4 Determinants of household perceptions

Results of the estimated marginal effects of the multinomial logit models are presented in Tables 8 to 10. The results show that most of the explanatory variables have statistically significant explanatory power at less than 10, 5 or 1% probability level and discussed subsequently.

### 3.1.5 Household characteristics

Literacy of the head of the household significantly affected perceptions of trends in seasonal total rainfall, daily rainfall intensity and number of extreme warm days among others. Being literate significantly decreased the probability of perceiving increased seasonal total rainfall, daily rainfall intensity and number of extreme warm days by 9.2, 9 and 7.8%, respectively and significantly increased the likelihood of increased perception of the number of warm days by 10.5%. These are in line with the recorded trends especially for the pastoral and agro-pastoral environment and imply that education enables household heads to be aware of changes in climate extremes as expected.

**Table 5.** Comparison of recorded and perceived trends of rainfall and temperature extremes for three distinct seasons in a pastoral ecoenvironment.

| Climate extreme event | Recorded | Perceived | | |
|---|---|---|---|---|
| | | Increasing (%) | Decreasing (%) | No change (%) |
| **Major rainy season** | | | | |
| Seasonal total rainfall | Decreasing (-1.47* mm/year) | 6.2 | 93.8 | 0 |
| Daily rainfall intensity | Decreasing (-0.16* mm/day/year) | 24.7 | 75.3 | 0 |
| number of extreme warm days | Increasing ( 0.39* days/year) | 97.5 | 2.5 | 0 |
| number of extreme warm nights | Increasing (1.22* days/year) | 75.3 | 24.7 | 0 |
| number of extreme cool days | Decreasing(-0.23* days/year) | 8.6 | 91.4 | 0 |
| number of extreme cool nights | Decreasing (-0.29* days/year) | 32.1 | 67.9 | 0 |
| **Small rainy season** | | | | |
| Seasonal total rainfall | No change(-0.83 mm/year) | 4.9 | 91.4 | 3.7 |
| Intensity of Rainfall | Decreasing (-0.064* mm/day/year) | 23.5 | 74.1 | 2.4 |
| number of extreme warm days | Increasing (0.39* days/year) | 98.8 | 1.2 | 0 |
| number of extreme warm nights | Increasing (0.70* days/year) | 72.8 | 27.2 | 0 |
| number of extreme cool days | Decreasing (-0.21* days/year) | 8.6 | 91.4 | 0 |
| number of extreme cool nights | Decreasing (-0.34* days/year) | 34.6 | 64.2 | 1.2 |
| **Dry season** | | | | |
| Seasonal total rainfall | No change(-0.99 mm/year) | 4.9 | 81.5 | 13.5 |
| Intensity of Rainfall | Decreasing (-0.092* mm/day/year) | 4.9 | 82.7 | 12.3 |
| number of extreme warm days | Increasing (0.43* days/year) | 96.3 | 2.5 | 1.2 |
| number of extreme warm nights | Increasing (0.89* days/year) | 72.8 | 27.1 | 0 |
| number of extreme cool days | Decreasing (-0.21* ) | 9.9 | 90.1 | 0 |
| number of extreme cool nights | Decreasing (-0.36* days/year) | 38.3 | 60.5 | 1.2 |

* Trends are significant P <5% probability level.

**Table 6.** Comparison of recorded and perceived trends of rainfall and temperature extremes for three distinct seasons in an ago-pastoral eco-environment.

| Climate extreme event | Recorded | Perceived | | |
|---|---|---|---|---|
| | | Increasing (%) | Decreasing (%) | No change (%) |
| **Major rainy season** | | | | |
| Seasonal total rainfall | No change (-0.40 mm/year) | 4.5 | 95.5 | 0 |
| Intensity of Rainfall | No change (-0.038 mm/day/year) | 9.1 | 88.6 | 2.3 |
| Number of extreme warm days | No change (0.13 days/year) | 68.2 | 29.5 | 2.3 |
| Number of extreme warm nights | No change (0.14 days/year) | 68.2 | 29.5 | 2.3 |
| Number of extreme cool days | Decreasing (-0.16* days/year ) | 29.5 | 65.9 | 4.6 |
| Number of extreme cool nights | Decreasing (-0.29* days/year) | 29.5 | 63.6 | 6.8 |
| **Small rainy season** | | | | |
| Seasonal total rainfall | No change (-0.93 mm/year) | 11.4 | 86.4 | 2.3 |
| Intensity of Rainfall | No change (-0.066 mm/day/year) | 9.1 | 86.4 | 4.6 |
| Number of extreme warm days | Decreasing (-0.17* days/year) | 72.7 | 27.3 | 0 |
| Number of extreme warm nights | No change (0.102 days/year) | 68.2 | 29.5 | 2.3 |
| Number of extreme cool days | No change (-0.069 days/year) | 29.5 | 65.9 | 4.5 |
| Number of extreme cool nights | No change (-0.053 days/year) | 25 | 68.2 | 6.8 |
| **Dry season** | | | | |
| Seasonal total rainfall | No change (-0.14 mm/year) | 25 | 50 | 25.0 |
| Intensity of Rainfall | No change (-0.046 mm/day/year) | 17.7 | 65.5 | 21.8 |
| Number of extreme warm days | No change (0.29 days/year) | 65.9 | 29.5 | 4.5 |
| Number of extreme warm nights | No change (0.054 days/year) | 65.9 | 25 | 9.1 |
| Number of extreme cool days | Decreasing (-0.16* days/year) | 38.6 | 52.3 | 9.1 |
| Number of extreme cool nights | No change (0.020 days/year) | 13.6 | 79.5 | 6.8 |

*Trends are significant at P <5% probability level.

**Table 7.** Comparison of recorded and perceived trends of rainfall and temperature extremes for three distinct seasons in an Mixed crop-livestock highland eco-environment.

| Climate extreme event | Recorded | Perceived | | |
|---|---|---|---|---|
| | | Increasing (%) | Decreasing (%) | No change (%) |
| **Major rainy season** | | | | |
| Seasonal total rainfall | Decreasing (-1.202* mm/year) | 47.8 | 51.1 | 1.1 |
| Intensity of Rainfall | No change (0.012 mm/day/year) | 48.9 | 51.1 | 0 |
| Number of extreme warm days | No change (0.051 days/year) | 77.2 | 14.1 | 8.7 |
| Number of extreme warm nights | Increasing (0.094* days/year) | 66.3 | 26.1 | 7.6 |
| Number of extreme cool days | No change (-0.032 days/year) | 18.5 | 77.2 | 4.4 |
| Number of extreme cool nights | No change (0.062 days/year) | 29.3 | 65.2 | 5.5 |
| **Small rainy season** | | | | |
| Seasonal total rainfall | No change (0. 147) | 47.8 | 47.8 | 4.4 |
| Intensity of Rainfall | No change (0.027 mm/year) | 28.3 | 67.4 | 4.4 |
| Number of extreme warm days | Increasing (0.109* days/year) | 65.2 | 29.3 | 5.4 |
| Number of extreme warm nights | No change (0.071 days/year) | 71.7 | 22.8 | 5.4 |
| Number of extreme cool days | Decreasing (-0.238* days/year) | 35.9 | 62.0 | 1.1 |
| Number of extreme cool nights | No change (0.070 days/year) | 30.4 | 66.3 | 3.3 |
| **Dry season** | | | | |
| Seasonal total rainfall | No change (-0.137 mm/year) | 13.0 | 45.7 | 41.3 |
| Intensity of Rainfall | No change (0.013 mm/day/year) | 16.3 | 44.6 | 39.2 |
| Number of extreme warm days | No change (0.016 days/year) | 90.2 | 6.5 | 3.3 |
| Number of extreme warm nights | Increasing (0.276* days/year) | 67.4 | 31.5 | 1.1 |
| Number of extreme cool nights | No change (-0.021 days/year) | 42.4 | 54.3 | 3.3 |
| Number of extreme cool days | No change (-0.038 days/year) | 27.2 | 69.6 | 3.3 |

*Trends are significant P<5% probability level.

**Table 8.** Marginal effects of explanatory variables from the multinomial logit perception models on seasonal total rainfall and intensity of rainfall.

| Explanatory variable | Seasonal total rainfall (TR) | | | Daily rainfall intensity(IR) | | |
|---|---|---|---|---|---|---|
| | Increase | Decrease | No change | Increase | Decrease | No change |
| Literacy | 0.0280 | -0.0915*** | 0.0635*** | 0.0240 | -0.0896** | 0.0240 |
| Land holding | -0.0055 | 0.0096 | -0.0041 | 0.0122 | -0.0026 | 0.0123 |
| Livestock ownership | 0.0025*** | -0.0025*** | -1.60E-05 | 0.0013** | -0.0017** | 0.0013** |
| Social/institutional responsibility | -0.0579*** | 0.1185*** | -0.0606** | -0.0431 | 0.1063** | -0.0431 |
| Distance from market | 0.0012 | -0.0021 | 0.0009 | -0.0005 | 0.0002 | -0.0005 |
| Access to ext. service | 0.0978* | -0.0789 | -0.0189 | -0.2040*** | 0.1905*** | -0.2040*** |
| Access to relief aid | -0.2355*** | 0.2272*** | 0.0083 | 0.1098** | -0.1258** | 0.1098** |
| Pastoral | -0.2143* | 0.2320** | -0.0177 | 0.1936** | -0.1476 | 0.1936** |
| Agro-pastoral | -0.4866*** | 0.5983*** | -0.1117** | 0.1064 | -0.0008 | 0.1063 |
| Major rainy season | -0.16456*** | -0.1623*** | 0.3268*** | -0.1682*** | -0.1505** | -0.1682*** |
| Small rainy season | -0.1323*** | -0.0414 | 0.1737*** | -0.0717** | -0.1062** | -0.0717** |

*, **, ***significant at < 10, <5 and < 1% P level, respectively.

**Table 9.** Marginal effects of explanatory variables from the multinomial logit perception models on number of extreme cool days and cool nights.

| Explanatory variable | Number of cool days (CD) | | | Number of cool nights (CN) | | |
|---|---|---|---|---|---|---|
| | Increase | Decrease | No change | Increase | Decrease | No change |
| Literacy | -0.0071 | -0.0106 | -0.0033 | 0.0317 | -0.0339 | 0.0022 |
| Land holding | 0.0178** | -0.0135 | -0.0075 | 0.0057 | -0.0025 | -0.0031 |
| Livestock ownership | -0.00013 | -0.0007 | -0.0005 | 0.00013 | -0.00056 | 0.00042 |
| Social/institutional responsibility | -0.0212 | -0.0019 | 0.0491* | -0.0517 | 0.0537 | -0.0020 |
| Distance from market | -0.0024** | 0.0029** | 0.0023** | -0.0022** | 0.00154 | 0.00065 |
| Access to ext. service | -0.0606 | 0.0694 | 0.0083 | -0.1886*** | 0.2149*** | -0.0263 |
| Access to relief aid | -0.1270*** | 0.1706*** | -0.0377 | -0.0609 | 0.0693 | -0.0084 |
| Pastoral | -0.2591** | 0.4812*** | -0.0682 | -0.2090** | 0.3324*** | -0.1234** |
| Agro-pastoral | -0.1682 | 0.4166*** | -0.1222 | 0.0545 | 0.0656 | -0.1201** |
| Major rainy season | 0.0317 | -0.0216 | 0.0663 | 0.0996** | -0.0925** | -0.0071 |
| Small rainy season | -0.0385 | 0.0358 | 0.0669 | 0.0812** | -0.0787** | -0.0025 |

*, **, *** significant at <10, 5 and 1% P level, respectively

**Table 10.** Marginal effect of explanatory variables from the multinomial logit perception models on number of extreme warm days and warm nights.

| Explanatory variable | Number of warm days (HD) | | | Number of warm nights (HN) | | |
|---|---|---|---|---|---|---|
| | Increase | Decrease | No change | Increase | Decrease | No change |
| Literacy | 0.1045** | -0.0777** | -0.0275 | 0.0302 | -0.0363 | 0.0061 |
| Land holding | -0.0099 | 0.0112 | -0.0012 | -0.0294*** | 0.0232** | 0.0062 |
| Livestock ownership | -0.0013 | 0.0016** | -0.00033 | -0.0017** | -0.00076 | 0.0025** |
| Social/institutional responsibility | -0.0029 | -0.0263 | 0.0291 | 0.0219 | -0.0586 | 0.0367 |
| Distance from market | 0.0024 | -0.00012 | -0.0023** | 0.00065 | 0.00024 | -0.00088 |
| Access to ext. service | -0.0477 | 0.0729 | -0.0252 | 0.1629*** | -0.1028* | -0.0602*** |
| Access to relief aid | 0.1123** | -0.0861** | -0.0261 | 0.0216 | 0.0104 | -0.0319 |
| Pastoral | 0.3607*** | -0.3171*** | -0.0436 | 0.4042*** | -0.3944*** | -0.0098 |
| Agro-pastoral | 0.3074** | -0.2172** | -0.0902** | 0.1185 | -0.0268 | -0.0917* |
| Major rainy season | 0.0878** | -0.0709** | -0.0169 | -0.0095 | 0.0342 | -0.0247* |
| Small rainy season | 0.1137** | -0.1156** | 0.0019 | -0.0266 | 0.0423 | -0.0157 |

*, **, *** Significant at < 10, 5 and 1% P level, respectively.

The size of farmland owned by households is related to perceptions on number of extreme cool days and warm nights. A unit increase in households' land holding size significantly increased the probability of increased perception of the number of extreme cool days by 1.9%. It significantly decreased the increased perception of the number of extreme warm nights by 2.9% and significantly increased the likelihood of decreased perception by 2.3%. However, these perceptions are not in line with the observed trends indicating that it is not the size of the farm, but the specific characteristics of the farm that may dictate household perceptions of changes in rainfall and temperature extremes trend.

A unit increase in livestock holding of the household significantly increased the probability to perceive increased seasonal total rainfall and daily rainfall intensity by 0.3 and 0.1%, respectively. It however, decreased the likelihood of decreased perception of seasonal total rainfall and daily rainfall intensity by 0.3 and 0.2%, respectively. Moreover, a unit increase in livestock ownership of the household significantly increased the probability of decreased perception of number of extreme warm days by 0.2% and significantly decreased the increased perception of the number of extreme warm nights by 0.2%. These perceptions are not supported by the observed trends. This might be due to better rainfall and temperature conditions in recent years that might result in better availability of pasture and water for livestock production.

### 3.1.6 Household access to institutional services

Having social and /or institutional responsibility significantly increased the probability of perceiving decreased seasonal total rainfall and daily rainfall intensity by 11.8 and 10.6%, respectively and increased the likelihood of perceiving decreased seasonal total rainfall by 5.8%. This indicates that as expectations social and/ or institutional responsibility enables farmers and herders to be more aware of climate extreme trends and that their perceptions on rainfall extremes may have been influenced by interactions with other peoples.

Contrary to expectations, a unit increase in distance from market center significantly increased the likelihood of decreased perception of number of extreme cool days by 0.3% and decreased the likelihood of increased perception by 0.2% and significantly decreased the probability of increased perception of the number of cool nights by 0.2%.This implies that though households far away from input centers have less access to information from market centers, they can perceive from experience the changes in number of extreme cool days and night better than those nearby to market centers.

As expected access to extension services significantly increased the probability of perceiving increased number of extreme warm nights by

16.3%. It significantly decreased the probability of increased perception of daily rainfall intensity and number of extreme cool nights by 20.0 and 18.9%, respectively. Moreover, access to extension services significantly decreased the likelihood of perceiving decreased number of extreme warm nights by 10.3%, respectively. It also significantly increased the probability of decreased perception of number of cool nights by 21.5%. However, unlike expectations access to extension services significantly increased the probability of perceiving increased seasonal total rainfall by 9.8% and decreased the likelihood of perceiving decreased daily rainfall intensity by 19.0%. Relief aid assistance also significantly increased households' likelihood of perceiving decreased seasonal total rainfall and number of extreme cool days by 22.7 and 17%, respectively. It significantly decreased the likelihood of increased perception of seasonal total rainfall and number of extreme cool days by 23.6 and 12.7%, respectively. Access to relief aid significantly increased the probability of perceiving increased number of warm days by 11%, but significantly decreased the likelihood of perceiving decreased number of extreme warm days by 8.6%. However, access to relief aid significantly increased the probability of perceiving increased daily rainfall intensity by 11.0% and decreased the likelihood of perceiving decreased daily rainfall intensity by 12.6%. This indicates that herding/farming households close to extension and relief aid services for information, advice and material benefits perceived as expected especially on number of extreme warm and cool days and nights. The results emphasize the importance of extension and relief aid services in climate change perception.

### 3.1.7 Eco-environmental factors

Compared to being in the mixed crop-livestock highland eco-environment, being in the pastoral eco-environment significantly increased the probability of perceiving decreased seasonal total rainfall, number of extreme cool days and number of extreme cool nights by 23.2, 48.0 and 33.0%, respectively. It also significantly decreased the likelihood of perceiving increased rainfall and number of extreme cool days by 21.4 and 25.9%, respectively. Moreover, being in the pastoral eco-environment significantly increased the probability of increased perception of intensity of rainfall, number of extreme warm days and number of extreme warm nights by 19.4, 36.0 and 40.0%, respectively. It significantly decreased the likelihood of increased perception of number of extreme cool nights by 20.9% and significantly decreased the likelihood of decreased perception of number of extreme warm days and number of extreme warm nights by 31.7 and 39.0%, respectively. Except for the intensity of rainfall, these are in line with the recorded trends and indicate that households in the pastoral eco-environments perceive climate extreme trends better than households in the mixed crop-livestock highland eco-environment. This might be related to the fact that pastoral areas are located in areas of high climate variability and frequent crop failures and livestock mortality to drought. As a result climate change is of major concern to them.

Similarly compared to being in the mixed crop-livestock highland eco-environment, households being in the agro-pastoral eco-environment significantly increased probability of perception of decreased number of extreme cool days by 41.7% and significantly decreased the likelihood of increased perception by 16.8%. It however, significantly increased the probability of perceiving increased rainfall and number of extreme warm days by 59.8 and 59.8%, respectively. It also significantly decreased the likelihood of perceiving decreased rainfall and number of warm days by 48.7 and 21.7%. This shows that unlike expectations agro-pastoralists are not better in perceiving recorded trends in climate extremes than the mixed crop-livestock highland farmers.

Compared to the dry season, the major rain season significantly decreased the probability of household perceptions of increased seasonal total rainfall and daily rainfall intensity by 16.0 and 16.8%, respectively. It significantly decreased the likelihood of perceiving decreased seasonal total rainfall by 16.0%, daily rainfall intensity by 15.0%, and number of extreme cool nights by 9.3%, and number of extreme warm days by 7.0%. It also significantly increased the probability of increased perception of number of extreme cool nights and number of extreme warm days by 10.0 and 8.8%, respectively.

Similarly compared to the dry season, the small rain season significantly decreased the likelihood of increased perception of seasonal total rainfall and daily rainfall intensity by 13 and 7% respectively. It also significantly increased the probability of increased perception of number of extreme cool nights by 8.2%; decreased the likelihood of decreased perception of intensity of rainfall, number of extreme cool nights and number of extreme warm days by 10.6, 7.9 and 11.6%, respectively. Moreover, it significantly increased the probability of perceiving increased number of extreme warm days by 11.0% and significantly decreased the likelihood of decreased perception of number of extreme warm days by 11.6%. These show that though seasons have agricultural significance, its effect on household perceptions are not clearly defined.

# 4. DISCUSSION

## 4.1 Household perception of extreme weather events vis-à-vis observed records

From the above results it is evident that households from the three eco-environments perceived increasing numbers of extreme warm days and warm nights, and conversely decreasing numbers extreme cool days and cool nights. The perceptions on rainfall extremes, however, were variable across season and eco-environments in such a way that majority of households in the pastoral and agro-pastoral eco-environments perceived decreasing total rainfall and intensity of rainfall, whereas the respondents were equally divided between perceiving increases and decreases for the major and small rainy seasons in the mixed crop-livestock highland eco-environment. The results are in line with other reports from similar environments (Dhaka et al., 2010; Acquah and Onumah, 2011). On the other hand, the relationship between household perceptions and station data were not systematic; in the pastoral eco-environment household perceptions were in line with the significant trends observed in recorded data, but in the agro-pastoral or mixed crop-livestock highland eco-environment the relationship was not clear. Maddison (2007) also found similar inconsistencies for rainfall and temperature between farmers' perceptions and recorded weather station data across many African countries. The results also revealed more inconsistencies among recorded data than household perception across eco-environments. This is in line with reports of Degefu and Bewket (2014) that showed considerably varied trends of climate extremes among neighboring stations within a given eco-environment.

## 4.2 Determinants of household perceptions

Herding/farming households' perceptions on rainfall and temperature extremes are significantly affected by a number of factors and the factors affecting each extreme variable may not be the same. Generally literacy, eco-environment, distance from market, social/institutional responsibility, access to extension and relief aid services caused/helped the majority of households to perceive increasing number of extreme warm days and warm nights, and conversely decreasing number of extreme cool days, cool nights, daily intensity of rainfall and seasonal total rainfall as expected. This shows that there is a need to consider these factors at policy level while planning community based climate change measures. So far the Ethiopian government has enacted various policy programs and strategies aimed at mitigating and adapting climate change. Some of this includes climate resilient green economy strategy, and agriculture and rural development polices and strategies which focuses on low/or no greenhouse gas emission, agricultural growth and energy development to reach the level of middle income by 2025.The implementation of these policies and strategies need the active participation of citizens at different capacities. As Ethiopia is a low income agrarian country, the grates emissions of greenhouse gases and the effects of the changing climate are both at small holder subsistence agricultural practices. Awareness and active participation of the rural population of the country engaged in small scale subsistence farming and herding is thus far important. The present study's finding of variation in households' perception of climate extremes trend indicates the need for eco-environment based policies and strategies to first make awareness on local and global scale changes in climate extremes. The policies and strategies should focus more on educating farming and herding households by incorporating climate change issues in both the formal and informal school curriculums and the adult education programs. The deployment of intermediate level trained agricultural extension workers close to farmers and herders also play vital role in information exchange with local communities. The extension workers, therefore, need to have regular access to information from the National Meteorological Agency of the country about their respective local weather report.

On the other hand, though they play an important role in climate change adaptation strategies (Deressa et al., 2011; Silvestri et al., 2012; Mandleni and Anim, 2011) farm land size, and livestock holding size of households could not help perceive changes in climate extremes. The implication is that households with large farmland size and livestock holding are better adapted to climate change and they may have less perception of climate extremes are changing. This shows climate extremes change is of less worry to better off households and there is a need to work on eco-environment based climate change adaptation strategies. The pastoral and agro pastoral eco-environment being under warm climate and low and erratic rainfall is dependent on livestock. In these areas, the policy programs need to focus on improving productivity of animals through better breeding, feeding and health care systems. On the other hand, in the mixed crop-livestock highlands the policy strategies need to focus on improving both crop and livestock productivity.

## 5. CONCLUSIONS

From the results of the present study it is apparent that herding/farming household perceptions of extremes of climate are similar for temperature extremes trend across eco-environments and seasons, whereas the perceptions on rainfall extremes vary with eco-environments and seasons. Moreover, household perceptions are sometimes at variance with recorded significant trends across eco-environments and seasons. Households in the pastoral eco-environment perceive changes in climate extremes better than households either in the agro-pastoral or mixed crop-livestock highland eco-environments. Household perceptions of the studied extreme events were significantly affected by a number of factors. Policy programs that enhance the literacy level of households and eco-environment based extension services may increase the level of awareness and understanding of climate change by households which help them better adapt to climate change.

## ACKNOWLEDGMENTS

The author acknowledges the German Academic Exchange Service (DAAD) for funding this work. They are also thankful to the National Meteorological Agency of Ethiopia for kindly providing the daily weather data of the studied stations. They also acknowledges herders and farmers whom were interviewed and development agents and experts who assisted in the course of the study

## REFERENCES

1. Abate T, Ebro A, Nigatu L (2010). Traditional resource utilization practices and pastoralists" perceptions on land degradation in the south-east Ethiopia. Tropical Grasslands 44:202-212.
2. Abebe D (2000). Pastoralism and pastoral production system. In: Pastoralism and Agro-pastoralism which way forward? Proceedings of the 8th annual conference of Ethiopian Society of Animal Production (ESAP) 24-26 August 2000, Addis Ababa, Ethiopia.
3. Acquah DA (2011).Public awareness and quality of knowledge regarding climate change in Ghana: a logistic regression approach. J. Sustain. Dev. Afr." 13(3):146-17.
4. Acquah H, Onumah EE (2011). Farmers Perception and Adaptation to Climate Change: An Estimation of Willingness to Pay. Agris on-line Papers in Economics and Informatics 3(4):31-39.
5. Akerlof K, Maibach EW, Fitzgerald D, Cedeno AY, Neuman A (2013). Do people personally experience"" global warming, and if so how, and does it matter? Glob. Environ. Change 23:81-91.
6. Angassa A, Oba G (2007). Relating long term rainfall variability to cattle population dynamics in communal rangelands and a government ranch in southern Ethiopia. Agric. Syst. 94(3):715-725.
7. Fosu-Mensah BY, Vlek PL, MacCarthy DS (2012). Farmers" perception and adaptation to climate change: a case study of Sekyedumase district in Ghana. Environ. Dev. Sustain. 14(4):495-505.
8. Bento S, Fatima D, Errahj M, Faysse N, Garin P, Richard-Ferroudji RJD, Dominique SL, Varanda M (2013). Farmers" relations to climate variability and changes: the case of groundwater users of coastal aquifers in France, Portugal and Morocco.

http://pascal.iseg.utl.pt/~socius/investigacao/Aquimed.ESA_Bentoet%20al_climate%20 and%20farmers.pdf. Accessed 20 June 2014

9. Bryan E, Ringler C, Okoba B, Roncoli C, Silvestri S, Herrero M (2013).Adapting agriculture to climate change in Kenya: Household strategies and determinants. J. Environ. Manage. 114:26-35.
10. Byrd TL, VanDerslice J, Peterson SK (2001). Attitudes and beliefs about environmental hazards in three diverse communities in Texas on the border with Mexico. Revpanam salud publica/pan.Am/ public health 9(3)154-160.
11. Coppock L (1994). The Borana Plateau of Southern Ethiopia: Synthesis of pastoral research, development and change, 1980-91 ILCA (International Livestock Center for Africa) Systems study 5. Addis Ababa, Ethiopia.
12. Degefu MA, Bewket W (2014) Variability and trends in rainfall amount and extreme event indices in the Omo-Ghibe River Basin, Ethiopia. Variability and trends in rainfall amount and extreme event indices in the Omo-Ghibe River Basin, Ethiopia. Reg. Environ. Change 14(2):799-810.
13. Deressa TT, Hassan RM, Ringler C, Alemu T, Yesuf M (2009). Determinants of farmers" choice of adaptation methods to climate change in the Nile Basin of Ethiopia", Glob. Environ. Change 19:248-255.
14. Deressa TT, Hassan RM, Ringler C (2011). Perception of and adaptation to climate change by farmers in the Nile basin of Ethiopia. J. Agric. Sci. 149:23-31.
15. Desta S (2000). Pastoralism and Natural Resource Management. In: Pastoralism and Agro-pastoralism which way forward? Proceedings of the 8th annual conference of Ethiopian Society of Animal Production ( ESAP) 24-26 August 2000, Addis Ababa, Ethiopia.
16. Dhaka BL, Chayal K, Poonia MK (2010). Analysis of Farmers" Perception and Adaptation Strategies to Climate Change. Libyan Agric. Res. Center J. Int. 1(6)388-390.
17. Enujeke EC, Ofuoku A (2012). Determinants of adaptation to climate change among arable crop farmers in Edo state, Nigeria and its implications for extension service. Int. J. Adv. Biol. Res. 22:220-227.
18. Frank E, Eakin H, L"opez-Carr D (2011). Social identity, perception and motivation in adaptation to climate risk in the coffee sector of Chiapas, Mexico. Glob. Environ. Change 21:66-76.
19. Gbetibouo GA (2009). Understanding Farmers' Perceptions and Adaptations to Climate Change and Variability: The Case of the Limpopo Basin, South Africa, Int. Food Policy Res. Institute (IFPRI), Discussion Paper 00849.
20. Gebru H (2001). Role of livestock in food security and food self sefishensi in the highland production system. In: Livestock in food security – Roles and contributions Proceedings of the 9th annual conference of the Ethiopian Society of Animal Production (ESAP) August 30-31 Addis Ababa, Ethiopia.
21. Greene WH (2000) Econometric analysis, Fourth edition, Prentice Hall, Mekasha et al. 65 New Jersey. Hansen J, Marx S, Weber E (2004).The Role of Climate Perceptions, Expectations, and Forecasts in Farmer Decision Making. The Argentine Pampas and South Florida. Final Report of an IRI Seed Grant Project. http://cred.columbia.edu/cred/files/ 2012/02/HansenMarxWeber_Clim.Perceptions_2004.pdf.Accessed 20 June 2012
22. Hasan Z, Akhter S (2011). Determinants of Public Awareness and Attitudes on Climate Change in Urban Bangladesh: Dhaka as a Case. European J. Soc. Sci. 21(1):154-162.
23. Hassan R, Nhemachena C (2008).Determinants of African farmers" strategies for adapting to climate change: Multinomial choice analysis. Afr. J. Agric. Resour. Econ. 2(1):83-104.
24. Henry AD (2000). Public perception of global warming. Res. Human Ecol. 7(1):25-30.
25. Hidalgo MC, Pisano I (2010). Determinants of risk perception and willingness to tackle climate change: A pilot study. Psyecol. 1(1):105-112.
26. Hulme M, Dessai S, Lorenzoni I, Nelson D (2009). Unstable climates: Exploring the statistical and social constructions of normal climate. Geoforum 40:197-206.
27. Ishaya S, Abaje IB (2008).„Indigenous people"s perception on climate change and adaptation strategies in Jema"a local government area of Kaduna state, Nigeria".Journal of geography and regional planning 1(8)138-143.

28. Lamma S, Devkota B (2009) Vulnerability of mountain communities to climate change and adaptation strategies. J. Agric. Environ. 10:65-71.
29. Maddison D (2007).The Perception of and Adaptation to Climate Change in Africa. Policy Research Working Paper 4308. The World Bank Development Research Group Sustainable Rural and Urban Development Team.https:// openknowledge. worldbank. org/bitstream/handle/10986/7507/wps4308.pdf?sequence=1.Accessed 20 June 2012
30. Mandleni B, Anim FDK (2011). Perceptions of Cattle and Sheep Farmers on Climate Change and Adaptation in the Eastern Cape Province of South Africa. J. Human Ecol. 34(2):107-112.
31. McSweeney C, New M, Lizcano G (2010). UNDP climate change profile for Ethiopia.http://country-profiles.geog.ox.ac.uk.Accessed 20 June 2014
32. Myers RH (1990).Classical and modern regression with applications, Duxbury Press, Belmont, CA.
33. New M, Hewitson B, Stephenson DB, Tsigna A, Kruger A, Manhique A, Gomez B, Coelho CAS, Masisi DN, Kululanga E, Mbambalala E, Adesina F, Saleh H, Kanyanga J, Adosi J, Bulane L, Fortunata L Mdoka ML, Lajoie R, (2006).„Evidence of trends in daily climate extremes over southern and West Africa. J. Geophys. Res. 111 D14102.
34. Nyong F, Adesina F, Ellasha BO (2007).The value of indigenous knowledge in climate change mitigation and adaptation strategies in the African Sahel. Mitigation and Adaptation Strategies Glob. Change 12:787-797.
35. O"brien RM (2007).A caution regarding rules of thumb for variance inflation factors.Quality & Quantity 41:673-690.
36. Ofuoku AU (2011). Rural farmers" perception of climate change in central agricultural Zone of Delta state, Nigeria. Indonesian J. Agric. Sci. 12(2):63-69.
37. Pasquaré FA, Oppizzi P (2012). How do the media affect public perception of climate change and geohazards?'An Italian case study. Glob. Planet. Change 90:152-157.
38. Piya L, Maharjan KL, Joshi NP (2012).Perceptions and Realities of Climate Change among the Chepang Communities in Rural Mid-Hills of Nepal. J. Contemporary India Studies: Space and Society Hiroshima University, 2:35-50.
39. R Development Core Team (2012). R: A language and environment for statistical computing. R Foundation for Statistical Computing, Vienna, Austria.ISBN 3-900051-07-0.http://www.R-project.org/. Accessed 20 June 2014
40. Sarkar S, Padaria RN (2010). Farmers" Awareness and Risk Perception about Climate Change in Coastal Ecosystem of West Bengal. Indian Res. J. Ext. Edu. 10(2):32-37.
41. Saroar M, Routray JK (2010). Why does climate awareness differ? Lessons learned from Bangladesh. 2nd International Conference: Climate, Sustainability and Development in Semi-arid Regions August 16 - 20, 2010, Fortaleza - Ceará, Brazil.
42. Seleshi Y, Zanke U (2004). Recent changes in rainfall and rainy days in Ethiopia". Int. J. Climatol. 24:973-983.
43. Semenza JC, Hall DE, Wilson DJ, Bontempo BD, Sailor DJ, George LA (2008). Public perception of climate change voluntary mitigation and barriers to behavior change. Am. J. Prevent. Med. 35(5):479-487.
44. Silvestri S, Bryan E, Ringler C, Herrero M, Okoba B (2012). Climate change perception and adaptation of agro-pastoral communities in Kenya. Reg. Environ. Change 12:791-802.
45. Stehr N, Storch H (1995).The social construct of climate and climate change. Clim. Res. 5:99-105
46. Storch H (2011). Regional climate knowledge for society. A paper presented on ESPI Workshop Remote Sensing regional climate change potentials and options to adapt. Wien 26-27.
47. Tank AMGK, Zwiers FW, Zhang X (2009). Guidelines on analysis of extremes in a changing climate in support of informed decisions for adaptation. Climate data and monitoring WCDMP-No.72.
48. Tansey J, O'riordan T (1999). Cultural theory and risk: a review. Health, Risk Soc. 1(1):71-90.

49. Tesso G, Emana B, Ketema M (2012). Econometric analysis of local level perception, adaptation and coping strategies to climate change induced shocks in North Shewa, Ethiopia. Int. Res. J. Agric. Sci. Soil Sci. 2(8)347-363.
50. Thornton PK, Jones PG, Owiyo TM, Kruska RL, Hererro M, Kristjanson P, Notenbaert A, Bekele N, Omolo A, Orindi V, Otiende B, Ochieng A, Bhadwal S, Anantram K, Nair S, Kumar V, Kulkar U (2006). Mapping climate vulnerability and poverty in Africa. Rep. dep. Int. dev., ILRI, PO Box 30709, Nairobi 00100, Kenya. P 200.
51. Trenberth KE, Jones PD, Ambenje P, Bojarlu R, Easterling D, Klein A, Tank PD, Rahimzadeh F, Renwick JA, Rusticucci M, Soden B, Zhai P (2007).Observations: Surface and atmospheric climate change. In: Climate change (2007): The physical science basis. Contribution of working group I to the fourth assessment report of the intergovernmental panel on climate change {Solomon S, Qin D, Manning M, Chen Z, Marquis M, Averyt KB, Tignor M, Miller HL (eds.)] Cambridge University press, Cambridge, United Kingdom and New York, NY, USA.
52. Vedwan N (2006). Culture, Climate and the Environment: Local Knowledge and Perception of Climate Change among Apple Growers in Northwestern India. J. Ecol. Anthropol. 10:4-18.
53. Viscusi WK, Zekhauser RJ (2006).The perceptions and valuations of the risks of climate change: A rationale and behavioral blend. Clim. Change 77:151-177.
54. Weber EU (2010).What shapes perceptions of climate change? Clim. Change 1:332-342.

# CHAPTER 7

# Aging effect of two decade old technologies of bean stem maggot (Ophiomyia spp.) management options in Ethiopia: Baseline information generation for future research direction

**Mulatwa Wondimu[1], Emana Getu[2*], Sisay Kidanu[3] Amare Fufa[1]**

[1]*Melkassa Agricultural Research Center, EIAR, P. O.Box 2003 Addis Ababa, Ethiopia.*

[2]*Addis Ababa University, College of Natural and Computational Sciences, P. O. Box 1176, Ethiopia.*

[3]*Jimma Agricultural Research Center, EIAR, P. O. Box 2003 Addis Ababa, Ethiopia.*

## ABSTRACT

Haricot beans are the most important food and cash crop for most of the Ethiopian farmers. However, bean stem maggot (BSM) (Ophiomyia species) is limiting the production of the crop particularly in dry areas. Three species are involved in the infestation: Ophiomyia phaseoli, Ophiomyia spencerlla and Ophiomyia centrosematis. Two decades ago, two management options of BSM were developed and used for the last twenty years. However, there is a tendency of decline in effectiveness. Hence, the current experiment was conducted to see the age effect on the efficacy of

BSM resistant varieties (Beshbesh and Melkae) and high bean plant population density (400,000 plants ha-1) (narrow spacing). The experiments were conducted at Shalla (southern Ethiopia) and Omonada (Jimma). Randomized Complete Block Design (RCBD) in three replications was used as the experimental design. The plot size was 10 m × 10 m. The experiment was conducted in 2014 and 2015. The treatments were BSM resistant varieties (Beshbesh and Melkae), high bean plant population, standard check (Awash-1) and Imdalem (seed dressing chemical). For plant population treatment and Imdalem 70% WS seed treatment a bean variety Awash-1 was used as a planting material. Data collected include stand count, seedling percent mortality by BSM, vigorously score, number of larva per plant and grain yield. The data were analyzed using SAS software and mean separation was done by Tukey Standardized Range Test (p<0.05). In all parameters measured, Melkae bean variety become susceptible to BSM and cannot be used any more as a resistant variety. From the result of the experiment, it can be concluded that Beshebeshe bean variety, high plant population and Imdalem seed dressing can be used as integrated management of BSM in Ethiopia in general and study sites in particular.

**Keywords**: Bean stem maggot, Imdalem 70% WS (Imdachloprid), seed dressing, plant population, resistant variety, two decades, proven technology.

# 1. INTRODUCTION

Haricot bean is the most important crop grown by African farmers in general and Ethiopian farmers in particular (Ampofo et al., 1994). Haricot bean production in Africa suffers from various constraints including biotic and abiotic. From the biotic constraints, the damage caused by insect pest infestations under field condition is tremendous. The most common insect pests which attack common bean in the field are the bean stem maggots (BSM) (Ophiomyia species), African bollworm (Helicoverpa armigera) and Aphids (Aphis fabae). Of the insect pests attacking haricot beans, BSM is considered as a major pest of central and many sub-Saharan African countries especially under poor soils and dry conditions (Ampofo et al., 1994; Ampofo, 2009). Three species of BSM (Ophiomyia phaseoli, Ophiomyia spencerella and Ophiomyia centrosematis) were identified in Ethiopia. O. phaseoli and O. spencerella are the most widely distributed and abundant of the three species. O. centrosematis occurs rarely and represents less than 2% of the total BSM population. The incidence of BSM species is influenced by one or a combination of environmental factors and cultural practices including altitude, sowing date growth stage and type of the host plant. O. phaseoli and O. centrosematis are more prevalent in warmer areas mostly in south and south western part of the country at the elevation of <1800 m above sea level, whereas O. spencerella is dominant at higher altitudes (cooler and wetter conditions). Bean stem maggot attacks the bean plant as soon as the first pair of leaves begins to unfold (Odendo et al., 2005). It is most severe in late or off-season planted crops (Nderitu et al., 1990; Ochilo and Nyamasyo, 2011), hotter drier seasons and in plants under stress such as poor soil fertility (Ampofo et al., 1994; Ojwang et al., 2010). Severe BSM attack especially during the early stages of plant growth causes high yield losses. In Africa, yield losses ranging from 80 to 100% have been reported (Ochilo and Nyamasyo, 2011; Munyasa, 2013).

Various control methods have been recommended by different authors. These include cultural control practices such as adjustment of planting date (Muleke et al., 2013), crop rotation and associated cropping (Abate and Ampofo, 1996; Amoabeng et al., 2014). These practices enable the host plant to escape the peak period of pest population. Other methods such as earthing or hilling up soil around the stem of seedlings encourage the development of adventitious roots above the damaged stem resulting in the recovery of plants from the BSM damage (Forbes et al., 2009). Use of chemicals especially seed dressing has also been found effective against BSM (Abate, 1991; Williamson et al., 2008; Uburyo, 2016). Other control methods include mulching and fertilizer applications, which enhance plant tolerance (Ampofo and Massomo, 1994; Gogo et al., 2012), and use of resistant variety (Ampofo, 1993; Kiptoo et al., 2016 (Beshbesh and Melkae) have been identified as resistant to BSM (Tsedeke, 1990). In Ethiopia, as Integrated Pest Management options of BSM Beshbesh and Melkae as resistant bean varieties, high plant populations per hectare and seed dressing with Imdalem 70% WS were under use for the last twenty years. However, there was an indication that some of the options showed a tendency of decline in efficacy. Hence, the current project was initiated to see the aging effect of BSM proven technologies in the management of BSM at the moment.

# 2. MATERIALS AND METHODS

## 2.1 Description of the study site

A field experiment was conducted at Shalla in southern Ethiopia and Omonada in Jimma in 2014 and 2015 main season under rain fed condition.

Shalla district is situated in west Arsi zone about 270 km south west of Addis Ababa. The area is lowland with an altitude of 1550 m above sea level, with a latitude of 38° 27'10.9"E and Longitude of 7°17'08.6"N. The site has a mean maximum temperature of 29.2°C and mean minimum temperature of 14.4°C. The soil texture is sandy loam and the site receives 763 mm mean annual rainfall, but with much variation in distribution and amount of 70% which occurs between the months of May and September.

Omonada district is situated in Jimma zone of Oromia regional state located at about 223 to 258 km south west of Addis Ababa. This Wereda lies between latitudes 7°29′59.99" N and longitudes 37°14′60" E. The elevation of the district ranges from 1000 to 3340 m. a. s. l. The area experienced an annual average rainfall of 1000 mm for 8 to 10 months (Haile and Tolemariam, 2008). The main rainy season extends from May to September and the small rainy season takes place from February to April. The soil type of the area is ento soil. The major crops grown include coffee, maize, teff, sorghum, barely, pulses (beans and peas), root crops (enset-false banana and potato) and fruits (Avocado and Pineapple).

## 2.2 Treatments, experimental design, data collected and data analysis

BSM resistant varieties, Beshbesh and Melake were collected from Melkassa research center lowland pulse division. These two varieties were released by Tsedeke (1990) as BSM resistant and high yielding varieties. These varieties were multiplied in off season and 1 kg of each variety was allotted per plot for each study location.

Imdalem (Imadachlopride WS 70% group) was collected from Addis Ababa chemical company. It was formulated in powder form and dissolved in water at the rate of 500 to 700 ml water/100 g chemical powder. First, 100 g of Imdalem powder and 500 ml of water were prepared and thoroughly mixed with 1 kg of haricot bean seed until all the seeds get wet. The treated seeds were exposed to the sun to help adhere the chemical to the seed. Then the seeds were sown at the end of June. High plant population density (400,000 plants/ha) were recommended for the management of BSM by Tsedeke (1990) used for the experiment. The recommended spacing for the high plant population density was 40 cm between rows and 6 cm between plants.

The experiment was laid out in RCBD design in three replications. The plot size was 10 × 10 m. Percent seedling mortality was assessed by counting dead seedlings per plot and comparing it against the whole plant population per plot. Such exercise was done three times starting two weeks from plant emergence. Percent mortality was calculated using the following formula reported by Ampofo et al. (1987).

PD= NDP×100/(TSC/P)

where PD = Percent damaged plant, NDP = No. of damaged plant per plot, TSC/P = Total stand count per plot.

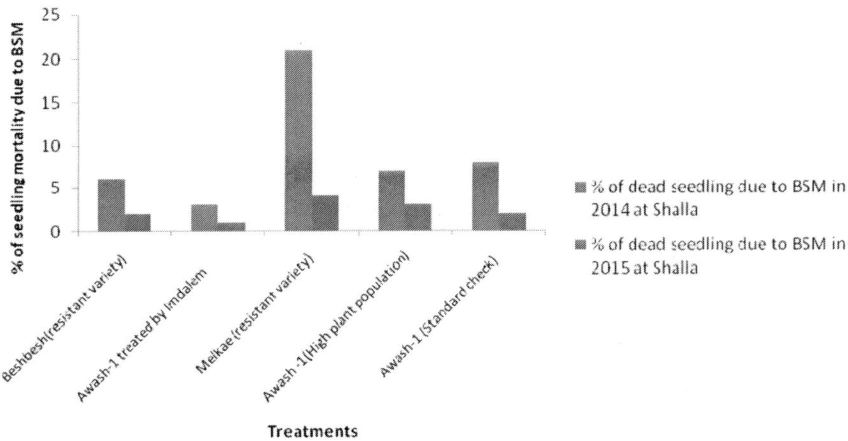


**Figure 1.** Effect of different BSM management options on percent of seedling mortality due to BSM in common bean on farmers' field at Shalla in 2014 and 2015.

Severity score were taken by visual observation using 0-5 scales where 0=no infestation, 1=less infested, 2=infested, 3=highly infested, 4=severly infested, 5=dead plants. This rating scale was adopted from work the work of Ampofo et al. (1987).

Ten plants per plot were randomly sampled and cut at ground level using knife. Then, the number of encountered BSM larvae and/or pupae was recorded and put in a petridish and taken to laboratory for further rearing. The pupal color was used for identification such that brown pupa was for O. phaseoli where as black pupa was for O. spencerella.

Vigorousity of the plant was also estimated by visual observation using 1-5 scale before plant maturity. Yield data (kg/ha) were taken from 20 plants from central rows after the seed dried to the optimum moisture content of 14%.

Data were analyzed using SAS program version 9.0. Percent mortality data were transformed to square root transformation before the analysis using one-way analysis of variance. Significant means ($p<0.05$) were separated using Tukeys studentized range test (HSD).

# 3. RESULTS AND DISCUSSION

## 3.1 Effect of different BSM management options at Shalla

Results obtained at Shalla are presented in Figures 1 and 2 and Table 1. Overall, the severity of BSM was higher in 2014 than 2015 at Shalla as percent seedling mortality ranges from 4 to 22% in 2014 as opposed to 2 to 13% in 2015. The highest percent seedling mortality was recorded on Melkae and the lowest on Awash-1 treated with Imdalem (Figure 1). Larval density per plant at Shalla was higher in 2014 than 2015. The highest larval density per plant was recorded on Melkae, while the lowest was on Awash-1 treated Imdalem. The level of bean plant vigorousity and yield per hectare at Shalla were given in Table 1. Vigorousity of bean plant was significantly lower in Melkae bean variety in both 2014 and 2015. Bean yield was also significantly lower in Melkae.

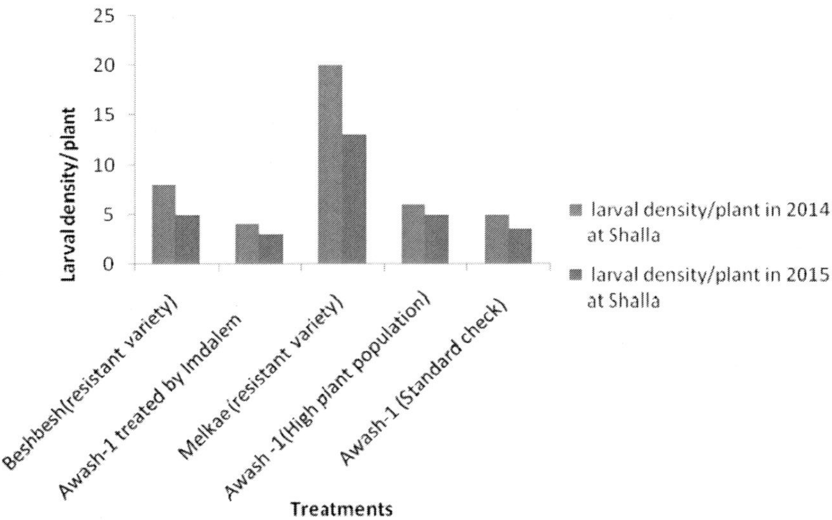


**Figure 2.** Effect of different BSM management options in larval density per-plant of BSM in common bean on farmers' field at Shalla in 2014 and 2015.

## 3.2 Effect of different BSM management at Omonada

Results obtained at Omonada are shown in Figures 3 and 4 and Table 2. Percent seedling mortality was significantly higher in Melkae. The highest percent seedling mortality of 15% was recorded in 2014 (Figure 3). BSM larval density per plant was recorded in 2014 on Melkae bean variety (Figure 4). Bean plant vigorousity and yield per hectare was significantly lower in Melkae bean variety (Table 2).

Results of both locations indicated that Melkae, the resistant variety lost its ability of overcoming the attack by BSM lost resistance due to age effect. The other technologies developed against BSM such as Beshebshe bean variety and high plant population remained effective in overcoming BSM damage. Ampofo (1993) and Tsedeke and Ampofo (1996) reported that some bean varieties which are currently resistant to some pests in general and BSM in particular can be susceptible at any time because of the fact that there are only few genes that control resistance which is known as vertical resistance. The fact that Melkae is a variety with vertical resistance as opposed to Beshebshe which looks horizontally, resistant variety became susceptible to BSM.

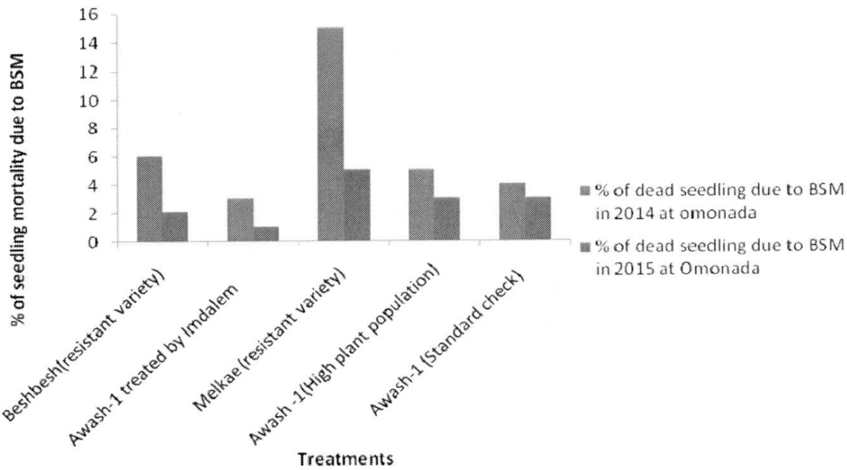


**Figure 3.** Effect of different BSM management options on percent of seedling mortality due to BSM in common bean on farmers' field at Omonada in 2014 and 2015.

**Table 1.** Effects of different BSM management options on the vigrousity (1-5 score) of bean plants and yield (kg/ha) on farmers' field at Shalla in 2014 and 2015.

| Treatments (Management options) | Vigorousity of bean plants | | Yield (kg/ha) | |
|---|---|---|---|---|
| | 2014 | 2015 | 2014 | 2015 |
| Beshbesh (resistance variety) | 5±0.58[a] | 5±0.20[a] | 252 9±0.45[a] | 2100±1.15[a] |
| Seeds dressed by Imdalem | 5±0.00[a] | 5±0.30[a] | 2492±0.23[a] | 2150±1.76[a] |
| Melkae (resistance variety) | 2±0.33[b] | 2±0.40[b] | 1781±0.56[b] | 1871±1.2[b] |
| High plant population | 5±0.57[a] | 5±0.34[a] | 2626±0.34[a] | 2133.3±2.52[a] |
| Awash-1 (standard check) | 4±0.18[a] | 4±0.24[a] | 2425±0.67[a] | 2000.0±0.07[a] |

Means followed by the same letters within a column are not significantly different from each other at 5%. Level of probability, Tukeys Studentized Range Test (HSD) (5%). *Vigrousity score (1-5) means, 1= vey week, 2= week, 3= good, 4= very good, and 5=excellent.

**Table 2.** Effect of different management options on percent dead seedling and severity score due to BSM at Omonada in 2014 and 2015.

| Treatments | Vigorousity of bean plants | | Yield (kg/ha) | |
|---|---|---|---|---|
| | 2014 | 2015 | 2014 | 2015 |
| Beshbesh (resistance variety) | 5±0.58[a] | 4±0.12[a] | 1757±12.50[a] | 1477±10.82[a] |
| Seeds dressed by Imdalem | 5±0.5 2[b] | 5±0.22[a] | 1412±10.60[a] | 1579±11.87[a] |
| Melkae (resistance variety) | 2±0.33[b] | 2±0.52[b] | 1167±7.70[b] | 1056±0.54[b] |
| High plant population | 4±0.12[a] | 5±0.27[a] | 1511±11.80[a] | 1573±11.87[a] |
| Awash-1 (standard check) | 4±0.11[a] | 4±0.0.13[a] | 1451±10.89[a] | 1563±11.78[a] |

Means followed by the same letters within a column are not significantly different from each other at 5% level (HSD).

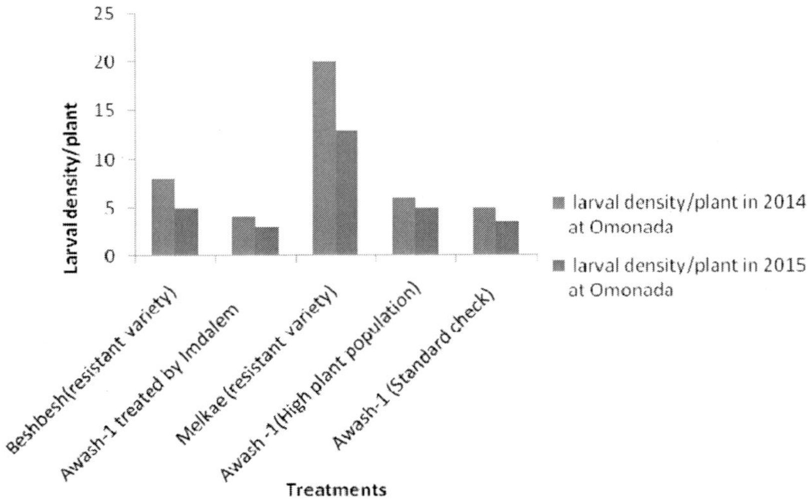


**Figure 4.** Effect of different BSM management options in larval density per-plant of BSM in common bean on farmers 'field at Omonada in 2014 and 2015.

# 4. CONCLUSION

From the current experiment, we came to know that Beshbesh bean variety, use of high plant population and Imdalem seed dressing can be used as integrated management of BSM in Ethiopia in general and study sites in particular.

# ACKNOWLEDGEMENT

The authors are very much grateful to Melkassa Research Center, Lowland Pulse and Crop Protection Teams for their both technical and fincial support for executing the experiment.

# REFERENCES

1. Abate T (1991). Seed dressing insecticides for Bean fly (Ophiomia Phaseoli (Tryon) Diptera: Agromyzidae) control in Ethiopia. Trop. Pest Manage (U.K) 37(4):334-337

2. Abate T, Ampofo JKO (1996). Insect pests of Beans in Africa: Their ecology management. Ann. Rev. Entomol. 41:45-73
3. Ampofo JKO (1993) Host plant resistance and cultural strategies for bean stem management. In: Ampofo, J.K.O (ed). Proceedings of the second meeting of the Pan-African working group on bean entomology. CIAT African workshop Ser. No 25. Cali, Colombia: CIAT.
4. Ampofo JKO, Massomo SMS (1994). Some cultural strategies for Bean fly (Diptera: Agromyzidae) management in Beans (Phaseolus Vulgaris) fields in Tanzania. Afr. Crop Sci. J. 6:351-356.
5. Ampofo K (2009). A Brief on Bean Stem Maggot (Bean fly) in Africa http://ipmnet.org/cicp/beanfly.html.
6. Amoabeng BW, Gurr GM, Gitau CW, Stevenson PC (2014) Cost benefit analysis of botanical Insecticide use in Cabbage: Implications for smallholder farmers in developing countries. Crop Prot. 57:71-76.
7. Forbes VE, Hommen U, Thorbek P, Heimbach F, Van den Brink PJ, Wogram J, Grimm V (2009). Ecological models in support of regulatory risk assessments of Pesticides: Developing a Strategy for the Future. Integrated Environ. Assessment Manage. 5:167-172.
8. Gogo EO, Saidi M, Itulya FM, Martin T, Ngouajio M (2012). Microclimate modification using eco-friendly nets for high quality Tomato transplant production by small-scale farmers in East Africa. Hort. Technol. 22:292-298.
9. Haile A, Tolemariam T (2008) The feed values of indigenous multipurpose trees for sheep: the case of Vrnonia amygdalina (Girawa), Buddelija polystachya (Anfare) and Maesa lanceolata Kelewa). Livest. Res. Rural Dev. 20(3):145-157.
10. Kiptoo M, Kinyua O, Kiplagat F, Wanjala ME, Kiptoo JJ, Cheboi JJ, Kimno SK, Rotich G, Ngurwe JK (2016) Evaluation of Common Bean (Phaseolus vulgaris L.) varieties for resistance to Bean Stem Maggot (Ophiomyia spp.) in Kenya Am. J. Exp. Agric. 12(3):1-7.
11. Munyasa AJ (2013) Evaluation of drought tolerance mechanisms in Meso american Dry Bean genotypes. University of Nairobi, Nairobi.
12. Muleke EM, Saidi M, Itulya FM, Martin T, Ngouajio M (2013). The Assessment of the use of eco-friendly nets to Ensure Sustainable Cabbage Seedling Production in Africa. Agron. 3:1-12.
13. Nderitu JH, Kayumbo HY, Mueke JM (1990). Bean fly infestation on common beans Phaseolus vulgaris in Kenya. Insect. Sci. App. 11:35-41.
14. Ochilo WN, Nyamasyo GH (2011) Pest Status of Bean Stem Maggot (Ophiomyia spp.) and Black Bean Aphid (Aphis fabae) in Taita District, Kenya [Situación de las plagas Del frijol: Gusano del Tallo. Trop. Subtrop. Agro Ecosyst. 13:91-97.
15. Odendo M, David S, Otsyula R (2005). Impact of Root-Rot resistant Bean varieties in Western Kenya: Application of impact diagramming. PABRA Millennium Workshop.
16. Ojwang PPO, Melis R, Songa JM, Githiri M (2010). Genotypic response of Common Bean genotypes to natural field populations of Bean fly (Ophiomyia phaseoli) under diverse environmental conditions. Field Crops Res. 117:139-145.
17. Tsedeke Abate (1990) Studies on genetics, cultural and insecticidal control against the bean fly, Ophiomyia phaseoli (Tryon) Diptera: Agromyzidae), in Ethiopia. PhD thesis, Simon Fraser University, Burraby, BC, Canada 177 p.
18. Uburyo I, Isaziyibish G (2016). Bean Stem Maggot control (Uburyo isazi yibishyimbo irwanywa) – EU pesticides database
19. Williamson S, Ball A, Pretty J (2008). Trends in Pesticide use and drivers for safer Pest management in four African Countries. Crop Prot. 27:1327-1334.

# CHAPTER 8

# Utilization of Beta vulgaris Agrowaste in Biodegradation of Cyanide Contaminated Wastewater

**E.A. Akinpelu, O.S. Amodu, N. Mpongwana, S.K.O. Ntwampe and T.V. Ojumu**

*1 Faculty of Engineering, Department of Chemical Engineering, Cape Peninsula University of Technology, Cape Town, South Africa*

*2 Bioresource Engineering Research Group, Cape Peninsula University of Technology, Cape Town, South Africa*

## 1. INTRODUCTION

Recent developments in biotechnology for agro- /agro-industrial waste utilization have identified a plethora of agricultural waste (agrowaste) that is suitable for microbial proliferation and production of a variety of high value biological products, which are useful in industrial as well as environmental applications. About 1.6 billion tons of agrowaste is reportedly generated globally per annum [25]. Considering the environmental degeneration caused by such waste, and the fact that they are readily available, research studies have been geared toward assessing the feasibility of converting such waste into value added products. Studies into the chemical and nutritional composition of agrowaste have equally identified some of them as suitable substrates for microbial cultivation [54, 40, 63, 69, 5].

In environmental bioremediation applications, microorganisms can be supported on solid agrowaste to provide the required macro- and micro-nutrients required for biofilm formation, which usually enhances the metabolic activities of the microorganisms for solubilization and biodegradation of contaminants, some of which are known to be potential human carcinogens [18, 22]. The paradigm shift from conventional substrates such as refined glucose, to unconventional substrates such as solid agrowaste or agro-industrial waste could be due to the fact that the latter mitigates operational costs, particularly for large-scale processes. Nutrients are considered the largest expense in industrial bioprocesses whereby the fermentation medium can account for a large proportion of fermentation costs [10, 39, 60]. Suitable agrowaste such as orange peel, apple pomace, wheat bran, sugar cane bagasse, wheat bran, soybean oil cake, jatropha curcas, whey waste, and Beta vulgaris, have been identified to support microbial growth and the synthesis of metabolites which can catalyze a number of reactions under suitable conditions[46, 42, 62, 5].

One of the most common wastewater pollutants is cyanide. It is usually released through various anthropogenic activities in the form of industrial effluent discharged from numerous industries. Another incessant anthropogenic source of cyanide deposition into the environment is through petroleum oil processing and its derivatives. Naturally, hydrocarbon oils such as petroleum contain cyano group compounds, which react with metals during thermal cracking operations to form metal cyanide complexes that culminate in wastewater [14]. Many of these cyanide complexes are known to be highly unstable, mainly due to thermal instability, thus releasing free cyanide into the environment under high temperature. It has been reported by Acheampong et al. [1] that, cyanide concentrations from facilities that serve industrialized areas could have cyanide concentration higher than 21.6 mg F-CN/L. Cyanide exposure is known to result in neurological disorders and thyroid abnormalities in humans [69, 55]; hence, a robust and economically feasible bioremediation process using renewable resources (agrowaste), i.e. an environmentally benign approach, is necessary to ensure a sustainable and an effective bioremediation process for cyanide deposited into the environment.

It is common to use oxidation methods for cyanide degradation and its complexes, such as the use of metal catalyzed hydrogen peroxide, and alkaline chlorination processes, including removal by ion-exchange resin [17]. This approach, though effective, has some drawbacks that are of major concern. The excess reagents used in the treatment tend to further pollute the environment, as well as increase operational costs. In addition, due to municipal regulations in some countries, the application of chemical methods on a large scale is not permissible. Considering that cyanide in wastewater is undesirable, if present, it must not exceed the discharge limit of 0.01 mg F-CN/L [23]. Thus, cyanide degradation using biotechnological processes is desirable.

It has been shown that several microorganisms such as algae, bacteria, and fungi, can produce enzymes that are capable of degrading free cyanide, cyanide complexes and by-products produced [3, 24, 33, 59]. Recently, studies have established sustainable cyanide biodegradation processes using various microorganisms such as Klebsiella sp., Pseudomonas sp., Acinetobacter sp., Bacillus sp., and many others [41, 58]. A fungal specie - Fusarium oxysporum, has equally been reported for its ability to produce enzymes such as nitrilase which readily hydrolyses cyano-compounds into a corresponding weak acid and ammonium-nitrogen, thus bioremediating the contaminated wastewater, with both the acid and ammonium-nitrogen produced being consumed for metabolic functions [37, 32]. Several agrowaste have also shown to be effective substrates for the cultivation of microorganisms and for the biodegradation of cyanic compounds [15, 30, 49].

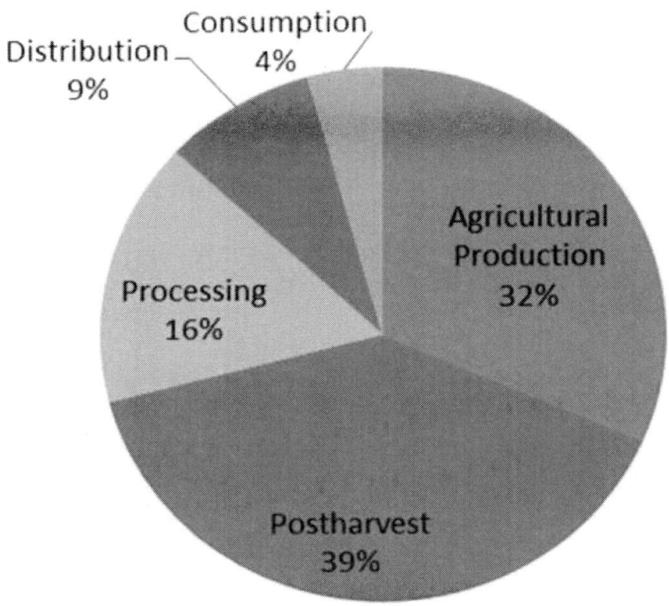


**Figure 1**. Classification of agrowaste production in South Africa [53]

The application of agrowaste as a substrate in cyanide biodegradation systems is particularly promising, as reported by Santos et al. [62]. Having a readily accessible waste material, microorganisms will be able to produce enzymes suitable for bioremediating contaminants in wastewater [59]. Besides their application as effective biosorbents, agrowaste can serve as a sole substrate for bioremediation purposes, on condition that it is compatible to the microbial community to be used [67, 45, 17, 16]. In South Africa, approximately 10 million tons of agrowaste is generated per annum [53], of which 96% is classified as pre-consumer waste (Figure 1). This is a large quantity of waste for a developing economy and should be put into profitable use to safe our environment.

## 2. APPLICATION OF FUSARIUM SP. AND BETA VULGARIS IN CYANIDE BIODEGRADATION

*Fusarium sp.* are widely distributed in environmental samples, particularly in soil. They can cause spoilage of agricultural produce and produce mycotoxins which contaminate cereal crops, affecting human and animal health, if the mycotoxins enter the food chain. *Fusarium sp.* has also been found useful in the hydrolysis of starch. The hydrolysed agricultural produce can be used to sustain the production of extracellular enzymes such as pectinase, cellulase, xylanase, amylase, and organic acids [43]. The fungus is also known for the production of cyanide hydratase and nitrilase including cyanidase. *Fusarium sp.* has been identified as having the ability to degrade cyanides through hydrolysis at varying temperature and pH, then metabolise the by-products as either nitrogen and carbon sources, respectively [52, 31]. The cyanide hydratase, converts the cyanide to amide products and ammonium-nitrogen while the nitrilase hydrolyse cyanide to produce a carboxylic acid [50]. Compared with other enzymes derived from bacteria, nitrilase and cyanide hydratase are of higher activity and can degrade various cyanides [59].

*Beta vulgaris* waste consists of water, carbohydrates, minerals and proteins which makes it a suitable substrate for microbial growth in the production of high value compounds [5, 45]. However, limited studies have shown its potential as a feed stock and solid support in a bioreactor for the biodegradation of cyanide in the presence of heavy metals [46]. Additionally, hydroxyl functional groups found in B. vulgaris waste can act as pseudo-catalysts for the conversion of cyanide to ammonium-nitrogen. Alhough the free hydroxyl functional group is a weak acid, they are able to deprotonate to produce alkoxides in the presence of a strong base like cyanides especially at high alkaline pH (Figure 2) [31, 62].

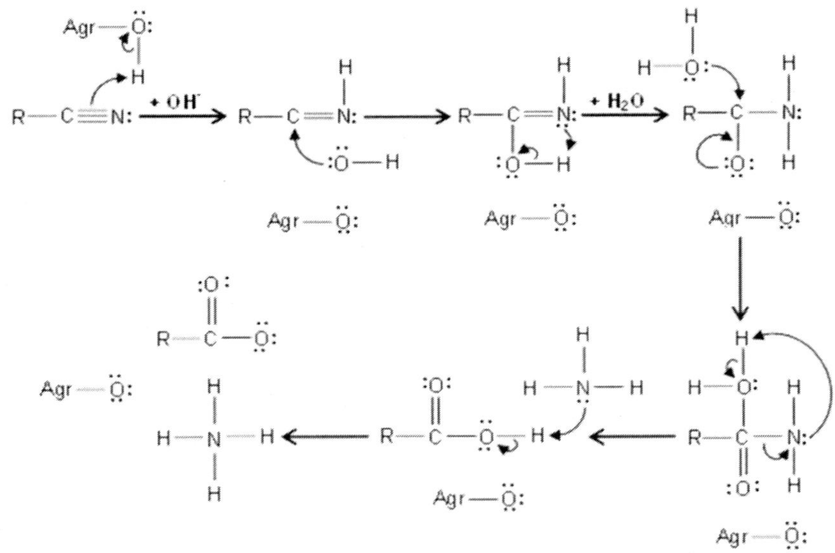

**Figure 2**. Pseudo-catalyst conversion of cyanide by free hydroxyl functional groups [61]

# 3. BIODEGRADATION OF CYANIDE BY FUSARIUM OXYSPORUM GROWN ON BETA *VULGARIS*

A number of different studies report on the application of cyanide degrading fungi. For instance, white rot fungi, Trametes versicolor, have been shown by Cabuk et al. [9] to tolerate cyanide concentration up to 130 mg F-CN/L, with complete degradation observed within 42 hours to produce minute quantities of ammonium-nitrogen (5.24 mg $NH_4^+$-N/L). Fourteen cyanide degrading fungi were examined by Pereira et al. [57] such as Fusarium sp. including Aspergillus sp. by Santos et al. [57, 62], and were found to tolerate cyanide concentration up to 520 mg F-CN/L. A list of other cyanide degrading species including degradation conditions are shown in Table 1.

There has been limited emphasis on the effect of carbon or nitrogen sources used in the biodegradation of cyanide. The viability of the agrowaste depends on the type of bioremediation required and the microorganism used. When the cultivating conditions are conducive, the minerals, proteins, carbohydrates and water in the agrowaste become easily accessible to the microorganisms [46]. Monosaccharides such as mannose, glucose and fructose present in the agrowaste can effectively support and/or enhance

**Table 1**. Cyanide degrading microbial species using different nutritional sources under different temperature and pH conditions

| Microorganism | C-source | N-source | Temperature (°C) | pH | Reference |
|---|---|---|---|---|---|
| *Fusarium oxysporum* | *Beta vulgaris* | KCN | 30 | 11 | This study |
| *Agrobacterium tumefaciens* | Starch | KCN | - | 7.2 | [58] |
| *Aspergillus awamori* | *Citrus sinesis* extract | KCN | 40 | 8.84 | [62] |
| *Baccillus pumilus* | Glucose | KCN | 40 | 8.5-9 | [64] |
| *Baccillus stearothermphilus* | - | NaCN | 27±2 | 7.8 | [6] |
| *Burkholderia cepacia* | Fructose, glucose, mannose | KCN | 30 | 10 | [2] |
| *Citrobacter sp., Pseudomonas sp.* | Sugarcane molasses, glucose | $[Cu(CN)_4]^{2-}$, $[Zn(CN)_4]^{2-}$ | 35 | 7.5 | [56] |
| *Cryptococcus humicolus* MCN2 | Glucose | KCN | 25 | 7.5 | [36] |
| *Escherichia coli* | Glucose | KCN | 30 | 9.2 | [26] |
| *Fusarium solani* | Glucose | $K_2Ni(CN)_4$, KCN | 25 | 7.0 | [8] |
| *Fusarium solani* | Yeast | KCN | 30 | 9.2-10.7 | [19] |
| *Fusarium oxysporum* | Glucose | KCN | 25 | 8.0 | [57] |

| | | | | | |
|---|---|---|---|---|---|
| *Fusarium oxysporum* immobilised on sodium alginate | Formamide | Cyanides | 25-30 | 8 | [11] |
| *Gloeocercospora sorghi, Stemphylium loti* | Glucose | KCN | 35, 28 | 5.3-5.7, 7.0 | [48] |
| *Klebsiella oxytoca* | Glucose | KCN | 30 | 7 | [34] |
| *Klebsiella oxytoca immobilised cell* | Alginate and cellulose triacetate | KCN | 30 | 7 | [13] |
| Mixed culture of bacteria | Glucose | $CN^-_{WAD}$ | 22 | 7.0 | [70] |
| Mixed culture of bacteria immobilised on ultrafiltration membranes | Phenol | Cyanides | 25 | - | [35] |
| *Pseudomonas fluorescens* | Glucose | Ferrocyanide | 25 | 5 | [21] |
| *Pseudomonas fluorescens* immobilised on calcium alginate | Glucose | Ferrocyanide | 25-35 | 4-7 | [20] |
| *Pseudomonas fluorescens* immobilised on zeolite | Zeolite | Tetra-cyano-nickelate (II) | 30 | - | [66] |

| | | | | | |
|---|---|---|---|---|---|
| *Pseudomonas pseudoalcaligenes CECT5344* | $CH_3COONa$ | NaCN | 30 | 9.5 | [38] |
| *Pseudomonas putida* BCN3 | Glucose | $[K_2[Ni(CN)_4]]$ | 30 | - | [63] |
| *Pseudomonas putida* immobilised on sodium alginate | NaCN | NaCN | 25 | 6.7 | [7] |
| *Pseudomonas putida* immobilised on sodium alginate | NaCN, sodium alginate | NaCN, Cyanates and thiocyanates | 25 | 7.5 | [12] |
| *Pseudomonas stutzeri* AK61 | - | KCN | 30 | 7.6 | [68] |
| *Pseudomonas sp.*(CM5, CMN2) | Glycerol | $CN^-_{WAD}$ | 30 | 9.2-11.4 | [4] |
| *Stemphilium loti* | | KCN | 25 | 6.5, 7.5 | [27] |
| *Trametes versicolor* | Citrate | KCN | 30 | 10.5 | [9] |
| *Trichoderma sp.* | Glucose | $CN^-$ | 25 | 6.5 | [24] |
| *Scenedesmus obliquus* | NaCN | NaCN | - | 10.3 | [33] |
| *Rhodococcus* UKMP-5M | Glucose | KCN | 30 | 6.6 | [41] |

microbial growth [2]. Other overriding factors which directly influence cyanide degradation include exposure to direct sunlight, temperature and pH. Cyanide compounds are soluble in water, thus dissociate and evaporate easily at low pH (i.e. pH<9) while under high salinity, the solubility decreases. Also at neutral pH, weak-acid dissociable (WAD) cyanides such as copper or zinc cyanide complexes, if present in a high concentration, dissociate, releasing a cyano group. Similarly, the reduction in temperature reduces the activity of microorganisms used in bioremediation. A number of studies have proven that, at low temperature (below 10oC), growth of microorganisms is inhibited, resulting in low removal rates of contaminants such as ammonium-nitrogen, nitrates and cyanide [72, 29, 73].

In this study, the biodegradation of cyanide in the presence of heavy metals (arsenic, copper, lead, iron and zinc), using Fusarium oxysporum grown on B. vulgaris waste as the sole carbon source, without any buffer solution, was investigated. The effect of temperature and pH on cyanide degradation with minimal ammonium-nitrogen production was studied using a response surface methodology.

# 4. MATERIALS AND METHODS

The experiments were carried out in batch cultures. *B. vulgaris* waste was milled to ≤ 100 μm. A broth of 0.5 g of milled waste in 10 mL distilled water was autoclaved at 116°C for 15 min to prevent thermal breakdown of reducing sugars [51]. To the waste broth, wastewater (20 mL) with 1 mL of a spore solution (2.25 x $10^6$ spore/mL) of Fusarium oxysporum was added to the *B. vulgaris* broth. The wastewater used had characteristics similar to the goldmine wastewater reported by Acheampong et al. [1] having metals such as arsenic, iron, copper, lead and zinc. The mixture was incubated for 48 hours in a rotary shaker at 70 rpm at the desired temperature and pH (- see Table 2). After this, KCN in distilled water, was added to make a final cyanide concentration of 500 mg CN-/L in the mixture. Thereafter, the mixture was incubated for a further 72 hours at 70 rpm at the desired temperature (- see Table 2). All experiments were carried out in duplicate in airtight multiport round bottom Erlenmeyer flasks (n = 28; final volume of 51 mL). Cyanide (CN) (09701) and ammonium-nitrogen ($NH_4^+$-N) (00683) test kits (MERCK®) were used to quantify the residual free cyanide and ammonium-nitrogen concentrations using a NOVA 60 spectroquant. Free cyanide volatilised was accounted for using the mass balance equations below:

$$CN^-_s - (CN^-_r + CN^-_v) = CN^-_b \quad (1)$$

**Table 2**. Experimental variation of pH and temperature

| Run | Temperature (°C) | pH |
|---|---|---|
| 1 | 19.5 | 8.5 |
| 2 | 9 | 11 |
| 3 | 19.5 | 8.5 |
| 4 | 30 | 11 |
| 5 | 30 | 6 |
| 6 | 19.5 | 8.5 |
| 7 | 9 | 6 |
| 8 | 19.5 | 8.5 |
| 9 | 19.5 | 12.04 |
| 10 | 34.35 | 8.5 |
| 11 | 4.65 | 8.5 |
| 12 | 19.5 | 4.96 |
| 13 | 19.5 | 8.5 |
| 14 | 19.5 | 8.5 |

$$CN^-_v = CN^-_{vo} - CN^-_{vf} \quad (2)$$

where $CN^-_s$ is the initial free cyanide concentration in the culture broth; $CN^-_r$ is the measured residual free cyanide after incubation; $CN^-_v$ is the volatilised free cyanide during incubation; $CN^-_b$ is the bioremediated free cyanide; $CN^-_{vo}$ is the initial free cyanide in control cultures (500 mg F-CN/L); and $CN^-_{vf}$ is the final free cyanide in control cultures. The control was prepared under the same conditions as other cultures without the *Fusarium oxysporum*.

The response surface methodology was used for the statistical design of the experiments to assess the influence of temperature and pH for optimal degradation of cyanide. A central composite design was used for the determination of optimal operating conditions with a minimum residual ammonium-nitrogen as one of the objectives. Design Expert software® version 6.0.8 (Stat-Ease Inc., USA) was used to generate the experimental runs.

The results (Table 3) indicated a variation in responses measured. There was appreciable degradation of cyanide in Runs 9, 4, 1, 3, 6, 8, 13, and 14, with the highest cyanide degraded being 263 mg F-CN/L (Run 9) and the lowest (83 mg F-CN/L) being observed for Run 11. However, both cases had a high residual ammonium-nitrogen of 210 mg $NH_4^+$-N/L and 120 mg $NH_4^+$-N/L, respectively. Both Runs 9 and 11 were axial points. Run 9 with an extremely high pH resulted in high residual ammonium-nitrogen while Run 11 with an extremely low temperature was observed to have minimal microbial activity despite the presence of a suitable quantity of B. vulgaris used as a carbon source. A similar scenario had earlier been reported by Zilouei et al. [72] and Zou et al. [73], whereby a low temperature was found to inhibit the growth of microorganisms, thus resulting in low removal of contaminants (ammonium-nitrogen, nitrate and nitrite). On the other hand, Runs 1, 3, 4, 6, 7, 8, 13 and 14 had up to 99% correlation with the predicted values for cyanide degradation which indicated a high accuracy of the model (Equation 4) used for predicting cyanide degradation. However, only Runs 4 and 7, which showed minimal residual ammonium-nitrogen presence, can be used for optimisation for a pilot scale process.

# 5. STATISTICAL MODEL ANALYSIS

The statistical model summary clarifies the fitness of the mean and quadratic models for the two responses based on the Sequential Model Sum of Squares and Lack of Fit Test. The responses were analysed using ANOVA to assess the significance of the variables in the model. A quadratic model was found to give the best fit for the experimental results.

**Table 3**. Coded experimental design variables and the corresponding response

| | | | Experimental value | Predicted value | Experimental value | Predicted value |
|---|---|---|---|---|---|---|
| 1 | 0 | 0 | 239 | 238.86 | 210 | 219.14 |
| 2 | -1 | 1 | 229 | 196.59 | 100 | 183.50 |
| 3 | 0 | 0 | 239 | 238.86 | 210 | 219.14 |
| 4 | 1 | 1 | 250 | 250.29 | 40 | 83.43 |
| 5 | 1 | -1 | 135 | 167.62 | 320 | 222.79 |
| 6 | 0 | 0 | 239 | 238.86 | 210 | 219.14 |
| 7 | -1 | -1 | 127 | 126.92 | 128 | 70.86 |
| 8 | 0 | 0 | 239 | 239.14 | 210 | 200.86 |
| 9 | 0 | 1.414 | 263 | 285.75 | 210 | 117.41 |
| 10 | 1.414 | 0 | 196 | 172.77 | 100 | 135.11 |
| 11 | -1.414 | 0 | 83 | 106.02 | 120 | 98.52 |
| 12 | 0 | -1.414 | 201 | 178.03 | 30 | 136.30 |
| 13 | 0 | 0 | 239 | 239.14 | 210 | 200.86 |
| 14 | 0 | 0 | 239 | 239.14 | 210 | 200.86 |

A and B represent coded level of variables.

**Table 4.** ANOVA for F-CN Reponse Surface Quadratic Model

| Factor | Coeff. Estimate | DF | Standard Error | 95% CL Low | 95% CL High | F Value | Prob > F | Significance |
|---|---|---|---|---|---|---|---|---|
| Intercept | 239 | 1 | 10.03 | 215.27 | 262.73 | 11.41 | 0.0029 | S |
| A | 23.6 | 1 | 8.69 | 3.05 | 44.50 | 7.37 | 0.0300 | S |
| B | 38.09 | 1 | 8.69 | 17.54 | 58.63 | 19.21 | 0.0032 | S |
| $A^2$ | -49.87 | 1 | 9.05 | -71.26 | -28.49 | 30.40 | 0.0009 | S |
| $B^2$ | -3.62 | 1 | 9.05 | -25.01 | 17.76 | 0.16 | 0.7005 | NS |
| AB | 3.25 | 1 | 12.29 | -25.81 | 32.31 | 0.07 | 0.7991 | NS |

S = significant; NS = Not significant; CL = Confidence Level; DF = Degree of freedom; "Prob > F" less than 0,05 indicates the model term is significant while values greater than 0.1 indicates the model term is not significant; Std. Dev. = 24.58; $R^2$ = 0.8907; Adj. $R^2$ = 0.8127; Pred. $R^2$ = -0.1858; Adeq. Precision = 10.341

The predicted response (Y) for the biodegradation of free cyanide in terms of the coded values was:

$$Y = 239 + 23.6A + 38.09B - 49.87A^2 - 3.62B^2 + 3.25AB \quad (3)$$

where *A* and *B* are the coded values of temperature and pH, respectively. When coefficients with significant effects were considered, Eq. (3) became;

$$Y = 239 + 23.6A + 38.09B - 49.87A^2 \quad (4)$$

A model reduction was appropriate since there were many insignificant model terms. Excluding these terms improved the model. The Model F-value of 11.41 for the cyanide biodegradation was significant; therefore, there was only a 0.29% chance that a "Model F-Value" this large could occur due to noise for the quadratic model. Statistically, an adequate ratio greater than 4 is desireable for measuring a signal to noise ratio; therefore, the adequate precision of 10.341 observed in this study indicates a passable signal that can be used to further navigate the design space. Figure 3 further justifies the fitness of the model with normality in the error term.

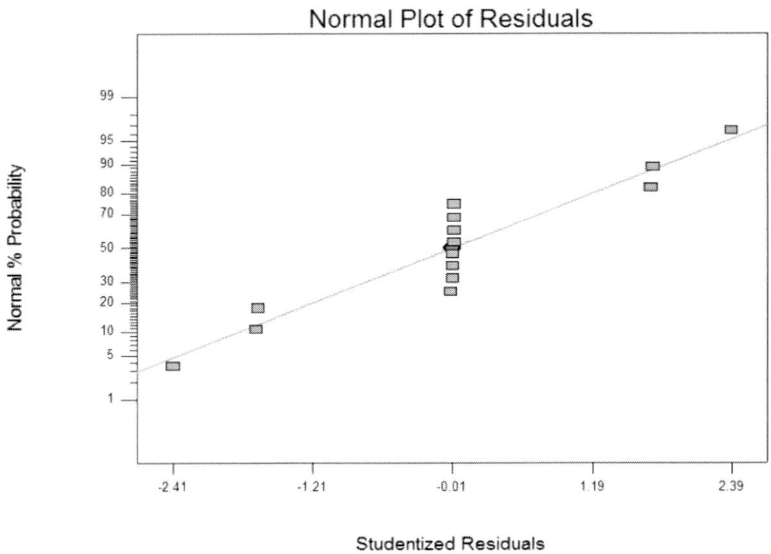


**Figure 3**. Normal probability plot of the residual F-CN

# 6. REPRESENTATION OF THE RESPONSE SURFACE MODEL

The interaction between independent variables can be studied by plotting three dimensional (3-D) curves of the response against the variables. It allows for the interpretation of experimental results and determination of optimal conditions. Elliptical contour shows the interaction between the independent variables is perfect while a circular contour indicates the variables are non-interactive [44, 47].

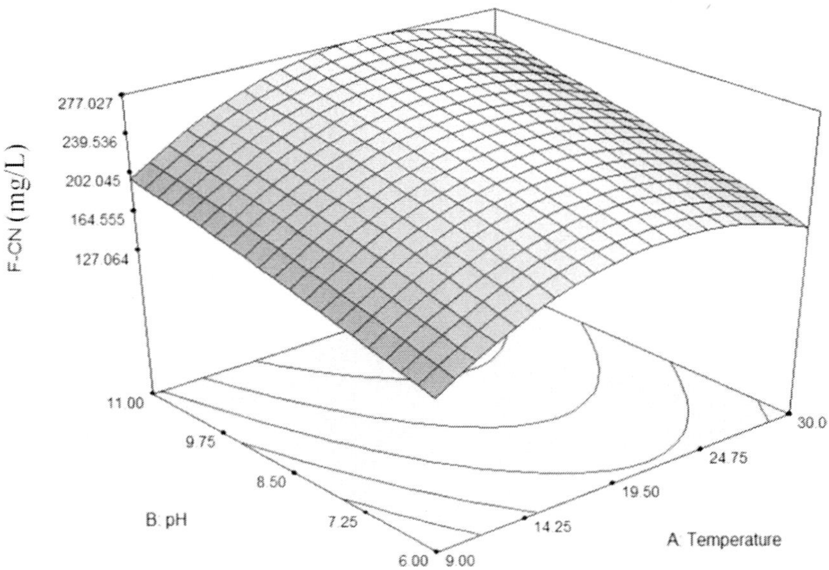


**Figure 4**. D plot showing interaction of independent variables on cyanide degradation

# 7. CYANIDE BIODEGRADATION OPTIMISATION

The optimisation was done using the Design-Expert software® numerical optimisation option where input factors were selected to achieve a desired perfomance. The numerical optimisation can maximise, minimise or achieve a targeted value: a single response; a single response subjected to upper and/or lower boundaries on other responses; and combinations of two or more responses. The desired goal for each variable and response is selected and the weight is chosen to show the degree of importance of individual goals. In this analysis, temperature and pH were set within range, cyanide degradation response was set at maximum while ammonium-nitrogen formation response was set at a minimum. The software gave three different solutions for this criteria with different desirability. The optimum point with the highest desirability was selected as shown in Fig.6 and 7. The optimal point with the maximum cyanide degradation of 250.436 mg F-CN/L and minimum ammonium-nitrogen formation of 74.285 mg $NH_4^+$-N /L was found to be at temperature of 30oC and pH of 11.

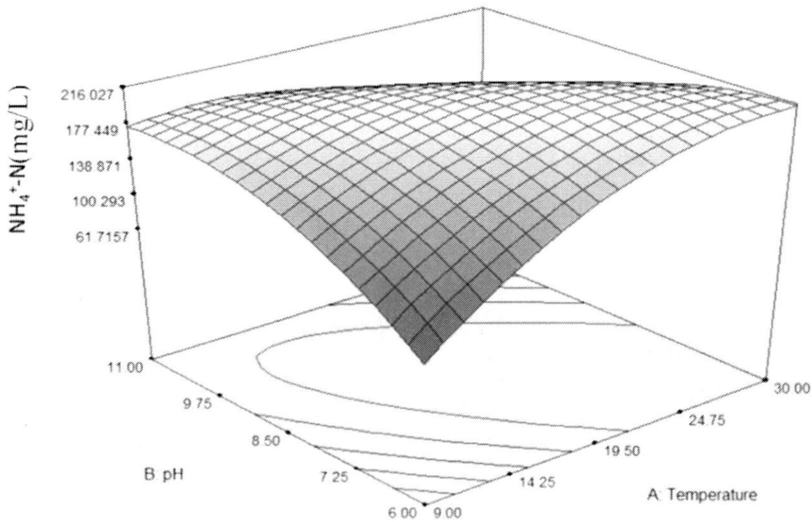


**Figure 5**. D plot showing interaction of independent variables on ammonium-nitrogen formation

# 8. CONCLUSION

*Fusarium oxysporum* cultures were grown on B. vulgaris waste to facilitate the biodegradation of cyanide, with the initial concentration of the cyanide being 500 mg CN-/L. The wastewater used was similar to the effluent discharged into ponds by goldmines having metals such as arsenic, copper, lead, iron and zinc.

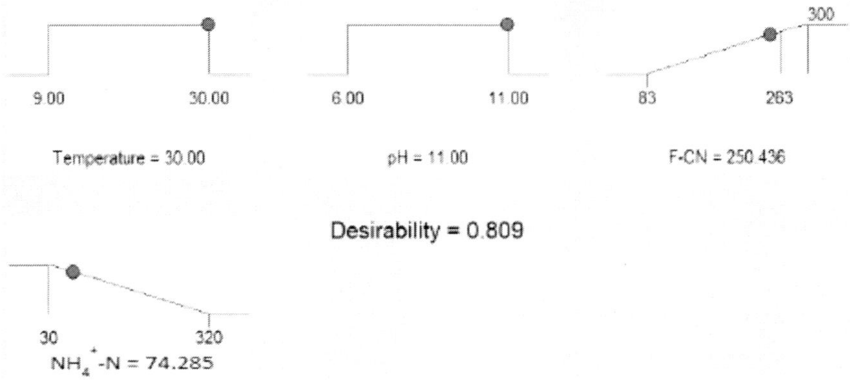


**Figure 6**. Desirability ramp for the numerical optimisation of cyanide degradation and ammonium-nitrogen formation

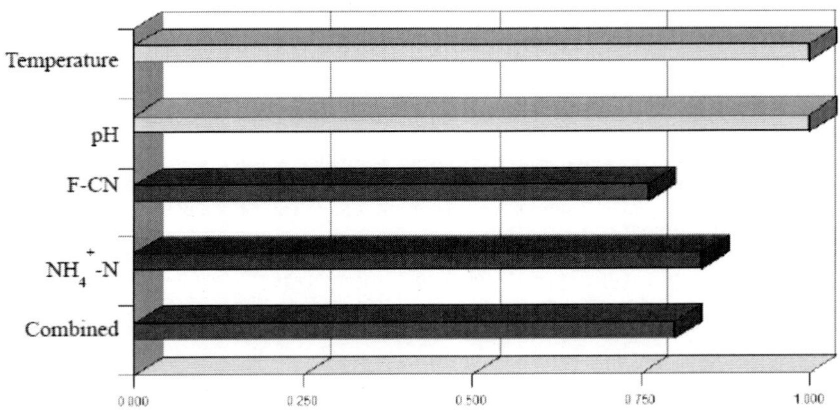


**Figure 7**. Desirability histogram for numerical optimisation of cyanide degradation and ammonium-nitrogen formation

The response surface plot identified temperature as a more significant factor affecting both the cyanide degradation and ammonium-nitrogen formation. The ammonium-nitrogen produced can be used as a nitrogen source by the fungus.

The optimum condition for maximum cyanide degradation and minimum ammonium-nitrogen formation was found at temperature 30oC and pH of 11 where cyanide of 250.436 mg F-CN/L was degraded and ammonium-nitrogen of 74.285 mg $NH_4^+$-N/L was formed.

# ACKNOWLEDGEMENTS

The authors acknowledge the CPUT University Research Fund RK 16 for funding this research.

# REFERENCES


1. Acheampong M, Paksirajan K, Lens PL. Assessment of the effluent quality from a gold mining industry in Ghana. Environmental Science and Pollution Research 2013; 20(6) 3799-3811.
2. Adjei MD, Ohta Y. Factors affecting the biodegradation of cyanide by Burkholderia cepacia strain C-3. Journal of Bioscience and Bioengineering 2000; 89(3) 274-277.
3. Akcil A. Destruction of cyanide in gold mill effluents: biological versus chemical treatments. Biotechnology Advances 2003; 21(6) 501-511.
4. Akcil A, Karahan AG, Ciftci H, Sagdic O. Biological treatment of cyanide by natural isolated bacteria (Pseudomonas sp.). Minerals Engineering 2003; 16(7) 643 - 649.
5. Amodu OS, Ntwampe SK, Ojumu TV. Emulsification of Hydrocarbons by Biosurfactant: Exclusive Use of Agrowaste. BioResources 2014; 9(2) 3508-3525.
6. Atkinson A. Bacterial cyanide detoxification. Biotechnology and Bioengineering 1975; 17(3) 457-460.
7. Babu GRV, Wolfram J, Chapatwala K. Conversion of sodium cyanide to carbon dioxide and ammonium-nitrogen by immobilized cells of Pseudomonas putida. Journal of Industrial Microbiology 1992; 9(3-4) 235-238.
8. Barclay M, Day JC, Thompson IP, Knowles CJ, Bailey MJ. Substrate-regulated cyanide hydratase (chy) gene expression in Fusarium solani: the potential of a transcription-based assay for monitoring the biotransformation of cyanide complexes. Environmental Microbiology 2002; 4(3) 183-189.
9. Cabuk A, Unal A, Kolankaya N. Biodegradation of cyanide by a white rot fungus, Trametes versicolor. Biotechnology Letters 2006; 28(16) 1313-1317.
10. Cameotra S, Makkar R. Synthesis of biosurfactants in extreme conditions. Applied Microbiology and Biotechnology 1998; 50(5) 520-529.
11. Campos MG, Pereira P, Roseiro JC. Packed-bed reactor for the integrated biodegradation of cyanide and formamide by immobilised Fusarium oxysporum CCMI 876 and Methylobacterium sp. RXM CCMI 908. Enzyme and Microbial Technology 2006; 38(6) 848-854.
12. Chapatwala KD, Babu GRV, Vijaya OK, Kumar KP, Wolfram JH. Biodegradation of cyanides, cyanates and thiocyanates to ammonium-nitrogen and carbon dioxide by immobilized cells of Pseudomonas putida. Journal of Industrial Microbiology and Biotechnology 1998; 20(1) 28-33.
13. Chen CY, Kao CM, Chen SC. Application of Klebsiella oxytoca immobilized cells on the treatment of cyanide wastewater. Chemosphere 2008; 71(1) 133-139.
14. Coty VF, Doll BE, Lifland PP (1980). Cyanide removal from wastewaters, Google Patents.
15. Cramer CS. Breeding and genetics of Fusarium basal rot resistance in onion. Euphytica 2000; 115(3) 159-166.
16. Das N, Chandran P. Microbial degradation of petroleum hydrocarbon contaminants: an overview. Biotechnology research international 2010; 2011.
17. Dash RR, Gaur A, Balomajumder C. Cyanide in industrial wastewaters and its removal: A review on biotreatment. Journal of Hazardous Materials 2009; 163(1) 1-11.
18. Dubey SK, Holmes DS. Biological cyanide destruction mediated by microorganisms. World Journal of Microbiology and Biotechnology 1995; 11(3) 257-265.

19. Dumestre A, Chone T, Portal J, Gerard M, Berthelin J. Cyanide degradation under alkaline conditions by a strain of Fusarium solani isolated from contaminated soils. Applied and Environmental Microbiology 1997; 63(7) 2729-2734.
20. Dursun AY, Aksu Z. Biodegradation kinetics of ferrous(II) cyanide complex ions by immobilized Pseudomonas fluorescens in a packed bed column reactor. Process Biochemistry 2000; 35(6) 615-622.
21. Dursun AY, Çalık A, Aksu Z. Degradation of ferrous(II) cyanide complex ions by Pseudomonas fluorescens. Process Biochemistry 1999; 34(9) 901-908.
22. EPA U (1999). Integrated Risk Information System (IRIS), EPA.
23. EPA. (1994). Treatment of cyanide heap leaches and tailings. EPA/530/R-94/037. USA.
24. Ezzi MI, Lynch JM. Biodegradation of cyanide by Trichoderma sp. and Fusarium sp. Enzyme and Microbial Technology 2005; 36(7) 849-854.
25. FAO (2014). "Food wastage: Key facts and figures." from http://www.fao.org/news/story/en/item/196402/icode/.
26. Figueira MM, Ciminelli VST, Andrade MCd, Linardi VR. Cyanide degradation by an Escherichia coli strain. Canadian Journal of Microbiology 1996; 42(5) 519-523.
27. Fry WE, Millar RL. Cyanide degradion by an enzyme from Stemphylium loti. Archives of Biochemistry and Biophysics 1972; 151(2) 468-474.
28. Gong J-S, Lu Z-M, Li H, Shi J-S, Zhou Z-M, Xu Z-H. Nitrilases in nitrile biocatalysis: recent progress and forthcoming research. Microbial Cell Factories 2012; 11 142.
29. Gu S, Wang S, Yang Q, Yang P, Peng Y. Start up partial nitrification at low temperature with a real-time control strategy based on blower frequency and pH. Bioresource technology 2012; 112(0) 34-41.
30. Gullicks H, Hasan H, Das D, Moretti C, Hung Y-T. Biofilm fixed film systems. Water 2011; 3(3) 843-868.
31. Gupta N, Balomajumder C, Agarwal V. Enzymatic mechanism and biochemistry for cyanide degradation: a review. Journal of Hazardous Materials 2010; 176(1) 1-13.
32. Gupta V, Nayak A. Cadmium removal and recovery from aqueous solutions by novel adsorbents prepared from orange peel and Fe2O3 nanoparticles. Chemical Engineering Journal 2012; 180 81-90.
33. Gurbuz F, Ciftci H, Akcil A. Biodegradation of cyanide containing effluents by Scenedesmus obliquus. Journal of Hazardous Materials 2009; 162(1) 74-79.
34. Kao CM, Liu JK, Lou HR, Lin CS, Chen SC. Biotransformation of cyanide to methane and ammonium-nitrogen by Klebsiella oxytoca. Chemosphere 2003; 50(8) 1055-1061.
35. Kowalska M, Bodzek M, Bohdziewicz J. Biodegradation of phenols and cyanides using membranes with immobilized microorganisms. Process Biochemistry 1998; 33(2) 189-197.
36. Kwon HK, Woo SH, Park JM. Degradation of tetracyanonickelate (II) by Cryptococcus humicolus MCN2. FEMS Microbiology Letters 2002; 214(2) 211-216.
37. Liang S, Guo X-Y, Feng N-C, Tian Q-H. Effective removal of heavy metals from aqueous solutions by orange peel xanthate. Transactions of Nonferrous Metals Society of China 2010; 20 s187-s191.
38. Luque-Almagro V, Blasco R, Huertas M, Martínez-Luque M, Moreno-Vivián C, Castillo F, Roldán M. Alkaline cyanide biodegradation by Pseudomonas pseudoalcaligenes CECT5344. Biochemical Society Transactions 2005; 33(Pt 1) 168-169.
39. Makkar R, Cameotra S. An update on the use of unconventional substrates for biosurfactant production and their new applications. Applied Microbiology and Biotechnology 2002; 58(4) 428-434.
40. Makkar RS, Cameotra SS, Banat IM. Advances in utilization of renewable substrates for biosurfactant production. AMB Express 2011; 1(5) 1-19.
41. Maniyam MN, Sjahrir F, Ibrahim AL. Bioremediation of cyanide by optimized resting cells of Rhodococcus strains isolated from Peninsular Malaysia. International Journal of Bioscience Biochemistry and Bioinformatics 2011; 1 98-102.
42. Mekuto L, Jackson VA, Ntwampe SKO. Biodegradation of free cyanide using Bacillus sp. consortium dominated by Bacillus safensis, Lichenformis and Tequilensis strains: A

bioprocess supported solely with whey. Bioremediation and Biodegradation 2013; s18(004) 1-7.
43. Mitchell DA, Berovič M, Krieger N (2006). Solid-state fermentation bioreactor fundamentals: introduction and overview, Springer.
44. Montgomery DC (2008). Design and analysis of experiments, John Wiley & Sons.
45. Mulligan CN. Environmental applications for biosurfactants. Environmental Pollution 2005; 133(2) 183-198.
46. Mussatto SI, Ballesteros LF, Martins S, Teixeira JA. Use of Agro-industrial waste in solid-state fermentation processes. Industrial waste. Croatia: InTech 2012; 121-140.
47. Myers RH, Montgomery DC, Anderson-Cook CM (2009). Response surface methodology: process and product optimization using designed experiments, John Wiley & Sons.
48. Nazly N, Knowles CJ, Beardsmore AJ, Naylor WT, Corcoran EG. Detoxification of cyanide by immobilised fungi. Journal of Chemical Technology and Biotechnology 1983; 33(2) 119-126.
49. Negi R, Satpathy G, Tyagi YK, Gupta RK. Biosorption of heavy metals by utilising onion and garlic waste. International Journal of Environment and Pollution 2012; 49(3) 179-196.
50. Nolan LM, Harnedy PA, Turner P, Hearne AB, O'Reilly C. The cyanide hydratase enzyme of Fusarium lateritium also has nitrilase activity. FEMS Microbiology Letters 2003; 221(2) 161-165.
51. Ntwampe SK, Santos BA. Potential of agro-waste extracts as supplements for the continous bioremediation of free cyanide contaminated wastewater. International Journal of Agricultural, Biosystems Science and Engineering 2013; 7(7) 285- 289.
52. O'Reilly C, Turner PD. The nitrilase family of CN hydrolysing enzymes – a comparative study. Journal of Applied Microbiology 2003; 95(6) 1161-1174.
53. Oelofse SH, Nahman A. Estimating the magnitude of food waste generated in South Africa. Waste Management and Research 2013; 31(1) 80-86.
54. Ojumu TV, Solomon BO, Betiku E, Layokun SK, Amigun B. Cellulase Production by Aspergillus flavus Linn Isolate NSPR 101 fermented in sawdust, bagasse and corncob. African Journal of Biotechnology 2003; 2(6) 150-152.
55. Osuntokun BO (1994). Chronic cyanide intoxication of dietary origin and a degenerative neuropathy in Nigerians. International Workshop on Cassava Safety 375.
56. Patil Y, Paknikar K. Development of a process for biodetoxification of metal cyanides from waste waters. Process Biochemistry 2000; 35(10) 1139-1151.
57. Pereira PT, Arrabaça JD, Amaral-Collaço MT. Isolation, selection and characterization of a cyanide-degrading fungus from an industrial effluent. International Biodeterioration and Biodegradation 1996; 37(1–2) 45-52.
58. Potivichayanon S, Kitleartpornpairoat R. Biodegradation of Cyanide by a Novel Cyanide-degrading Bacterium. Biodegradation 2014; 1 1362-1365.
59. Rao M, Scelza R, Scotti R, Gianfreda L. Role of enzymes in the remediation of polluted environments. Journal of Soil Science and Plant Nutrition 2010; 10(3) 333-353.
60. Rodrigues L, Teixeira J, Oliveira R. Low-cost fermentative medium for biosurfactant production by probiotic bacteria. Biochemical Engineering Journal 2006; 32(3) 135-142.
61. Santos BAQ Continous bioremediation of electroplating effluent. Unpubished MTech thesis, Cape Peninsula University of Technology, Cape Town, South Africa 2013; 98.
62. Santos BAQ, Ntwampe SKO, Doughari JH. Continuous biotechnological treatment of cyanide contaminated waters by using a cyanide resistant species of Aspergillus awamori. Environmental Biotechnology–New Approaches and Prospective Applications, M. Petre, Ed. Croatia: InTech 2013; 123-146.
63. Silva-Avalos J, Richmond MG, Nagappan O, Kunz DA. Degradation of the metal-cyano complex tetracyanonickelate (II) by cyanide-utilizing bacterial isolates. Applied and Environmental Microbiology 1990; 56(12) 3664-3670.
64. Skowronski B, Strobel GA. Cyanide resistance and cyanide utilization by a strain of Bacillus pumilus. Canadian Journal of Microbiology 1969; 15(1) 93-98.

65. Sobrinho HBdS, Luna JMd, Rufino RD, Porto ALF, Sarubbo LA. Assessment of toxicity of a biosurfactant from Candida sphaerica UCP 0995 cultivated with industrial residues in a bioreactor. Electronic Journal of Biotechnology 2013; 16(4) 4-4.
66. Suh Y-J, Park JM, Yang J-W. Biodegradation of cyanide compounds by Pseudomonas fluorescens immobilized on zeolite. Enzyme and Microbial Technology 1994; 16(6) 529-533.
67. Van Hamme JD, Singh A, Ward OP. Recent advances in petroleum microbiology. Microbiology and Molecular Biology Reviews 2003; 67(4) 503-549.
68. Watanabe A, Yano K, Ikebukuro K, Karube I. Cyanide hydrolysis in a cyanide-degrading bacterium, Pseudomonas stutzeri AK61, by cyanidase. Microbiology 1998; 144(6) 1677-1682.
69. Way JL. Cyanide intoxication and its mechanism of antagonism. Annual Review of Pharmacology and Toxicology 1984; 24(1) 451-481.
70. White DM, Schnabel W. Treatment of cyanide waste in a sequencing batch biofilm reactor. Water Research 1998; 32(1) 254-257.
71. Zhu Z, Zhang F, Wei Z, Ran W, Shen Q. The usage of rice straw as a major substrate for the production of surfactin by Bacillus amyloliquefaciens XZ-173 in solid-state fermentation. Journal of Environmental Management 2013; 127(0) 96-102.
72. Zilouei H, Soares A, Murto M, Guieysse B, Mattiasson B. Influence of temperature on process efficiency and microbial community response during the biological removal of chlorophenols in a packed-bed bioreactor. Applied Microbiology and Biotechnology 2006; 72(3) 591-599.
73. Zou S, Yao S, Ni J. High-efficient nitrogen removal by coupling enriched autotrophic-nitrification and aerobic-denitrification consortiums at cold temperature. Bioresource Technology 2014; 161 288-296.

# CHAPTER 9

# Microalgal Biotechnology: Prospects and Applications

**Soha S.M. Mostafa**

*Microbiology Department, Soil, Water and Environment Research Institute, Agricultural Research*

## 1. INTRODUCTION

There is a current worldwide interest in finding new and safe antioxidants from natural sources such as plant material to prevent oxidative deterioration of food and to minimize oxidative damage to living cells [1]. Microalgae are photosynthetic microorganisms that are able to rapidly generate biomass from solar energy, $CO_2$ and nutrients in bodies of water. This biomass consists of important primary metabolites such as sugars, oils and lipids, for which process path-ways exist for the production of high-value products including human and animal feed supplements, transport fuels, industrial chemicals and pharmaceuticals. Algal biomass and algae-derived compounds have a very wide range of potential applications, from animal feed and aquaculture to human nutrition and health products. Some algae are considered as rich sources of natural antioxidants. Although macroalgae have received much attention as potential natural antioxidants [2]. Furthermore, the qualities of the microalgal cells can be controlled, so that they contain no herbicides and pesticides, or any other toxic substances, by using clean nutrient media for growing the microalgae. The value of microalgae as a source of natural antioxidants is further enhanced by the relative ease of purification of target compounds. Reports on the antioxidant activity of microalgae are limited. Because cyanobacteria are largely unexplored, they represent a rich opportunity for discovery; the expected rate of rediscovery is far lower than for other better- studied groups of organisms Li et al. 2007 [3]. In this chapter, we focus on many desirable chemicals are the products of secondary metabolism triggered under conditions not conducive to fast growth. For those chemicals to be produced by microalgae, one needs to develop new strains (faster growth, higher substrate tolerance, etc.) by classical selection or genetic manipulation so microalgal biomass can be produced consistently. Highlight the role of dietary antioxidants and their potential benefits in health and disease directly or indirectly by the plant nutrition and animal feed to produce healthy organic food. Investigate the different biological activities of algae and the relations with its biochemical composition, pigments and different constituents which may vary with salt stressed culture conditions and describe the antioxidant characteristics of algae.

## 2. WHAT ARE MICROALGAE?

Microalgae are prokaryotic or eukaryotic photosynthetic microorganisms that produce carbohydrates, proteins and lipids as a result of photosynthesis. They can grow rapidly and live in harsh conditions due to their unicellular or simple multicellular structure. Examples of prokaryotic microorganisms are Cyanobacteria (Cyanophyceae) and eukaryotic microalgae are for example green algae (Chlorophyta) and diatoms (Bacillariophyta). Microalgae are present in all existing earth ecosystems, not just aquatic but also terrestrial, representing a big variety of species living in a wide range of environmental conditions. It is estimated that more than 50,000 species exist, but only a limited number, of around 30,000, have been studied and analyzed [4]. Sunlight, water, nutrients and arable land are the major requirements for growing algae. Micro algae have the ability to fix $CO_2$ using solar energy with efficiency 10 times greater than that of the terrestrial plants with numerous additional technological advantages. Algae are more efficient at utilizing sunlight than terrestrial plants, consume harmful pollutants, have minimal resource requirements and do not compete with food or agriculture for precious resources [5].

## 3. ALGAL METABOLITES

Metabolites are the intermediates and products of metabolism. The term metabolite is usually restricted to small molecules. A primary metabolite is directly involved in the normal growth, development, and reproduction. A secondary metabolite is not directly involved in those processes, but usually has important ecological function. The induction of secondary metabolism is linked to particular environmental conditions or developmental stages. Secondary metabolites are those chemical compounds in organisms that are not directly involved in the normal growth, development or reproduction of organisms.The exploration of these organisms for pharmaceutical purposes has revealed important chemical prototypes for the discovery of new agents, stimulating the use of sophisticated physical techniques and new syntheses of compounds with biomedical application. In this regard, both secondary and primary metabolisms have been studied as a prelude to future rational economic exploitation (Figure 1).The secondary metabolism is of restricted distribution, while the primary metabolism furnishes intermediates for the synthesis of essential macromolecules [6].

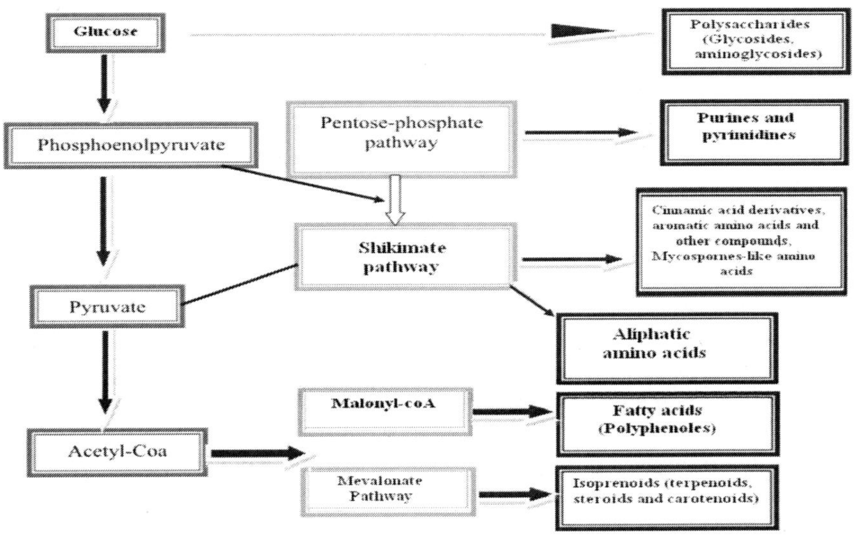


**Figure 1.** Main pathways of some secondary and primary metabolites biosynthesis modified fromBurja et al. [7]

# 4. WHAT ARE PHYTOCHEMICALS?

"Phyto" is the Greek word for plant. The term "phytochemicals" refers to a wide variety of compounds produced by plants. Phytochemicals are chemical compounds formed during the plants normal metabolic processes. There are many "families" of phytochemicals and they help the human body in a variety of ways. Phytochemicals may protect human from a host of diseases. These chemicals are often referred to as "secondary metabolities" of which there are several classes including alkaloids, flavonoids, coumarins, glycosides, gums, polysaccharides, phenols, tannins, terpenes and terpenoids Phytochemicals are naturally occurring, nonnutritive chemicals. They appear to work alone and in combination, and perhaps in conjunction, with vitamins [8].

# 5. MICROALGAL BIOACTIVE COMPOUNDS

Microalgae are significant resource for bioactive metabolites, particularly cytotoxic agents with applications in cancer chemotherapy. From the marine microalgae such as from the blooms of *Phaeocystis* sp., antibiotic substances were listed. *Phaeocystis pouchetii* is reported to produce chemicals such as acrylic acid, which constitutes about 7.0% of the dry weight. The antibiotic substances thus produced are transferred throughout the food chain and found in the digestive tract of *Antartic penguins.* Production of ß carotene and vitamins by the halotolerant alga *Dunaliella* sp. is documented. These compounds have much importance for the Mariculture activities [9]. Cyanobacteria have been identified as one of the most promising group of organisms from which novel and biochemically active natural products are isolated. Cyanobacteria such as *Spirulina,Anabaena*, *Nostoc* and *Oscillatoria* produce a great variety of secondary metabolites. Cyanobacteria produce a wide variety of bioactive compounds, which include 40% lipopeptides, 5.6% amino acids, 4.2% fatty acids, 4.2% macrolides and 9% amides. Cyanobacterial lipopeptides include different compounds like cytotoxic (41%), antitumor (13%), antiviral (4%), antibiotics (12%) and the remaining 18% activities include antimalarial, antimycotics, multi-drug resistance reversers, antifeedant, herbicides and immunosuppressive agents [7]; besides the immune effect, blue green algae improves metabolism. Cyanobacteria are also known to produce antitumor, antiviral, antifungal compounds and have a cholesterol-lowering effect in animals and humans [10]. Many of the pharmaceutically interesting compounds in cyanobacteria are peptides, including cyanobacterial toxins and important candidates for anti-cancer drugs. Peptide synthetases are common in cyanobacteria and responsible for the production of cyanobacterial hepatotoxins and other peptides. Polyketide synthetases are also involved in the biosynthesis of certain cyanobacterial bioactive compounds (e.g. microcystins). A number of extracts were found to be remarkably active in protecting human lymphoblastoid T-cells from the cytopathic effects of HIV infection. Active agents consisting of sulfolipids with different fatty acid esters were isolated from *Lyngbya lagerheimii* and *Phormidium tenue*. Cyanovirin is a protein isolated from an aqueous cellular extract of *Nostoc elipsosporum* prevents the in vitro replication and citopathicity of primate retroviruses.Cryptophycin 1, an active compound isolated from *Nostoc* strain, exerts antiproliferative and antimitotic activities by binding to the ends of the microtubules, thus blocking the cell cycle at the metaphase of mitosis. Research has been focused on its potent antitumor activity and a synthetic analogue, cryptophycin-52, is at present in Phase II clinical trials. Sulfated polysaccharide, calcium spirulan. A novel water soluble extracts of cyanobacteria have found to be an antiviral agent. This compound appears to be selectively inhibiting the penetration of enveloped viruses into host cells, thereby preventing the replication. The effect was described for many different viruses like herpex simplex, measles, and even HIV-1. Among eukaryotic microalgae, a glycoprotein prepared from *Chlorella vulgaris* culture supernatant exhibited protective activity against tumor metastasis and chemotherapy-induced immunosuppression in mice [11]. Hereafter, a brief discuss of the commercial application of the most explored compounds from algae and the biosynthetic pathways of fatty acids, steroids and carotenoids.

## 5.1. Fatty acids (fa)

Microalgae include essential fatty acids (EFAs) such as linoleic, arachidonic, linolenic, ?-linolenic acids etc. that must be in diet for healthy growth. These acids cannot be synthesized fast enough by body to meet needs [12]. Fatty acids are structural components of many lipids, and the types and amounts of fatty acids vary considerably among algae. In recent years, fatty acids compositions in large scale production of microalgae including marine algae have created considerable interest among researchers. This is mainly because of the health benefit of mono and polyunsaturated fatty acids (MUFA and PUFA) that can be found in plants including microalgae. Moreover, polyunsaturated fatty acids (PUFAs) play key roles in cellular and tissue metabolism, including the regulation of membrane fluidity, electron and oxygen transport, as well as thermal adaptation [13]. The biosynthesis of EPA occurs through a series of reactions that can be divided into two distinct steps. First is the de novo synthesis of oleic acid (18:1 $\omega$9) from acetate, followed by conversion to linoleic acid (18:2 $\omega$-6) and α-linolenic acid (18:3 $\omega$-3). The subsequent stepwise desaturation and elongation steps form an $\omega$-3 PUFA (Fig. 2). Inside the cell, EPA is normally esterified (by cyclooxygenase and lipooxygenase activities) to form complex lipid molecules and plays an important role in higher animals and humans as the precursor of a group of eicosanoids, hormone-like substances such as prostaglandins, thromboxanes and leucotrienes that are crucial in regulating developmental and regulatory physiology (Figure 2) [14]. Consumption of n-3 PUFAs from both seafood and plant sources may reduce coronary heart disease (CHD) risk as reported by Mozaffarian et al. [15] in a cohort study of 45,722 men. Thus, many health supplement stores now sell preparation of microalgae such as *Spirulina* and *Chlorella* packed in capsule or caplets, or even in food and beverages known to have therapeutic values in treating hypercholesterolemia, hyperlipidaemia and atherosclerosis [16]. The fatty acid contents of microalgae are influenced by the environmental and cultural condition selected for its growth [17]. Some of the environmental conditions include heterotrophic, photoautotrophic and nitrogen deprivation or stimulation. Although some microalgae species are cultivated as sources of these fatty acids, transgenic algae engineered to produce EPA, like transgenic oilseed crops, could provide an alternative sustainable source of oil for human consumption [18]. However, the possibility for deploying transgenic organisms nutritionally enhancedwith EPA is currently limited by continued consumer antipathy to transgenic food products. One alternative would be to use EPA from transgenic algae as a high potential food source in aquaculture. In this way, the significant health benefits of these fatty acids could be delivered into the human diet, without the requirement of direct ingestion of genetically modified food.

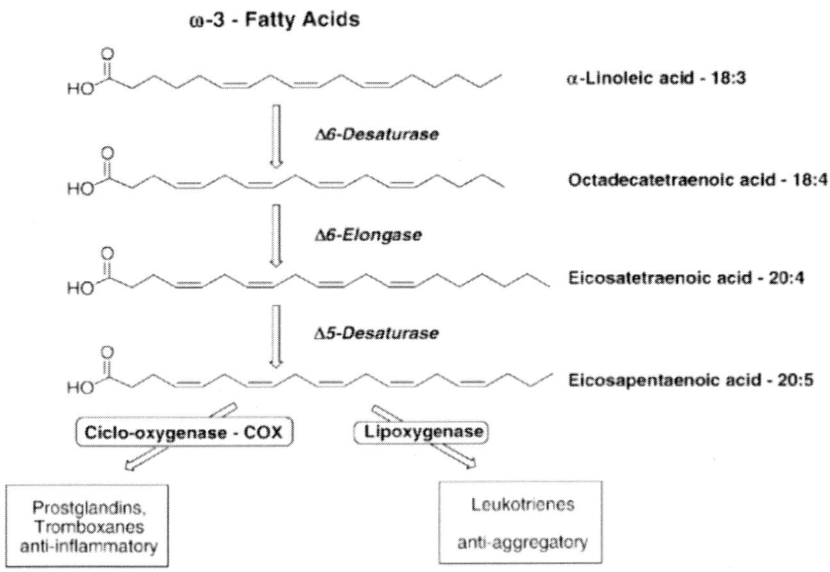


**Figure 2.** A simplified biosynthesis scheme of eicosapentaenoic acid and eicosanoid (prostaglandins, thromboxanes, leukotrienes) modified from Sayanova and Napier [19].

## 5.2. Sterols

Sterols are one of the most important chemical constituents of microalgae [20]. Sterols are the main component of eukaryote organisms and different classes of organisms have divergent sterols patterns. It is because of this that sterols act as a fingerprint for organic matter input into an aquatic environment. Furthermore, sterols have a relatively high resistance to degradation when settled in anoxic sediments and persist in the environment for a longer period of time. Of all the sterol compounds, cholesterol is the most abundant and ubiquitous one in the environment, which is due to it having a variety of sources [21]. Most biologically produced sterols are planar 3β-hydroxy tetracyclic structures commonly containing a methyl- or ethyl- substituted C7-C11 hydrocarbon side chain, and exhibiting a range of methyl-substitution (C4, C14) patterns on the polycyclic nucleus with varying degrees and positions of unsaturation (C5, C7, C8). The rigid structure of the sterols (Figure 3), caused by the fused ring system, provides the cell membrane integrity and stability thus, holds the membrane together. In general, there is not a specific sterol that can be uniquely linked to one algal source. Many of the sterols previously discussed are also found in other groups of algae [22].

## 5.3. Pigmentation in aquacultures

Astaxanthin (Figure 4) is a red pigment common to several aquatic organisms including microalgae, seagrasses, shrimp, lobsters and fish such as salmon and trout. Crustaceans are unable to synthesize carotenoids de novo and require astaxanthin (or appropriate precursors) in their diet in order to acquire the adequate color for seafoodmarket acceptance [23]. Several natural sources–such as the algae *Dunaliella salina* and *Spirulinamaxima*–or synthetic β-carotene, canthaxanthin and astaxanthin have been used for this purpose. Astaxanthin is, in fact, one of the most expensive components of salmon farming, accounting for about 15%of total production costs [48]. Among the several natural sources of astaxanthin applied in aquaculture, the green unicellular freshwater alga *Haematococcus pluvialis* has been explored by biotechnology companies [24].

β-Carotene is one of the important members of the family of carotenoids; a group of natural fat-soluble stereoisometric pigments. β-Carotene shows pro-vitamin A activity and as such it plays an important role in the human body[25]. β-Carotene can be also used as a coloring agent. Therefore, β-carotene has several applications in food, pharmaceuticals and cosmetics. The great demand of β-carotene has been met by industry, mainly by synthetic production. Increasing demand for natural carotenoids has resulted in growing interest in extracting β-carotene from different natural sources. *Dunaliella salina* is the main source for the natural β-carotene in the market. The estimated market size for natural β-carotene is 10-100 tonnes.year-1 and its price is >750 €.Kg-1 [26]. In addition, β-carotene (like other carotenoids) is a strong antioxidant, scavenging potentially harmful oxy radicals, which are commonly associated with the induction of certain cancers (Leach et al., 1998) and there is an inverse relation between the consumption of certain carotenoids and the risk of cancer [25]. The demonstrated antioxidant activity of carotenoids is the basis of the protective action of these compounds against oxidative stress in many organisms and situations. Effects of carotenoids on human health are, in general, associated with their antioxidant properties. Notwithstanding, not all of the biological activities ascribed to carotenoids must be necessarily linked to their ability to prevent accumulation of free radicals and reactive oxygen species. The halophilic green biflagellate microalga *Dunaliella salina* has since long been recognized as an efficient biological source of this carotenoid. Many epidemiological studies suggest that humans fed on a diet high in β-carotene from *Dunaliella*, which maintains higher than average levels of serum carotenoids, have a lower incidence of several types of cancer and degenerative diseases [27]. The xanthophyll astaxanthin has many applications in nutraceuticals, cosmetics, and food and feed industries. Recently, a variety of additional potential applications of this carotenoid, mainly related to human health and nutrition properties, have been claimed [28]. Lutein is among the most important carotenoids in foods and human serum and, together with zeaxanthin, is the essential component of the pigment present in the macula lutea (or yellow spot) in the eye retina and in the eye lens. Lutein is used as food dyes and especially as feed additives in aquaculture and poultry farming. During the last few years, additional applications for lutein have received considerable interest, especially those related to human health. Mainly on the basis of epidemiological studies, lutein is currently considered as effective agent for the prevention of a variety of human diseases. The microalga *Muriellopsis* sp. and other chlorophycean species are able to accumulate lutein as a part of their biomass. An established commercial system for the production of lutein from microalgae does not exist yet, although the basis for outdoor production of lutein-rich cells of strains of Muriellopsis and *Scenedesmus* at a pilot scale has alreadybeen set up [29].

## 5.4. Mycosporine-like amino acids

A remarkable group of marine natural products are the mycosporine-like amino acids (MAAs). An outstanding characteristic of these compounds is their high UV absorption with molar absorptivities (ε) of around 40 000 l $mol^{-1}$ $cm^{-1}$ (e.g. Takano et al. [30]). MAAs are water-soluble, low molecular-weight (generally <400) compounds composed of either an aminocyclohexenone or an aminocyclohexenimine ring, carrying nitrogen or amino alcohol substituents [31]. They are found in a wide variety of marine, freshwater and to a smaller degree in terrestrial organisms. There is limited evidence that MAAs are derived from early steps of the shikimate pathway. However, the biochemical pathway of MAA synthesis is still largely unknown, as well as its genetic base. The most primitive organisms capable of MAA synthesis are cyanobacteria [32].

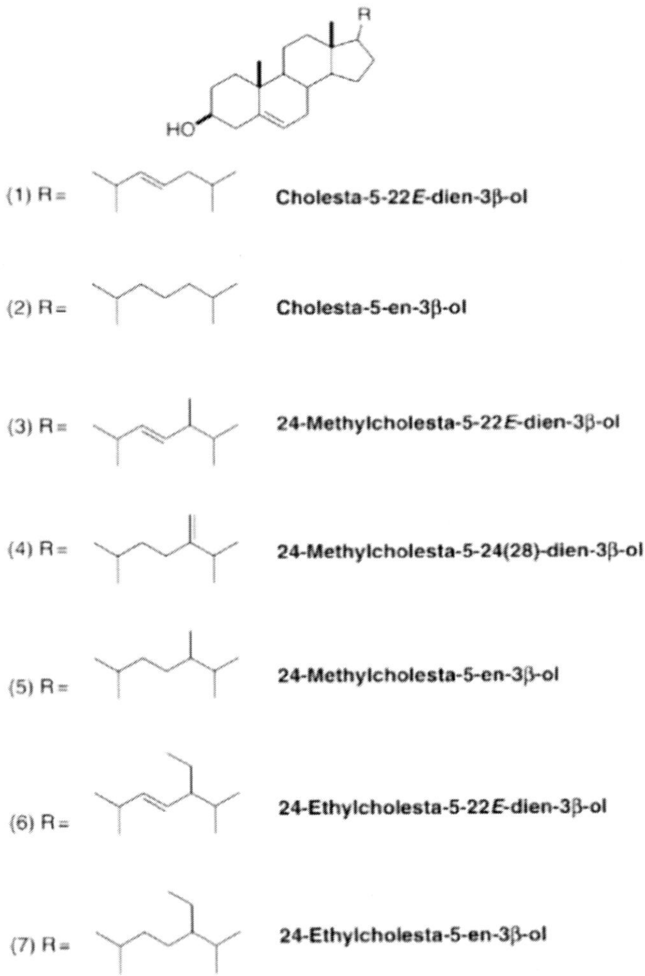


**Figure 3.** Some sterols found in marine and freshwater microalgae (modified from Ponomarenko et al. [34]

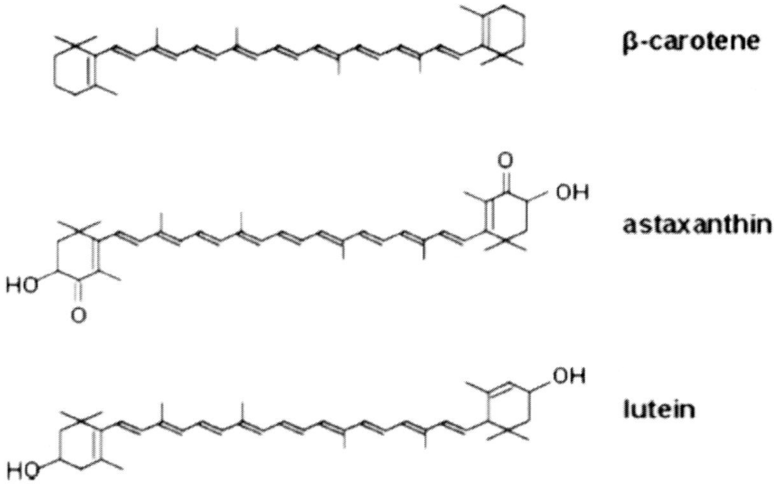


**Figure 4.** Chemical structure of β-carotene and of the xanthophylls astaxanthin and lutein, main carotenoids from microalgae with commercial interest

# 6. BIOLOGICAL ACTIVITY OF MICROALGAE

Many of the microalgal metabolites have chemical structure and possess interesting biological activity. Microalgae are a unique source of therapeutic substances, particularly from cyanobacteria. Among cyanobacteria *Spirulina* sp. has undergone numerous and rigorous toxicological studies that have highlighted its potential therapeutic applications in the area of immunomodulation, anticancer, antiviral, and cholesterol reduction effects.

## 6.1. Antioxidant activity

Hydrogen peroxide is a product of microalgae and plants through of photosynthesis, photorespiration, respiration and other metabolic processes, as result from the enzymatic activity of glycolate oxidase, urate oxidase and amino acid oxidase. However, major pathway for production of $H_2O_2$ is conversion from superoxide ($O_2^{-}$) produced through the transfer of an electron from ferredoxin of photosystem I (PSI) to $O_2$ (Mehler reaction) by the action of Superoxide Dismutase (SOD). However, it is suspected that those antioxidants are responsible for some side effects such as liver damage and carcinogenesis. Antioxidants can involve with the oxidation process by scavenging free radicals, chelating catalytic metals and by acting as oxygen scavengers [33]. Recently many researchers are interested in finding any natural antioxidants having safety and effectiveness, which can be substituted for current and commercial synthetic antioxidants, BHA and BHT. Microalgae have become good candidates for sources of natural antioxidants, as revealed by a number of recent studies [34-35]. Algae contain several enzymatic and nonenzymatic antioxidant defense systems to maintain the concentration of ROS ($O_2^-$ and $H_2O_2$) to protect cells from damage [36]. The main cellular components susceptible to damage by these ROS are lipids (peroxidation of poly-unsaturated fatty acids in membranes), proteins (denaturation), carbohydrates and nucleic acids. The essential for ROS detoxification during normal metabolism and particularly during stress, are antioxidant defenses system [37]. The primary scavenging enzymatic defenses system include SOD, calalase (CAT) and glutathione peroxidase, (GPX) and peroxiredoxin (PrxR) [38]. These enzymic detoxification system involving the action of SOD and reductase, either quench toxic compounds or regenerate antioxidants with the help of reducing power provided by photosynthesis [39]. However, at low levels, $H_2O_2$ resulted in induction of defense genes such as glutathione S-transferase and glutathione peroxidase. The hydrophilic antioxidants AA and GSH effectively scavenge oxygen radicals. Carotenoids and TOH remove ROS directly from the pigment bed [40]. Also, Foyer and Noctor [41] reported that the changes in ROS, fluctuations in the antioxidants concentrations in photosynthetic cells might have important consequences not only for defense metabolism but also for the regulation of genes associated with adaptive responses. Several bioactive metabolites produced by cyanobacteria and algae have been discovered by screening programs, employing target organisms quite unrelated to those for which the metabolites evolved [42]. Shanab et al. [43] studied the antioxidant activity of aqueous extracts of nine microalgal species namely, *Nostoc muscorum, Anabaena flos aquae, Anabaena oryza*e, *Nostoc humifusum*, *Oscillatoria* sp., *Spirulina platensis, Phormediumfragile, Wollea saccata* and *Chlorella vulgaris*. Antioxidant activity of the algal extracts was performed using 2,2 diphenyl-1-picrylhydrazyl (DPPH) test and 2,2/-azino-bis (ethylbenzthi-azoline-6-sulfonic acid (ABTS) radical action assay which revealed higher antioxidant activity than DPPH mehod. Concerning DPPH, the antioxidant activity of nine tested algal species ranged between 30.1 and 72.4% comparing with the standarad antioxidant BHT (80.2%). Using ABTS method, which was more sensitvie than the DPPH method (Figure 5), the antioxidant activity ranged between 31.2 and 75.9% (Standarad BHT showed 85.6%); *Spirulina platensis, Oscillatoria* sp, *Anabaena flous-aqua* and *Nostoc muscorum* recorded the highest (75.9, 75.6, 73.6 and 72.8%, respectively) antioxidant activity which is could be attributed to the extracellular and intracellular secondary metabolits content (Total phenolic content, terpenoids and alkaloids) of these microalgae (Tables 1,2).

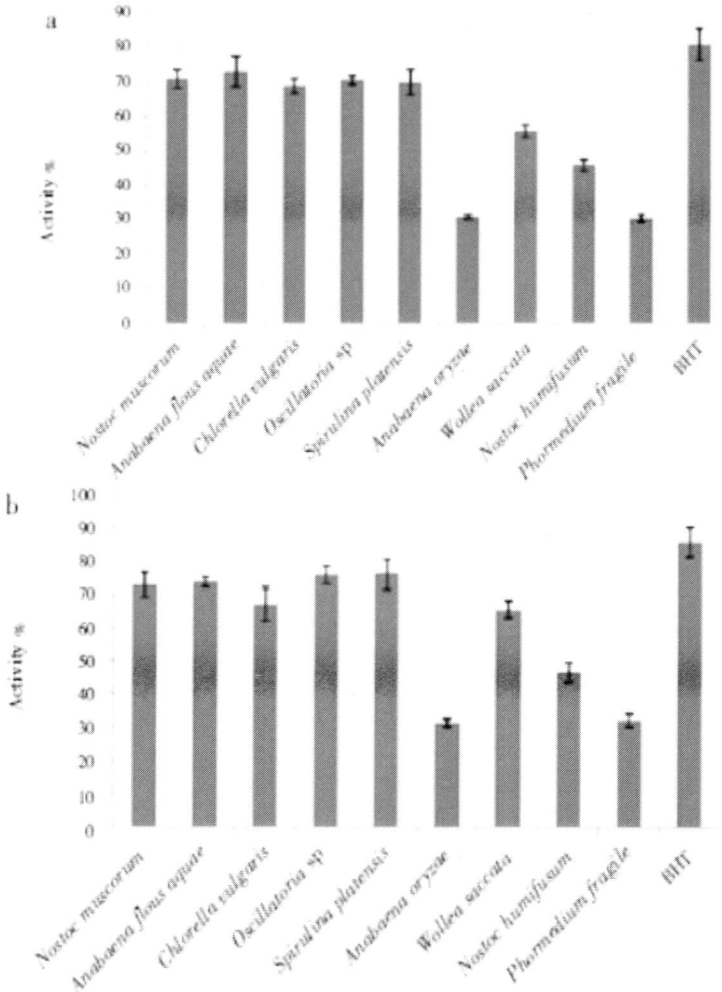


**Figure 5.** Antioxidant activity of algal wter extracts at 100 g/ml against DPPH (a) and ABTS (b) radicals [43]

The extractracellular phytochemicals metabolites (%) released in the algal cultures show large variability. *Anabaena oryza, Phomidium fragile* and *Wallea saccate* (Table 1) recorded the highest extracellular total phenolic compounds (0.0085, 0.0078 and 0.0074% respectively). The highest terpenoids contents were achieved by *Phormidium fragile, Spirulina platensis* and *Wollea saccata* (0.0055. 0.0050 and 0.0049%, respectively).

**Table 1.** Secondary metabolites (as mg/100 g) in algal filterates (extracellular) [43]

| Metabolites | *Nostoc muscorum* | *Anabaena flous aquae* | *Chlorella vulgaris* | *Oscillatoria* sp | *Spirulina platensis* | *Anabaena oryzae* | *Wollea saccata* | *Nostoc humifusum* | *Phormedium fragile* |
|---|---|---|---|---|---|---|---|---|---|
| *T. phenolic* | 1.91±0..02 | 4.10±0.51 | 2.90±0.05 | 4.40±0.65 | 4.80±0.74 | 8.50±1.87 | 7.40±0.84 | 3.10±0.055 | 7.80±0.96 |
| Terpenoids | 1.80±0.01 | 2.10±0.16 | 2.10±0.27 | 2.50±0.68 | 5.00±0.52 | 8.10±0.65 | 4.90±0.64 | 2.10±0.0 | 5.50±0.32 |
| Alkaloids | 35.00±1.65 | 39.00±2.98 | 36.00±1.64 | 49.00±2.65 | 68.00±3.61 | 75.00±4.95 | 59.00±3.94 | 42.00±1.66 | 68.00±3.74 |
| LSD 0.05 | 2.60 | 2.61 | 2.61 | 2.64 | 2.60 | 1.423 | 1.36 | 2.60 | 3.93 |

**Table 2.** Secondary metabolites (as %) in algal cells (inracellular) [43]

| Metabolites | *Nostoc muscorum* | *Anabaena flous aquae* | *Chlorella vulgaris* | *Oscillatoria sp* | *Spirulina platensis* | *Anabaena oryzae* | *Wollea saccata* | *Nostoc humifusum* | *Phormedium fragile* |
|---|---|---|---|---|---|---|---|---|---|
| *T. phenolic* | 0.61±0.06 | 0.32±0.05 | 0.20±0.00 | 0.55±0.0 | 0.71±0.14 | 0.40±0.00 | 0.10±0.00 | 0.34±0.0 | 0.36±0.00 |
| Terpenoids | 0.10±0.01 | 0.15±0.02 | 0.09±0.00 | 0.20±0.0 | 0.14±0.00 | 0.12±0.00 | 0.14±0.00 | 0.10±0.0 | 0.08±0.00 |
| Alkaloids | 2.30±0.20 | 1.9±0.03 | 2.45±0.05 | 2.62±0.15 | 3.02±0.06 | 1.60±0.14 | 1.50±0.40 | 1.65±0.16 | 1.80±0.25 |
| LSD 0.05 | 0.066 | 0.105 | 0.0036 | 0.018 | 0.029 | 0.067 | 0.031 | 0.031 | 0.066 |

**Table 3.** Phycobilins pigments (mg/ml) in different aqueous filtrate of some microalgae [43]

| Algal species | CPC | APC | CPE | Total phycobilins |
|---|---|---|---|---|
| *Nostoc muscorum* | 0.089±0.010 | 0.140±0.020 | 0.0±0.0 | 0.229±0.020 |
| *Anabaena flous aquae* | 0.000±0.000 | 0.017±0.000 | 0.0±0.0 | 0.017±0.000 |
| *Chlorella vulgaris* | ND | ND | ND | ND |
| *Oscillatoria sp* | 0.021±0.000 | 0.001±0.000 | 0.0±0.0 | 0.022±0.000 |
| *Spirulina platensis* | 0.030±0.000 | 0.113±0.000 | 0.0±0.0 | 0.143 ±0.000 |
| *Anabaena oryzae* | 0.009±0.000 | 0.002±0.000 | 0.0±0.0 | 0.011±0.000 |
| *Wollea saccata* | 0.002±0.000 | 0.033±0.000 | 0.0±0.0 | 0.035±0.000 |
| *Nostoc humifusum* | 0.012±0.000 | 0.019±0.000 | 0.0±0.0 | 0.031±0.000 |
| *Phormedium fragile* | 0.130±0.010 | 0.030±0.000 | 0.0±0.0 | 0.160±0.030 |
| LSD 0.05 | 0.005 | 0.0057 | — | 0.003 |

The maximum values of the extracellular alkaloids were recorded by *Anabaena oryza*, *Phomidium fragile*, *Anabaena oryza*, *Spirulina platensis* and *Phomidium fragile* (0.075, 0.068 and 0.068%, respectively). While, the highest percentages of these metabolities complaining with the released extracellularly metabolities, *Spirulina platensis*, *Nostoc muscourum* and *Oscillatoria* sp. recorded the greatest intracellular total phenolic compounds (0.71, 0.6 and 0.55% respectively), while *Wollea saccata* showed the least content (0.1%) as shown in Table (2).

Concerning terpenoids, *Anabaena flos aquae*, *Spirulina platensis* and *Wollea saccata*, recorded the highest contents (0.15, 0.14 and 0.14% respectively). Alkaloids determination in algal cultures showed that *S. platensis*, *Oscillatoria* sp. and *Chlorella vulgaris* showed the highest contents (3.02, 2.6 and 2.45% respectively). Phycobiliprotein pigments (Table 3) were determined in water extracts of the tested algal species. Normally, phycobilin pigments in cyanobacteria comprised phycocyanin, allophycocyanin and phycoerytherin (the blue, gray and red colors respectively). *Phormidium fragile* recorded the higher C-phycocyanin (CPC) content (0.13 mg/ml) while *Anabaena oryzae* and *A. flos aquae* recorded the least and absence of phycocyanin (0.0089 and 0.0 mg/l respectively).

*Nostoc muscorum* contained the greatest allophycocyanin (APC) pigment (0.14 mg/l), while, *Oscillatoria* sp have the least content (0.001 mg/l). Concerning total phycobilin pigments (phycocyanin and allophycocyanin), *Nostoc muscrum*, *Phormedium fragile* recorded the highest contents (0.229 and 0.16 mg/l) followed by that of S. platensis (0.143 mg/ml), while *Anabaena oryzae* showed the least total phycobilin pigment content (0.011 mg/l).*Chlorella vulgaris* is green alga which have other pigments (Chlorophyll and carotenoids) than the phycobilins (absent).

Aqueous extracts of the tested algal species showed wide range of colours (green, blue, violet, pink, ligh-blue) in spite of the fact that eight of the tested algae were of cyanobacteria and only one species was a green alga, their water extracts showed highly variable colors (Figure 6) which may be attributed in part to their phycobiliprotein constituents (ratios of phycocyanin to allophycocyanin and approximately absence of phycoerytherin pigments), and in part to the produced major polar secondary metabolites. All these substances may not only caused the alteration of the pH values of the algal aqueous extracts, but also the induced biological activities which may be attributed to the synergistic effects of these compounds. The aqueous extract of the tested algal species (8 cyanobacteria and one green alga) have variable colors (Figure 6) ranging from green, violet, blue, light blue and pink color, which can be used as an additive coloring agents to different food products (natural, non toxic) instead of the synthetic coloring substances which may be carcinogenic [43].

**Figure 6.** Color of aqueous extracts of the different microalgae (1. Nostoc muscorum, 2. Anabaena flos aquae, 3.Chlorella vulgari, 4. Oscillatoria sp., 5. Spirulina platensis, 6. Anabaena oryzae, 7. Wollea saccat, 8. Nostoc humifusum and 9. Phormedium fragile [43]

Phycobiliprotein pigments were known by its antioxidant activity [44], increasing of these pigments production as a result of doubling nitrate concentration in the growth culture media, led to a progressive increase in the antioxidant activity recorded by both DPPH and ABTS assays in the two cyanobacteria under investigation.Keeping in mind that, synergetic effect occurred between the polar secondry metabolites especially the phenolic compounds and the polysaccharides in antioxidant activity. Increasing nitrate concentrations in the culture media of both cyanobacteria species

(*N. muscorum* and *Oscillatoria* sp) led to a marked enhancement in phycobiliprotein production (Table 4) which was translated in an obvious increase in antioxidant activity (by DPPH and ABTS) in both species under study while decreasing the nitrate content, phycobilin pigments production were consequently decreased in both species and its complete absence was recorded on nitrogen starvation especially in case of *Oscillatoria* sp., and no allophycocyanin pigments were produced as found by Shanab et al.[43]. They also investigated that increasing nitrate conc. and the consequent increase in phycobilin pigments production, have the major role in enhancing the antioxidant activity may be attributed. The decrease in nitrate conc. was followed by an obvious decrease in phycobilin pigment and even an absence of one of its constituents on nitrate starvation. The antioxidant activity in both species (by both assays) was apparently not affected comparing with the control (1.5 g/l nitrate). Under stress conditions, it was known that, deviation in metabolic pathways may occur. In presence of nitrate, nitrogenous compounds, including the phycobilin pigments were increasingly produced leading, together with other antioxidant active secondary metabolites (as phenolics), to a marked increase in biological activity.

The decrease in nitrate content induced a stress condition and not only a decrease in nitrogen skeleton compounds as phycobilin pigment production, but an increase in the carbon skeleton compounds (as phenolics) as a result of metabolic alterations under these stress conditions. So on decreasing nitrate content, the antioxidant activity remain at a level comparable or even higher than the control due to the synergistic effect of the phycobilin pigment and the phenolic compounds produced in excess under stress nitrate condition which have high redox potentials. On nitrogen starvation the recorded antioxidant activity (Comparable to those in presence of high nitrate content (6-9 g/l) was largely due to the high production of the carbon skeleton compounds (phenolic compounds) which show potent antioxidant activity [45].Shalaby et al. [46] stated that cultivation of *Spirulina platensis* under salt stress conditions (0.02 M as control), 0.04 and 0.08 M NaCl led to a remarkable alteration of algal metabolism as well as an enhancement or induction of biologically active compounds. Biochemical analysis of salt stressed algal revealed that lipid content was slightly increased together with certain saturated and unsaturated fatty acids especially the polyunsaturated ones (γ-inolenic acid, omega 3 fatty acid).

**Table 4.** Antioxidant activity of the nitrogen stressed promising algal species using DPPH and ABTS radicals [43]

| Treatment | *Oscillatoria* sp. | | *Nostoc muscorum* | |
|---|---|---|---|---|
| | DPPH | ABTS | DPPH | ABTS |
| Control (1.5 g/L $NaNO_3$) | 59.80±0.95 | 69.80±1.45 | 69.80±1.22 | 72.60 |
| 3 g/L | 60.20±1.58 | 70.00±0.95 | 70.60±1.00 | 75.30±2.30 |
| 6 | 61.50±0.88 | 71.60±0.63 | 70.20±0.98 | 76.10±3.00 |
| 9 | 68.00±3.60 | 73.60±2.80 | 72.90±0.51 | 80.30±1.65 |
| 0.75 | 60.30±1.50 | 69.80±4.31 | 71.50±0.64 | 70.00±0.58 |
| 0.37 | 62.50±2.45 | 73.60±2.55 | 70.00±1.60 | 74.30±0.47 |
| 0.0 | 66.80±1.67 | 73.00±1.60 | 74.00±2.78 | 76.80±0.61 |
| LSD 0.05 | 0.968 | 0.956 | 0.987 | 1.001 |

## 6.2. Anticancer activity

Today cancer is the largest single cause of death in men and women, and chemoprevention has been a promising anticancer approach aimed at reducing themorbidity andmortality of cancer by delaying the process of carcinogenesis. A variety of compounds fromnature sources have been shown to be beneficial for the inhibition of cancer, such as flavonoids, phenolic acids, carotenoids, etc.; the mechanisms which suppress tumorgenesis often involve inhibition of tumor cell mediated protease activity, attenuation of tumor angiogenesis, promotion of cell cycle arrest, induction of apoptosis and immunostimulation, etc. In addition, Chinery et al. [47] also reported their use with the chemotherapy agents 5-fluorouracil and antioxidants could cause complete remissions in colorectal cancer, where only partial remission is possiblewith chemotherapy agents only; therefore, antioxidants have been proposed to have potential for the prevention and treatment of diseases associated with active oxygen species, especially in cancer diseases. Moreover, experimental and epidemiological evidence suggests that anti-inflammatory drugs may also decrease the incidence of mammary cancer, tumor burden, and tumor volume [48].The medicinal value of cyanobacteria was appreciated as early as 1500 Bc, when strains of Nostoc were used to treat gout, fistula and several forms of cancer. Cyanobacteria are a rich source of potentially useful natural products. Over 40 different Nostocales species, the majority of which are Anabaena and Nostoc spp. Produce over 120 natural products (Secondry metabolities) having activities such as anti-HIV anticancer, antifungal, antimalarial and antimicrobial. Cyanovirin (CV-N, cyanoviorin-N), a 101 amino acid protein extracted from *Nostoc ellipsosporum* was found to have potent activity against all human immunodeficiency viruses such as HIV-1, M and T tropic strains of HIV-1, HIV-2, SIV (Simian), and FIV (Feline) [7]. The cosmopolitan distribution of cyanobacteria indicates that they can cope with a wide spectrum of global environmental stress, such as heat, cold, desiccation, salinity, nitrogen starvation, photooxidation, anaerobiosis and osmotic stress. They have developed a number of mechanisms by which cyanobacteria defend themselves against environmental stressors. Important among them are the production of photoprotective compounds such as mycosporine-like amino acids (MAAs) and Scytonemin enzymes such as superoxide dismutase, catalase and peroxidases repaire of DNA damage and synthesis of shock proteins [49]. Shanab et al. [43] investigated anticancer efficiency of the algal water extracts against Ehrlich Ascites Carcinoma cell (EACC) and Human hepatocellular cancer cell line (HepG2). Anticancer efficiency of the algal water extracts was investigated against Ehrlich Ascites Carcinoma cell (EACC) and Human hepatocellular cancer cell line (HepG2). The anticancer efficiency of the algal aqueous extracts illustrated in Figure (7) and using EACC and HepG2 cell lines, recorded that the anticancer activity ranged between 15.68 to 87.25 % in case of EACC cell line and from 9.5 to 89.4% using HepG2 cell line *Nostoc muscorum* aqueous extracts recorded the highest anticancer activity in both cell lines (87.25% in case of EACC and 89.4% in case of HepG2), followed by *Oscillatoria* sp. (67.40 and 77.8% in EACC and HepG2 respectively). In case of *N. muscorum*, the anticancer activity against EACC cell line ranged between 83.0 and 90.4% at all nitrate concentrations (increase and decrease) compared to the control (85.9%). Comparable anticancer activity was recorded at both the highest nitrate conc and starvation (90.4 and 89.9% respectively). The anticancer activity against HepG2 cell line recorded more or less comparable activities were recorded at most nitrate conc compared to the control (85.6, 86.9, 88.7 and 88.6 % at 3, 6, 9 and 1.5 g/l). At nitrate starvation the highest anticancer activity against HepG2 cell line was recorded (92.3%). In case of *Oscillatoria* sp., the anticancer activity against EACC and HepG2 recorded an increase in activity on both increasing and decreasing nitrate conc comparing with the control. Higher activity was recorded against both cell lines at higher nitrate conc (82.6 and 75.9% in case of EACC and HepG2 respectively) and at nitrate starvation (82.9 and 82.0% respectively) compared to the control (68.3 and 70.4 % against EACC and HepG2 respectively). Water extracts of the tested promising algal species demonstrated higher anticancer efficiencies against both EACC and HepG2 cell lines (87.25 and 89.4% respectively) in case of *N. muscorum* and 67.40 and 77.8% in case of *Oscillatoria* sp.). Under stress nitrogen conditions, these two cyanobacteria species recorded higher anticancer activities on exess limitation or starvation of nitrate comparing with its normal content in growth media.

The recorded maximum activity in both species against both cell lines at the highest nitrate content (9 g/L) may be attributed mainly to the higher content of the phycobiliprotein pigments produced under excess nitrate contents. Nitrate limitation and starvation, in spite of the caused decrease in phycobilin pigment production due to metabolic alteration expected under stress conditions, the carbon skeleton compounds as phenolic may replace phycobilin shortage in inducing similar anticancer activity of or even higher efficiency caused by great phycobilin contents at higher nitrate supplementation [43]. These results demonstrated that the compounds responsible for anticarcinogenic activity was highly polar as the phycobilins, phenolic compounds and polysaccharides which induced apoptosis of the cancer cells as reported by Aboul-Enein et al [50], which go parallel with these results coincides with the results obtained by Wang et al [51] who reported that the aqueous extract of red algae mainly contain c-phycocyanin, exhibited higher antipraleferation inducing apoptosis body formation. The authors explained that phycocyanin interact with membrane associated B-tubulin and glyceraldehydes-3-phosphate dehydrogenase (GAPDH), caused polymerization of microtubules and actins filaments leading to arrested the cell cycle at G0/G1 phase. As these aqueous extracts exhibited antioxidant and anticancer activities, its effect as coloring agent is amplified by these biological efficiencies which are very important for human health. Also, it can be used for the manufacture of pharmaceutical drugs (antioxidant and anticancer).

Table (5) recorded the anticancer efficiency of nitrate stressed *N. muscorum* and *Oscillatoria* sp. against EACC and HepG2 cell lines. The promising cyanobacterial species *N. muscorum* and *Oscillatoria* sp. induced both the highest antioxidant (by DPPH and ABTS methods) and anticancer activities (using EACC and HepG2 cell lines) which may be attributed to their large content in total phycobiliprotein pigments together with the higher secondary metabolites content (phenolic compounds, terpenoids, alkaloids [43].

The antimicrobial activity of microalgae has been attributed to compounds belonging to several chemical classes –including indoles, terpenes, acetogenins, phenols, fatty acids and volatile halogenated hydrocarbons [52] for instance, the antimicrobial activity of supercritical extracts obtained from the microalga *Chaetoceros muelleri* were related to its lipid composition [53]. However, the antimicrobial activity detected in several pressurized extracts from *Dunaliella salina* may be explained not only by several fatty acids, but also by such compounds as - and -ionone, - cyclocitral, neophytadiene and phytol. Efforts to identify the compounds directly responsible for those antimicrobial features –e.g. chlorellin [54] have been on the run, but are still relatively incipient owing the some new classes of compounds found. Microalgal cell-free extracts are already being tested as additives for food and feed formulation, in attempts to replace antimicrobial compounds of synthetic origin currently in use – including subtherapeutical doses of antibiotics employed as prophilatic measure in animal breeding [55].

Recall, in this regard, the growing resistance of some bacterial strains arising from the widespread and essentially unrestricted use of antibiotics in cattle handling, and by domestic consumers use via self-prescription [56]. However, a key factor for their eventual economic feasibility is the possibility of operating large photobioreactors under aseptic conditions, which are able to produce biomass and metabolites to sufficiently high levels [57].

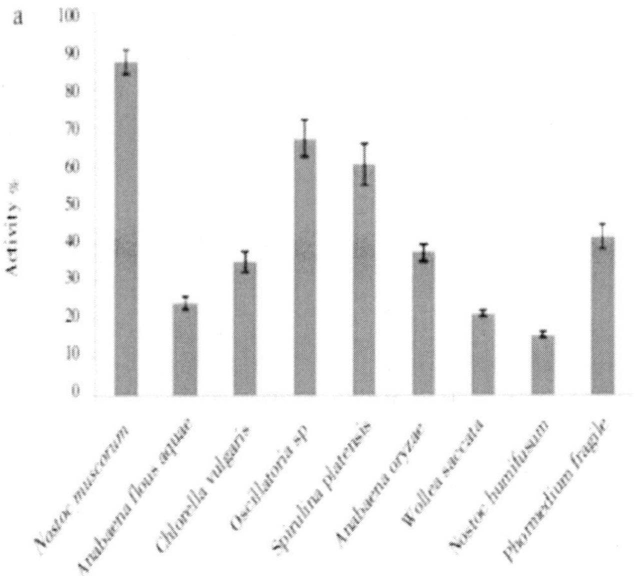


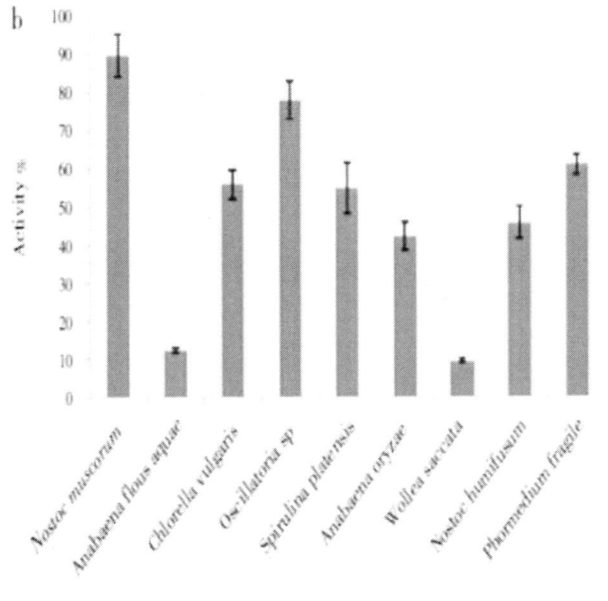


**Figure 7.** a, b). Anticancer activity of algal water extracts at 100 μg/ ml against EACC (a) and HepG2 (b) cell lines [43]

**Table 5.** Anticancer activity of the nitrate stressed promising algal species using EACC and HepG2 cell lines [43]

| Treatment | *Oscillatoria* sp. | | *Nostoc muscorum* | |
|---|---|---|---|---|
| | HepG2 | EACC | HepG2 | EACC |
| Control (1.5 g/L $NaNO_3$) | 70.40±1.87 | 68.30±0.58 | 88.60±2.96 | 85.90±3.60 |
| 3 g/L | 72.10±2.60 | 69.70±1.64 | 85.60±2.30 | 84.60±4.00 |
| 6 | 70.60±0.80 | 72.60±2.34 | 86.90±1.80 | 85.60±2.98 |
| 9 | 75.90±2.61 | 82.60±4.65 | 88.70±0.95 | 90.40±4.82 |
| 0.75 | 72.60±3.05 | 68.30±0.72 | 86.70±1.00 | 83.00±1.64 |
| 0.37 | 74.80±2.50 | 78.30±5.96 | 88.00±2.70 | 84.00±0.50 |
| 0.0 | 82.00±4.85 | 82.90±3.40 | 92.30±1.65 | 89.90±1.96 |
| LSD 0.05 | 0.575 | 0.907 | 1.63 | 0.505 |

## 6.3. Antimicrobial activity

### 6.1.1. Antiviral activity

A number of infectious diseases caused by viruses have emerged (and re-emerged) in recent years. Although several antiviral drugs have been specifically developed, drug-resistant mutations are constantly occuring – so new antiviral active principles are necessary, especially those from sources that do not constitute (or are exposed to) viral pools. This is why microalgae have received a strong attention as potential suppliers of antiviral agents [58]; Viral growth is generally divided into three stages, and antiviral action may take place at a single or more stages: Stage I, which consists on adsorption and invasion of cells; Stage II, or eclipse phase, during which the cell is forced to synthesize multiple copies of said virus; and Stage III, or maturity and release of virus particles. For instance, the anti-HSV activity of the antiviral compound acyclovir® is expressed at stage II, but the anti-HSV factor from *Dunaliella* sp. inactivates the viral function at stage I. Sulphated exopolysaccharides from marine microalgae have been claimed to interfere with Stage I of some enveloped viruses they offer competitive advantages because of their broad antiviral spectrum against e.g. HSV and HIV-1 [59]. Apparently, their inhibitory effect arises from interaction with the positive charges on the virus or on the cell surface – which prevents penetration of the former into the host cells; they may also selectively inhibit reverse transcriptase in the case of HIV, thus hampering production of new viral particles after infection yet the exact step during viral replication when they act remains to be elucidated. Antiviral highly sulfated polysaccharides from several species of red microalgae consist mainly of xylose, glucose and galactose [60]; they are unusually stable when exposed to extreme pH and temperature [61]. Despite their successful antiviral performance, the metabolic pathways leading to sulfated polysaccharides are still poorly known. Their secretion by unicellular red algae was originally characterized via radiolabeling – which howed biosynthesis of the carbon chain, and sulfation of the resulting polysaccharide to occur in the Golgi apparatus [62]; these findings were confirmed in *Porphyridium* sp. [63] and other red microalgae [64]. More recently, [65] used 14C pulse-chase experiments and ultrastructural microscopy to conclude that brefeldin A–a membrane-traffic inhibitor of the Golgi apparatus, decreases the contents of the bound and the soluble forms of polysaccharides, while inhibiting cell-wall binding of polysaccharides to a greater extent than its soluble counterpart (in both actively growing and resting cells). Discovery of small molecules that can specifically disrupt a particular protein-protein interface remains a challenge-but is of a particular interest in virology, since the antiviral drugs currently available target only viral proteins.

### 6.1.2. ANTIBACTERIAL ACTIVITY

Most efforts were devoted to the study of antibiotic resistance in bacteria for several reasons: (i) bacterial infections are responsible for most community-acquired and nosocomial infections; (ii) the large and expanding number of antibacterial classes offers a more diverse range of resistance mechanisms; and (iii) the ability to move bacterial resistance determinants into standard, well-characterized bacterial strains facilitates more detailed studies of the underlying molecular mechanisms [66]. Pratt et al. [67] isolated the first antibacterial compound from a microalga, *Chlorella*; a mixture of fatty acids, viz. chlorellin, was found to be responsible for that inhibitory activity against both Gram+ and Gram- bacteria. Research aimed at identifying antibacterial active principles produced by microalgae has meanwhile boomed [68]. This realisation arose e.g. from the risk associated with several multidrug-resistant *Staphylococcus aureus* (MRSA) strains, which have been causing an increased concern in healthcare institutions worldwide – since they are not susceptible to most conventional antibiotics. Hence, discovery of novel antibacterial compounds following distinct biochemical mechanisms of action is urged. Antibiotics are typically less effective against Gram- bacteria because of their complex, multilayered cell wall structure – which makes it more difficult for the active compound to penetrate them [69]; this justifies why the antibacterial activity of the supernatant (and methanolic extracts) is more potent against Gram+ than Gram- bacteria [68,70]. The exact mechanism of action of fatty acids remains unknown: they may act upon multiple cellular targets, even though cell membranes are the most probable ones – as membrane damage will likely lead to cell leakage and reduction of nutrient uptake, besides inhibiting cellular respiration; conversely, Desbois [71] claimed a peroxidative process.Furthermore, compounds synthesized by *Scenedesmus costatum*, and partially purified from its organic extract, exhibited activity against aquaculture bacteria because of their fatty acids longer than 10 carbon atoms in chain length – which apparently induce lysis of bacterial protoplasts. The ability of fatty acids at large to interfere with bacterial growth and survival has been known for quite some time, but recent structure-function relationship studies suggest that said ability depends on both their chain length and degree of unsaturation. Such compounds as cholesterol can antagonize antimicrobial features [53] so both composition and concentration of free lipids should be taken into account [72]. Among microalgal-derived oxylipins, the antibacterial activities of polyunsaturated aldehydes deserve a special ention. Such compounds are synthesized by diatoms, e.g. *S. costatum* and *Thalassiosira rotula*. One illustrative example is decadienal–probably derived from (the polyunsaturated) arachidonic acid (C20:4 n-3), which exhibits a strong activity against such important human pathogens as MRSA and *Haemophilus influenza* – with MIC values of 7.8 and 1.9 μg/mL, respectively, and well as against *E. coli* and *Pseudomonas aeruginosa*, and *S. aureus and Staphylococcus epidermidis* (Gram- and Gram+ bacteria, respectively). Furthermore, it impairs growth of diverse marine bacteria, such as (the Gram-) *Aeromonas hydrophila, L. anguillarum, Alteromonas haloplankti, Photobacterium phosphoreum* and *Psychrobacter immobilis*, and the (Gram+) *Planococcus citreus* and *Micrococcus luteus* [73].

### 6.1.3. Antifungal activity

Algae are one of the chief biological agents that have been studied for the control of fungi plant pathogens [74]. Various strains of cyanobacteria are known to produce intracellular and extracellular metabolites with diverse biological activities such as antibacterial, antifungal and antiviral activity [75]. These biologically active compounds include antibiotics and toxins [76]. Hussien *et al.* [77] screened the effect of culture filtrates of nine algal strains (*Anabaena flosaquae, Anabaena oryzae, Chlorella vulgaris, Nostoc muscorum, Nostoc humifusu, Oscillatoria* sp., *Phormedium fragile, Spirulina platensis* and *Wollea saccata*) at concentrations of 10, 20, 30 and 40% on mycelium growth of the plant pathogenic fungi *Cercospora beticola* causing leaf spot disease in sugar beet comparing with different concentrations of the synthetic fungicide Topsin M70 (100, 200, 300 and 400 ppm). They found that generally, all the algal culture filtrates reduced the fungal mycelium growth but the highest fungal mycelium growth inhibition percentages were achieved by *Spirulina platensis, Oscillatoria* sp. and *Nostoc muscorum* (Figure 8) at the concentrations of 30% (100, 100 and 82%, respectively) and at 40% (100, 100 and 100%, respectively). While, the chemical synthetic fungicide Topsin M70 completely inhibited the fungal mycelium growth at the concentrations of 200, 300 and 400 ppm. Fungal spores production (number of spores) was completely inhibited by the previous three algal culture filtrates at 300 and 400 ppm particularly at the concentration of 40%. Same results were obtained by Topsin M70. The antifungal activity of the algal culture filtrates has been attributed to the presence of bioactive compounds i.e. total phenolic compounds, total saponins and alkaloids in the algal culture filtrates.

## 6.2. Nematicidal activity

Root-knot nematodes, *Meloidogyne* spp., are among the most damaging nematodes in agriculture, causing an estimated US$ 100 billion loss/year worldwide [78]. The application of chemical nematicides has been found as an effective measure for controlling nematodes but they have toxic residual effect on the environment particularly on non-target organisms and human health. In addition, the use of chemical nematicides is prohibited in organic farming. Nematicidal potential of cyanobacteria has remained unexplored except for a few reports, which suggest that endospores of Microcoleus and *Oscillatoria* spp. killed nematodes [79].

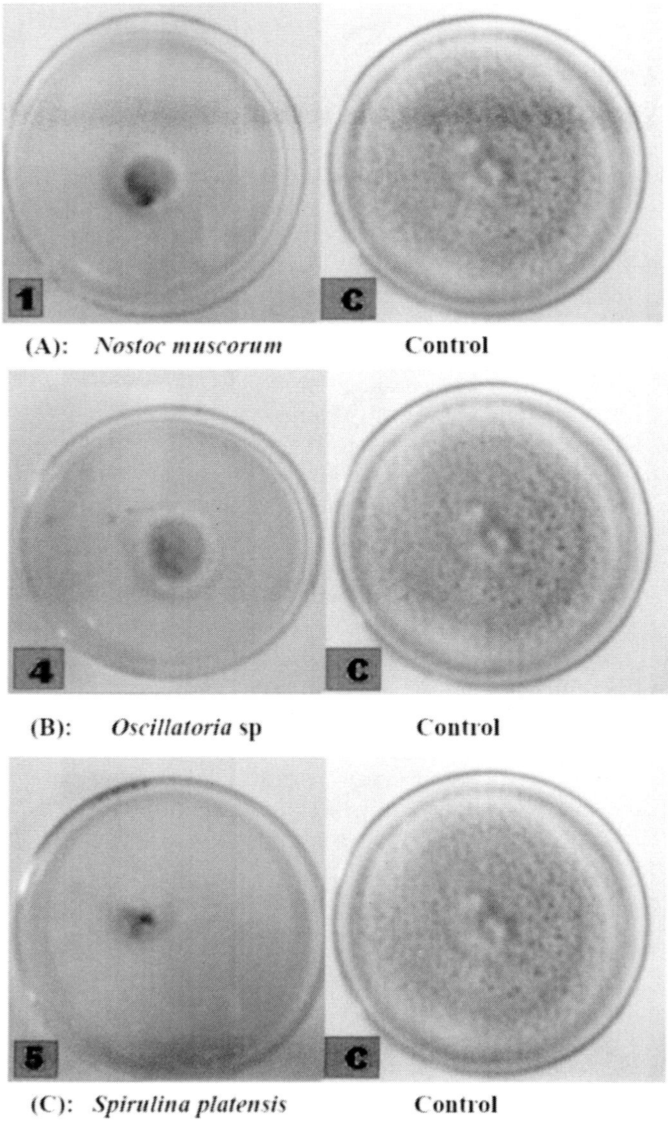


**Figure 8.** The efficacy of (A) *Nostoc muscorum,* (B) *Oscillatoria* sp and (C) *Spirulina platensis* culture filtrates (40 $ml.L^{-1}$) in suppressing the fungal mycelia growth diameter of *Cercospora beticola* [77]

Culture filtrates of Microcoleus vaginatus inhibited hatching of *Meloidogyne incognita* eggs and killed second stage juveniles. Microalgal metabolites have attracted attention, because they are a resource for toxins, and potential new drugs [80].Shawky et al. [81] studied the nematicidal effect of nine culture filtrates of algal strains (*Nostoc muscorum, Anabaena flos aquae, Anabaena oryzae, Chlorella vulgaris, Wolleasaccata, Phormedium fragile, Oscillatoria* sp., *Nostoc humifusum* and *Spirulina platensis*), *Azolla pinnata* aqueous extract filtrate (1:2w/v) and compost watery extract filtrate (1:5w/v) in controlling the population of the root knot nematode,*Meloidogyne incognita* in cucumber under both laboratory and greenhouse conditions. Laboratory experiment revealed that high juvenile mortality percentage occurred during all the exposure periods of all treatments, the best results were after 72 hr exposure. Only five cyanobacterial strains, namely, *Spirulina platensis, Oscillatoria* sp., *Anabaena oryzae, Nostoc muscorum*and *Phormedium fragile*, followed by compost watery extract, significantly increased juveniles mortality over 70% at the highest concentration of 1:10 (84.3, 80.4, 78.9, 75.4, 72.5 and 70.1%, respectively). A*zolla pinnata* aqueous extract filtrate achieved 69.8% at the same concentration while, *Anabaena flos aquae* and *Chlorella vulgaris* recorded the lowest effect on mortality percentage (52.1 and 40.1%, respectively) at the concentration of 1:10. In the greenhouse experiment, the combination of mixing five algal culture filtrates of *S. platensis, Oscillatoria* sp., *A. oryzae, N. muscorum* and *P. fragile* with *A. pinnata* aqueous extract filtrate and compost extract achieved the highest reduction in the number of the 2ndstage juveniles in soil, the numbers of galls, developmental stages, females, egg masses, egg numbers/egg mass in roots of cucumber plants comparing with the individual treatment and the non treated control. In addition, all combinations significantly improved fresh weight of roots and shoots and increased the yield of cucumber plants.

## 6.3. Mollscicidal activity

The snail intermediate hosts of schistosomiasis are the sites of intense multiplication of this parasite, thus their control strategies are considered a priority of the reduction of schistosomiasis transmission [82]. Although chemical molluscicides are to certain extent quite successful in curbing the disease concerned. However, in view of their side effects, interest in environmentally friendly approaches and use of biological control agents have been revived [83]. Mostafa and Gawish [84] stated that the algal culture filtrate of *Spirulina platensis* was proved for its mollscicidal activity against *Biomphalaria alexandrina* snails the intermediate host of *Schistosoma mansoni* in Egypt that accommodates their survival and fecundity and also for its effect on hatchability of snail's eggs and viability of the free living larval stages of the parasite (miracidia and cercariae). The culture filtrate proved to have a lethal effect on snails with 90 LC 0.23% of the filtrate. *B. alexandrina* snails stopped egg laying after one week of continuous exposure to the sublethal concentrations 0.19 and 0.17%, while those exposed to 0.07% laid few ones.It could be due to the phytochemical constituents of the culture filtrate i.e. total phenolic compounds, alkaloids and total saponins. This was confirmed by histological examination that showed a severe damage in the hermaphrodite gland cells of snails (Figure 9) exposed to these concentrations. *B. alexandrina* eggs of 3 and 7 days old failed to hatch post exposure to 0.1% of algal filtrate, while 22% and 10% only hatched after exposure to 0.05% of this filtrate. Free cell culture filtrate shows also marked miracidicidal and cercaricidal activities as 2% of this filtrate killed most of these organisms within 15 minutes of exposure. It is concluded that the byproducts of the blue-green alga *Spirulina platensis* has a lethal effect against adult *B. alexandria* snails, reduced or stopped their oviposition, hence minimize the snail populations available for the parasite transmission. Therefore it may be a potential source of effective compounds for control of *Schistosoma mansoni*.

# 7. OTHER APPLICATIONS AND PRODUCTS FROM MICROALGAE

Microalgae have found commercial applications as natural sources of valuable macromolecules, including carotenoids, long-chain polyunsaturated fatty acids, and phycocolloids. As photoautotrophs, their simple growth requirements make them attractive for bioprocesses aimed at producing high added-value compounds that are in large demand by the pharmaceutical market. The productivity and biochemical composition of microalgae depend strongly on the mode of cultivation, medium composition, and nutrient profile. Consequently, numerous efforts aimed at elucidating the practical impacts of the aforementioned parameters have been developed [56]. Thus, there is a growing interest in the area of research on the positive effect of algae on human health and other benefits.

## 7.1. Food

The first use of microalgae by humans dates back 2000 years to the Chinese, who used*Nostoc* to survive during famine. Microalgae for human nutrition are nowadays marketed in different forms such as tablets, capsules and liquids. They can also be incorporated into pastas, snack foods, candy bars or gums, and beverages [85-86].

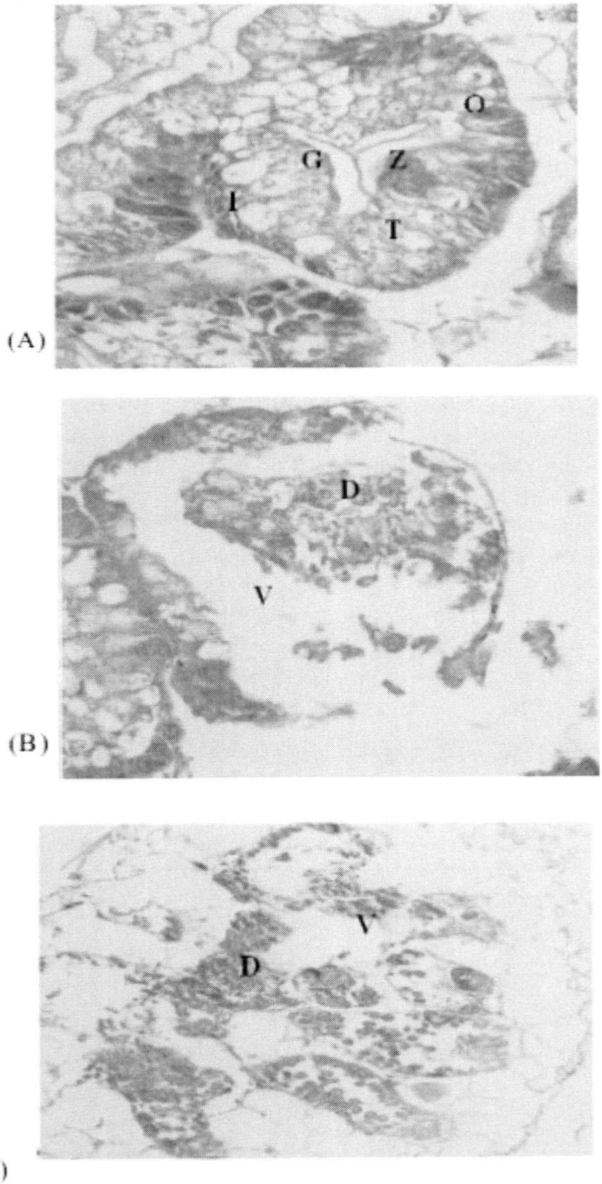


**Figure 9.** Section in the hermaphrodite gland of *Biomphalaria alexandrina* snail: control (A), exposed to 0.07% (B) and to 0.17% (C) concentrations of *Spirulina platensis* culture filtrate. G=Spermatogonia, T=Spermatid, Z=Spermatozoa, I= Oogonia, O=Oocyte V= vacules D=degeneration [84]

Owing to their diverse chemical properties, they can act as a nutritional supplement or represent a source of natural food colorants. The commercial applications are dominated by four strains: *Arthrospira, Chlorella, Dunaliella salina* and *Aphanizomenon flos-aquae*. *Arthrospira* is used in human nutrition because of its high protein content and its excellent nutritive value [87-88]). In addition, this microalga has various possible healthpromoting effects: the alleviation of hyperlipidemia, suppression of hypertension, protection against renal failure, growth promotion of intestinal *Lactobacillus*, and suppression of elevated serum glucose level [85-86]. A significant amount of *Arthrospira* production is realized in China and India. *Chlorella* can also be used as a food additive owing to the taste- and flavour-adjusting actions of its coloring agent [85]. *D. salina* is exploited for its β-carotene content that can reach 14% of dry weight [89]. For human consumption, Cognis Nutrition and Health, the world's largest producer of this strain, offers *Dunaliella* powder as an ingredient of dietary supplements and functional foods. The last major commercial strain application is *A. flos-aquae*. According to many research studies, used alone or in combination with other nutraceuticals and natural food products, *A. flos-aquae* promotes good overall health [85,90].The consumption of *Arthrospira (Spirulina)* by the Kanembu was also reported by Delisle et al. [91] in a survey of household food consumption and nutritional adequacy in Wadi zones of Chad. Table (6) presents a comparison of the general compositions of human food sources with that of different microalgae.

**Table 6.** General composition (%) of dry matter of different human food sources and algae [92].

| Commodity | Protein | Carbohydrate | Lipid |
|---|---|---|---|
| Bakers' yeast | 39 | 38 | 1 |
| Meat | 43 | 1 | 34 |
| Milk | 26 | 38 | 28 |
| Rice | 8 | 77 | 2 |
| Soybean | 37 | 30 | 20 |
| *Anabaena cylindrica* | 43-56 | 25-30 | 4-7 |
| *Chlamydomonas rheinhardii* | 48 | 17 | 21 |
| *Chlorella vulgaris* | 51-58 | 12-17 | 14-22 |

| | | | |
|---|---|---|---|
| *Dunaliella salina* | 57 | 32 | 6 |
| *Porphyridium cruentum* | 28-39 | 40-57 | 9-14 |
| *Scenedesmus obliquus* | 50-56 | 10-17 | 12-14 |
| *Spirulina maxima* | 60-71 | 13-16 | 6-7 |
| *Synechococcus sp.* | 63 | 15 | 11 |

## 7.2. Feed

Microalgae can be incorporated into the feed for a wide variety of animals ranging from fish (aquaculture) to pets and farm animals. In fact, 30% of the current world algal production is sold for animal feed applications [92] and over 50% of the current world production of *Arthrospira* is used as feed supplement [85]. In 1999, the production of microalgae for aquaculture reached 1000 t (62% for molluscs, 21% for shrimps, and 16% for fish) for a global world aquaculture production of $43x10^6$ t of plants and animals [93]. The importance of algae in this domain is not surprising as they are the natural food source of these animals. The main applications of microalgae for aquaculture are associated with nutrition, being used fresh (as sole component or as food additive to basic nutrients) for coloring the flesh of salmonids and for inducing other biological activities. The most frequently used species are *Chlorella, Tetraselmis, Isochrysis, Pavlova, Phaeodactylum, Chaetoceros, Nannochloropsis, Skeletonema* and *Thalassiosira* [85,94]. Many nutritional and toxicological evaluations have proved the suitability of algal biomass as feed supplement [92]. *Arthrospira* is largely used in this domain and concerns many types of animal: cats, dogs, aquarium fish, ornamental birds, horses, cows and breeding bulls. Algae positively affect the physiology (by providing a large profile of natural vitamins, minerals, and essential fatty acids; improved immune response and fertility; and better weight control) and their external appearance (resulting in healthy skin and a lustrous coat) of animals. In poultry rations, algae up to a level of 5-10% can be used safely as partial replacement for conventional proteins. Prolonged feeding of algae at higher concentrations produces adverse effects. The yellow color of broiler skin and shanks as well as of egg yolk is the most important characteristic that can be influenced by feeding algae [92].

## 7.3. Agricultural purposes

Humans have practiced agriculture for more than 10,000 years, but only in the past 50 years or so have farmers become heavily dependent on synthetic chemical fertilizers and pesticides. It contributes to numerous forms of environmental degradation, including air and water pollution, soil depletion and diminishing biodiversity. Synthetic chemical pesticides and fertilizers are polluting soil, water, and air, harming both the environment and human health. Soil is eroding much faster than it can be replenished—taking with it the land's fertility and nutrients that nourish both plants and those who eat them. Chemical fertilizers can gradually increase the acidity of the soil until it begins to impede plant growth. Chemically fertilized plots also show less biologic activity in the soil food web (the microscopic organisms that make up the soil ecosystem) than do plots fertilized organically with manure or other biologic sources of fertility [95]. The best way, however, is to use as much as possible microbial products, functional bio-fertilizers and bio-controllers and reduce the amount of the use of chemical fertilizers or pesticides. Heterocystous cyanobacteria and several nonheterocystous cyanobacteria are known for their ability to fix atmospheric nitrogen. The fertility of many tropical rice field soils has been mainly attributed to the activity of nitrogen-fixing cyanobacteria. An estimation showed that more than 18 kg N $ha^{-1}$ $year^{-1}$ was added to the soils by cyanobacteria. Inoculation of cyanobacteria to increase the fertility of soils has been successfully attempted. Recently, nitrogen-fixing cyanobacteria have been reported to dominate desert crusts worldwide. This is believed to contribute significantly to the fertility of desert soils and may eventually facilitate vegetation of deserts [96]. Algae as biofertilizers are a promising alternative to avoid soil pollution caused by agrochemicals. Also, they recover the nutrients content to soil as they secrete exo-polysaccharides that improve soil structure and bio-active substances that enhance the plant growth. Algae are known to be one of the most promising sources as bio-control agents of any residues, thereby having positive impact on human health [97]. Microorganisms play an important role in various chemical transformations of soils and thus, influence the availability of major nutrients like nitrogen, phosphorus, potassium and sulphur to the plants. Cyanobacteria and phosphate-solubilizing bacteria were used as biofertilizers to increase crop production [98].The cyanobacterial ability to mobilize insoluble forms of inorganic phosphates is evident from the finding of kleiner and Harper [99] who reported more extractable phosphates in soils with cyanobacterial cover than in nearby soils without cover. Cyanobacteria can fix about 25 kg N/ha/season. Apart from nitrogen fixation, inoculation with cyanobacteria is also reported to reduce considerably the total sulphides and ferrous iron content of the soil. Blue-green algae constitute an important group of microorganism capable of nitrogen fixation. Most of the species possess nitrogen fixation ability to the order Nostocales and Stigonematales. Over 100 species of blue-green algae are known to fix atmospheric nitrogen. These have been found to be very effective on the rice and banana plantation. In field condition, overall increase in the gram yield of rice is amounted to about 586 kg/ha. In case of crops other than rice, algalization increased nearly 34 per cent yield. India is one of the countries where agro-chemical conditions appear to be favourable where blue-green algae technology has been put forward. In some parts of the country, production of BGA inoculants has been commercialised. Producing inoculum in artificially controlled conditions is well defined, but relatively expensive. On the other hand open-air soil culture is simpler, less expensive and easily adaptable by the farmers. Field scale production of algae biofertilizer is also possible. 20-25 kg dry algae can be obtained on 40 m field. Adopting this method, 15 t/ha of wet BGA can be obtained by the farmers. Farmers can also produce algae for countryard of the house [96]. Blue-green algal extracts comprise a great number of bioactive compounds that influence plant growth and development. They mostly contain growth phyto-regulators like gibberellins, auxin, cytokinin, ethylene and abscisic acid [100]. This group of microorganisms have been reported to benefit plants by producing growth promoting regulators resemble gibberellin and auxin, vitamins, amino acids, polypeptides, antibacterial and antifungal substances that exert phytopathogen biocontrol and polymers especially exopolysaccharides that were reported to enhance growth and productivity of plants like *Daucus carota* [101], *Santalum album* [102], *Oryzae sativa* [103], *Lilium alexandrae* hort [104] and *Beta vulgaris* L.[100]. Non-nitrogen fixing cyanobacteria can enrich phosphorus and potassium contents in soils, laying indirect major role in plant growth promotion [105]. Cyanobacteria also enhance the soil biological activity in terms of increasing the total bacterial, total cyanobacterial counts, $CO_2$ evolution, dehydrogenase and nitrogenase activities. Many researches suggested that up to 50% of the recommended dose of the mineral nitrogen fertilizers could be saved by using some species of nitrogen fixing cyanobacteria. The obtained results emphasized the prospects and potentials of using cyanobacteria biofertilizers as renewable natural nitrogen resources for many crops. They are none polluting, inexpensive, utilize renewable resources (inorganic nutrients and atmospheric $CO_2$) in addition to their ability in using free available solar energy, atmospheric nitrogen and water [106-108].

## 7.4. Mitigation of $co_2$: why algae for $co_2$ sequestration?

Many options that have been proposed and that are in use for capturing $CO_2$ can be seen as economically, socially and environmentally short-sighted. A common approach is taking measures to offset any immediate effects, often by simple relocation of the emissions. Injection of flue gases into oceanic or geological sinks is examples of such "end-of-pipe" solutions [109]. Algae cultivation can yield a broad range of useful end products, apart from biofuels. The sequestration of $CO_2$ into algal biomass can become profitable also through the production of high value products such as pigments and high-grade lipids, which are extractable from several species of algae. Brennan and Owende [110] also mention high value products such as animal feed supplements being extractable from the microalgae species *Chlorella*, *Scenedesmus* and *Spirulina*. The urgent need for substantive net reductions in $CO_2$ emissions to the atmosphere can be addressed via biological $CO_2$ mitigation, coupled with transition to more extensive uses of biofuel, nuclear and renewable energy sources. Microalgae have attracted a great deal of attention for $CO_2$ fixation and biofuel production because they can convert $CO_2$ (and supplementary nutrients) into biomass via photosynthesis at much higher rates than conventional biofuel crops can. This biomass may then be transformed into methane or hydrogen, using processes mediated by anaerobic bacteria; an integrated process for hydrothermal production of methane via microalgae has been discussed recently [111-112]. Of particular interest is the production of oils by microalgae because of the ease of their synthesis (a lack of a nitrogen source usually suffices to trigger this form of secondary metabolism). Lipid extraction and re-esterification are accomplished with short-chain alcohols and other by-products of secondary metabolism (i.e. polyunsaturated fatty acids, bcarotenes or polymers [111]. Upon extraction, such oils can be hydrolyzed and then re-esterified with methyl- or ethyl alcohol moieties to obtain biodiesel. Microalga-mediated $CO_2$ fixation and biofuel production can be rendered more sustainable by coupling microalgal biomass production with existing power generation and wastewater treatment infrastructures (Figure 10). Microalgae can utilize low-quality water, such as agricultural runoff or municipal, industrial or agricultural wastewaters, as a source of water for the growth medium as well as a source of nitrogen, phosphorus and minor nutrients [113].

## 7.5. Wastewater treatments

Wastewater nitrogen and phosphorous as microalgae nutrients aquaculture systems involving microalgae production and wastewater treatment (e.g. of amino acids, enzyme, or food industries wastewaters) seems to be quite promising for microalgae growth combined with biological cleaning. This allows nutrition of microalgae by using organic compounds (nitrogen and phosphorous) available in some manufactures wastewater, not containing heavy metals and radioisotopes. Additionally, microalgae can mitigate the effects of sewage effluent and industrial sources of nitrogenous waste such as those originating from water treatment or fish aquaculture and at the same time contributing to biodiversity. Moreover, removing nitrogen and carbon from water, microalgae can help reduce the eutrophication in the aquatic environment.

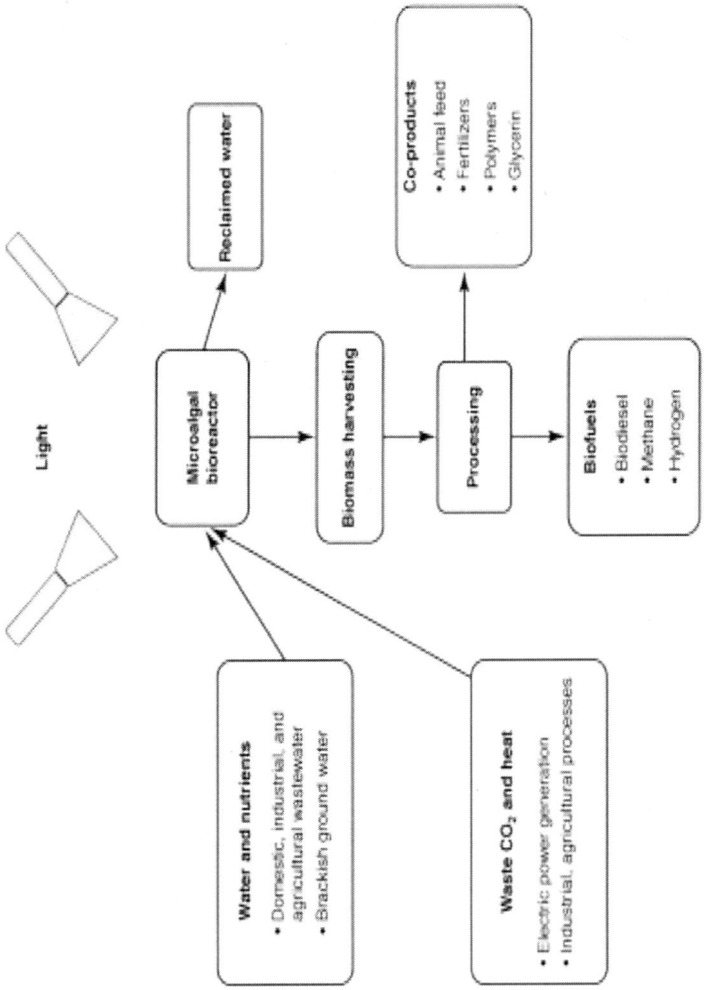


**Figure 10.** Integration of microalgal bioreactors into existing wastewater and power generation infrastructures. The overall process uses microalgae to capture industrially produced waste $CO_2$ in photobioreactors, coupled with treatment of nutrients in wastewater. $CO_2$ is converted into algal biomass by photosynthesis in the presence of light. After processing (biological, physical or thermochemical), the biomass generated can be used for production of biodiesel, methane or other fuels and co-products (e.g. animal feeds and polymers)

Aslan and Kapdan [114] used *C. vulgaris* for nitrogen and phosphorus removal from wastewater with an average removal efficiency of 72% for nitrogen and 28% for phosphorus (from 3 to 8 mg/L $NH_4^+$ and 1.5–3.5 mg/L $PO_{4}^{-3}$). Other widely used microalgae cultures for nutrient removal are *Chlorella* [115] and *Spirulina* species [116]. Nutrient removal capacities of *Nannochloris* [117]), *Botryococcus brauinii* [118] and cyanobacterium *Phormidium bohneri* have also been investigated [119-120]. Environmental applications production of biodiesel and other bio-products from microalgae can be more environmentally sustainable, cost-effective and profitable, if combined with processes such as wastewater and flue gas treatments. In fact various studies demonstrated the use of microalgae for production of valuable products combined with environmental applications [113,121].

## 7.6. Biofuel production

Microalgae can potentially be employed for the production of biofuels in an economically effective and environmentally sustainable manner. The production of these biofuels can be coupled with fuel gas $CO_2$ mitigation, wastewater treatment and the production of high-value chemicals. The efficiency is low but there is much room for improvement. The use of microalgae is seen as, at least, a partial solution to climate change and energy problem [122]. Many microalgae are exceedingly rich in oil which can be converted to biodiesel using existing technology. More than 50% of their biomass as lipids, sometimes even up to 80%, and oil levels of 20-50% are quite common [123].

Lipids production and biodiesel extraction from algae depend on algal species and extraction solvent system [124].There is a unique opportunity to both treat wastewater and provide nutrients to algae using nutrient-rich effluent streams. By cultivating microalgae, which consume polluting nutrients in municipal wastewater, and abstracting and processing this resource, then the goals of sustainable fuel production and wastewater treatment can be combined [174,125].The efforts span over many areas of "algae to fuels" technologies including production system development, algae harvest, algae strain development and genetic modification, algae products development, etc. Screening and genetic modification of algae strains will play an increasingly important role. Genetic engineering has the potential to improve the overall algal biomass yield and lipid yield. Discovery of new strains and genetically modified strains capable of secreting hydrocarbons to extracellular spaces will open some new opportunities; however, challenges with recovering the secreted liquids or volatiles remain. There is a need to develop high throughput screening and analysis methods. Current harvest and dewatering are still too energy intensive. New techniques and strategies must be devised to lower the costs. Direct conversions such as in situ transesterification and hydrothermal liquefaction offer the possibility to process wet algae. Fractionation of algal biomass, before or after oil extraction, deserves a closer look because it may play an important role in offsetting the costs. New techniques to disrupt algae cellular structures to improve oil extraction efficiency are needed [126].

## 7.7. Heavy metals and phycoremediation

Metals are directly or indirectly involved in all phases of microbial growth. Many metals such as sodium, potassium, iron, copper, magnesium, calcium, manganese, zinc, nickel and cobalt are vital for biological functions, while others such as aluminum, cadmium, silver, gold, mercury and lead are not known to have necessary biological functions. All these elements can interact with microbial cells and be accumulated as a result of different mechanisms [127]. Some of these mechanisms have biotechnological importance and can be applied for the bioremediation of metals from industrial effluents. The capability of some microbial species to adsorb some heavy metals on their surface [128-129] or accumulate them within their structure is a chief route for the removal of heavy metals from contaminated environment [130-132]. Another fashion for the detoxification of heavy metals by microorganisms is the chelation of these metals inside or outside their cells after converting them into other forms to reduce their toxicity. In 2007, Lefebvre et al. [133] working with some cyanobacterial strains (*Limnothrix planctonica*, *Synechococcus leopoldiensis* and *Phormidium limnetica*) demonstrated their ability to convert $Hg^{2+}$ into elemental mercury Hg° and meta-cinnabar (β-HgS) under pH controlled and aerated conditions. The transformation of mercury into β-HgS was attributed to the interaction with metal binding sulfhydryl protein as an intermediate step in metal sulfide synthesis. Moreover, some of the freshwater algae *Limnothrix planctonica* and *Selenastrum minutum* were recorded for their ability to bio-transform $Hg^{2+}$ into a form with the analytical properties of β-HgS under aerobic conditions due to the presence of some protein and non-protein thiol chelators [134]. Furthermore, Lengke et al. [135] investigated the gold bioaccumulation by cyanobacterium Plectonema boryanum from gold (III)-chloride solutions. They confirmed that the reduction mechanism of gold (III) to metallic gold by this organism involves the formation of an intermediate gold (I)-sulfide due to a chelation process via some thiol compounds. Recntly Essa and Mostafa [136] studied the effeicincy of three cyanobacterial isolates (*Spirulina platensis*, *Nostoc muscorum*, and *Anabaena oryzae*) individually or as a mixed culture to precipitate some heavy metals ($Hg^{2+}$, $Cd^{2+}$, $Cu^{2+}$ and $Pb^{2+}$) out of their solutions through using the culture biogas produced during their aerobic growth in a batch bioreactor. Variable capabilities of metal bioprecipitation were recorded by the three algal isolates. FT-IR studies showed the existence of –OH groups in the metal precipitate produced by the algal isolates while –NH groups were identified only in the metal precipitates produced by *N. muscorum*, and *A. oryzae*. This study highlighted a novel approach for heavy metals bioremediation through the transformation of these metals into nitrogen complexes and/or hydroxide complexes via using the culture biogas produced by some cyanobacterial species.

# 8. MICROALGAL PRODUCTION

Microalgae for human nutrition are nowadays marketed in different forms such as tablets, capsules and liquids. They can also be incorporated into pastas, snack foods, candy bars or gums, and beverages. In addition, this microalga has various possible healthpromoting effects: the alleviation of hyperlipidemia, suppression of hypertension, protection against renal failure, growth promotion of intestinal Lactobacillus, and suppression of elevated serum glucose level [86-87]. Owing to their diverse chemical properties, they can act as a nutritional supplement or represent a source of natural food colorants. The commercial applications are dominated by four strains: *Arthrospira, Chlorella, D. salina* and *Aphanizomenon flos-aquae*. *Arthrospira* is used in human nutrition because of its high protein content and its excellent nutritive value [87,88,137]. A significant amount of *Arthrospira* production is realized in China and India. The world's largest producer Hainan Simai Enterprising Ltd. is located in the Hainan province of China. This company has an annual production of 200 t of algal powder, which accounts for 25% of the total national output and almost 10% of the world output. The largest plant in the world is owned by Earthrise Farms and streches over an area of 440,000 $m^2$(located at Calipatria, CA, USA; Figure 11). Their production process is presented in Figure 12. Their Arthrospira-based products (tablets and powder) are distributed in over 20 countries around the world. Many other companies sell a wide variety of nutraceuticals made from this microalga. For example, the Myanmar *Spirulina* Factory (Yangon, Myanmar) sells tablets, chips, pasta and liquid extract, and Cyanotech Corp. (a plant in Kona, Hawaii, USA) produces products ranging from pure powder to packaged bottles under the name *Spirulina pacifica*. Cyanotech Corp. has developed an orginal process for drying the biomass in order to avoid the oxidation of carotenes and fatty acids that occurs with the use of standard dryers. The patented process employs a closed drying system that is kept at low oxygen concentrations by flushing with nitrogen and carbon dioxide. The process relies on a very cold ocean water crown from a depht of 600 m just offshore to provide dehumidification and actually dries microalgal products in less than 6 s (Figure 13). *Chlorella* is produced by more than 70 companies; Taiwan

Chlorella Manufacturing and Co. (Taipei, Taiwan) is the largest producer with 400 t of dried biomass produced per year. Significant production is also achieved in Klötze, Germany (130 – 150 t dry biomass per year) with a tubular photobioreactor. This reactor consists of compact and vertically arranged horizontal running glass tubes with a total length of 500,000 m and a total volume of 700 m3 (Figure 14). The world annual sales of *Chlorella* are in excess of US$ 38 billion [85]. The most important substance in *Chlorella* is β-1,3-glucan, which is an active immunostimulator, a free-radical scavenger and a reducer of blood lipids (9, Ryll et al., Abstr. Europ. Workshop Microalgal Biotechnol., Germany, p. 56, 2003). However, various other health-promoting effects have been clarified (efficacy on gastric ulcers, wounds, and constipation; preventive action against atherosclerosis and hypercholesterolemia; and antitumor action) [85,122,138].

**Figure 11.** Earthrise Farms Arthrospira production plant (Calipatria, CA, USA)

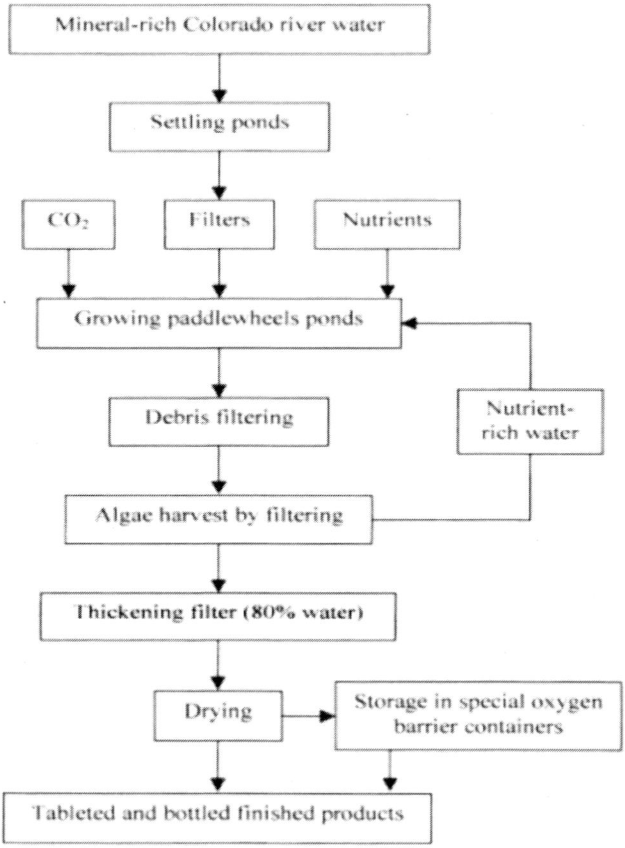


**Figure 12.** Earthrise Farms microalgal production process

# 9. CONCLUSION

Meeting the increasing water demands with limited resources advocates Egypt to find innovative and sustainable approaches for management. It is essential to maximize the benefits of the available resources and to minimize the wastes and losses, not only in water resources but also in all economical and social resources, and in an integrated framework believing that everything is related to everything. So would it not be possible to kill several birds with one stone, using algae for absorbing $CO_2$ at the same time as providing nutrient recovery from food industrial effluents and domestic wastewater and producing renewable energy (fuels), as well as other pharmaceutical products, food, feed and fertilizer from the biomass? In recent years, microalgal culture technology is a business oriented line owing to their different practical applications. Innovative processes and products have been introduced in microalgal biotechnology to produce vitamins, proteins, cosmetics, health foods and animal feed. For most of these applications, the market is still developing and the biotechnological use of microalgae will extend into new areas.With the development of algal cultures and screening techniques, microalgal biotechnology can meet the challenging demands of food, feed, pharmaceutical industries, fuels and biofertilizers. The general needs of the human society are continuously increasing. We need every new compound which may be useful for the human society. More food, new drugs, and other goods are highly necessary for the benefit of humankind. The only question is the existence of sufficient natural and technical resources to fulfill these demands. Fortunately, in the area of the research of bioactive microbial products it seems that the ever expanding scientific and technical possibilities are increasing together with the continuously widening needs of the human.

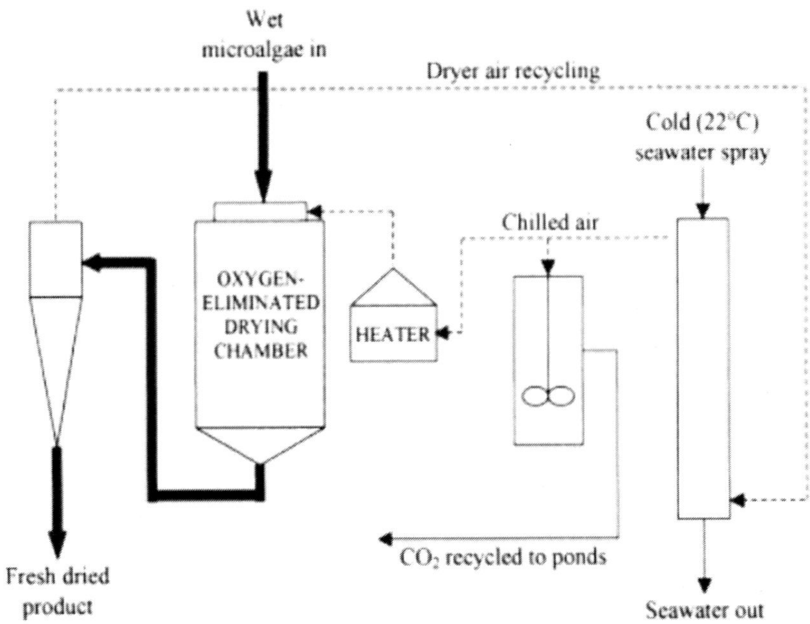


**Figure 13.** Cyanotech process for drying microalgae biomass

**Figure 14.** Glass tube photobioreactor (700 m3) producing Chlorella biomass (Klötze, Germany)

# REFERENCES


1. Pratt DE (1992) Natural antioxidants from plant material. In: Huang MT, Ho CT and Lee CY (Editors), Phenolic Compounds in Food and their Effects on Health. II.American Chemical Society. ACS Symposium Series, 507.Washington. 54-71.
2. Abou El alla FM and Shalaby EA (2009) Antioxidant Activity of Extract and Semi-Purified Fractions of Marine Red Macroalga. Gracilaria Verrucosa. Aust J. Bas. App. Sci.3:3179-85.
3. Li HB, Cheng KW, Wong CC, Fan KW, Chen F and Jiang Y (2007) Evaluation ofcapacity and total phenolic content of different fractions of selectedmicroalgae. Food Chemistry. 102:771–776.
4. Richmond A. (2004) Handbook of microalgal culture: biotechnology and appliedphycology. Blackwell Science Ltd.
5. Brennan L and Owened PAuthor Vitae (2010) Biofuels from microalgae-Areview oftechnologies for production, processing and extractions of biofuels and co-products.Renewable and Sustainable Energy Reviews. 14(2):557-577.
6. Dos Santos MD, Guaratini T, Lopes JLC, Colepicolo P and Lopes NP (2005) Plant celland microalgae culture. In:Modern Biotechnology in Medicinal Chemistry andIndustry. Kerala, India: Research Signpost.
7. Burja AM, Banaigs B, Abou-Mansour E, Burgess JG and Wright PC (2001) Marinecyanobacteria - a prolific source of natural products. Tetrahedron. 57:9347-9377.

8. Tyagi S, Singh G, Sharma A and Aggarwal G (2010) Phytochemicals as candidatetherapeutics: An over view. International Journal of Pharmaceutical Sciences Reviewand Research. 3(1):53-55.
9. Lipton AP (2003) Marine bioactive compounds and their application in mariculture.Marine Ecosystem, Univ. of Kerala, Kariavattom, 2(4): 695-581.
10. Iwata K, Inayama T and Katoh T (1990) Effect of *Spirulina platensis* on plasmalipoprotein lipase activity in fructose induced hyperlipidemia in rats. J. Nutr. Sci.Vitaminol. 36:165-171.
11. Barsanti L and Gualtieri P (2006) Algae and men. In: Algae: Anatomy, Biochemistry,and Biotechnology. Taylor and Francis Group LLC, CRC Press. 251–291.
12. Valeem EE and Shameel M (2005) Fatty acid composition of blue-green algae of Sindh,Pakistan. Int. Journal of Phycology and Phycochemistry. 1:83-92.
13. Funk CD (2001) Prostaglandins and leukotrienes: advances in eicosanoids biology.Science. 294:1871–1875.
14. Wen Z-Y and Chen F (2003) Heterothrophic production of eicosapentaenoic acid bymicroalgae. Biotechnol. Adv. 21:273–294.
15. Mozaffarian D, Ascherio A, Hu FB, Stampfer MJ, Willett WC, Siscovick DS and RimmEB (2005) Interplay between different polyunsaturated fatty acids and risk of coronaryheart disease in men. Circulation, 111(2):157-164.
16. Eussen S, Klungel O, Garssen J, Verhagen H, van Kranen H, van Loveren H,Rompelberg C (2010) Support of drug therapy using functional foods and dietarysupplements: focus on statin therapy. Br. J. Nutr. 103(9): 1260-1277.
17. Petkov G and Garcia G (2007) Which are fatty acids of the green alga *Chlorella*? Bioch.Systemat. Ecol. 35:281-285.
18. Abbadi A, Domergue F, Meyer A, Riedel K, Sperling P, Zank T and Heinz E (2001)Transgenic oilseeds as sustainable source of nutritionally relevant C20 and C22polyunsaturated fatty acids? European Journal of Lipid Science and Technology.103:106–113.
19. Sayanova OV and Napier JA (2004) Eicosapentaenoic acid: biosynthetic routs and thepotential for synthesis in transgenic plants. Phytochemistry. 65:147–158.
20. Ponomarenko LP, Stonik IV, Aizdaicher NA, Orlova TY, Popvskaya GI, Pomazkina GVand Stonik VA (2004) Sterols of marine microalgae Pyramimonas cf. cordata(prasinophyta), Atteya ussurensis sp. nov. (Bacollariophyta) and a spring diatom bloomform Lake Baikal. Comp. Biochem. Physiol., B 138, pp.65–70.
21. Froehner S, Martins RF and Errera MR (2008) Assessment of fecal sterols in BariguiRiver sediments in Curitiba, Brazil. Environment Monitoring Assessment.DOI:10.1007/s10661-008-0559-0.
22. Volkman JK, Kearney P and Jeffrey SW (1990) A new source of 4-methyl sterols and5a(H)-stanols in sediments: prymnesiophyte microalgae of the genus Pavlova. OrganicGeochemistry. 15:489-497.
23. Meyers SP, Latscha T (1997) Carotenoids. In: D'Abramo, L.R., Conklin, D.E., Akiyama,D.M. (Eds.), Crustacean Nutrition, Advances in World Aquaculture, vol. 6. WorldAquaculture Society, Baton Rouge, LA, 164–193.
24. Sommer TR, D'Souza FML and Morrissy NM (1992) Pigmentation of adult rainbowtrout, Oncorhynchus mykiss, using the green alga Haematococcus pluvialis.Aquaculture. 106: 63–74.
25. Chen BH, Chuang JR, Lin JH and Chiu CP (1993) Quantification of pro-vitamin Acompounds in Chinese vegetables by high-performance Liquid Chromatography. JFood Prot. 56 (1): 51-54.
26. Pulz O, Scheibenbogen K and Grob W (2001) Biotechnology with cyanobacteria andmicroalgae. In: Rehm, H. J., Reed, G. editors. A Multi-Volume Comprehensive TreatiseBiotechnology Germany. Wiley-VCH Verlag GmbH. 10:107-136.
27. Ben-Amotz A (1999) *Dunaliella* β-carotene: from science to commerce. In: Seckbach J(ed) Enigmatic microorganisms and life in extreme environments. Kluwer, Netherlands.401–410.

28. Higuera-Ciapara I, Félix-Valenzuela L, Goycoolea FM (2006) Astaxanthin: a review ofits chemistry and applications. Crit. Rev. Food Sci. Nutr. 46:185–196.
29. Del Campo JA, García-González M and Guerrero MG (2007) Outdoor cultivation ofmicroalgae for carotenoid production: current state and perspectives, Appl MicrobiolBiotechnol. 74:1163–1174.
30. Takano S, Nakanishi A, Uemura D and Hirata Y (1979) Isolation and structure of a 334nm UVabsorbing substance, porphyra-334 from the red alga Porphyra tenera Kjellman.Chem. Lett. 8:419-20.
31. Nakamura H, Kobayashi J and Hirata Y (1982) Separation of mycosporine-like aminoacids in marine organisms using reverse-phase high performance liquidchromatography. J. Chromatogr. 250:113-18.
32. Klisch M and Häder D (2008) Mycosporine-Like Amino Acids and Marine Toxins - TheCommon and the Different. Marine Drugs. 6: 147-163.
33. Büyükokurog˘lu ME, Gülçin I, Oktay M and Küfreviog˘lu ÖI (2001) In vitro antioxidantproperties of dantrolene sodium. Pharmacol. Res. 44:491-495.
34. Karawita R, Senevirathne M, Athukorala Y, Affan A, Lee YJ, Kim SK, Lee JB and Jeon YJ(2007) Protective effect of enzymatic extracts from microalgae against DNA damageinduced by H2O2. Mar. Biotech. 9: 479-490.
35. Lee SH, Karawita R, Affan A, Lee JB, Lee BJ and Jeon YJ (2008) Potential antioxidantactivites of enzymatic digests from benthic diatoms *Achnanthes longipes, Amphoracoffeaeformis*, and *Navicula* sp. (Bacillariophyceae). J. Food Sci. Nutr. 13: 166-175.
36. Abd El-Baky H Hanaa, El Baz FK and El-Baroty GS (2004) Production of antioxidant bythe green alga *Dunaliella salina*. Intl. J. Agric. Biol. 6(1): 49-57.
37. Suzuki N and Mittler R (2006) Reactive oxygen species and temperature stresses: Adelicate balance between signaling and destruction. Physiol. Plantarum, 126:45-51.
38. Mittler R, S Vanderauwera, M Gollery and F Van Breusegem (2004) Reactive oxygengene network of plants. Trends. Plant Sci. 9: 490-498.
39. Tausz RM, Soledad M and Grille D (1998) Antioxidative defence and photoprotection inpine needles under field conditions. A multivariate approach to evaluate patterns ofphysiological responses at natural sites. Physiol. Planta. 104:760-768.
40. Polle A and Rennenberg H (1994) Photooxidative stress in trees. In: Foyer, C.H. andP.M. Mullineaux (Eds.). Causes of photoxidative stress and amelioration of defenseSystems in Plants. Boca Raton, FL: CRC Press.
41. Foyer CH and Noctor G (2000) Oxygen processing in photosynthesis: Regulation andsignaling. Rev. New Phytol. 146:359-388.
42. Zhang J, Zhan B, Yao X, Gao Y, Shong J. Evaluation of 28 marine algae from theQingdao coast for antioxidative capacity, determination of antioxidant efficiency, totalphenolic content of fractions and subfractions derived from Symphyocladia latiuscula(Rhodomelaceae). J Appl Phycol 2007; 19: 97-108
43. Shanab SMM, Mostafa SSM , Shalaby EA and Mahmoud GI (2012) Aqueous extracts ofmicroalgae exhibit antioxidant and anticancer activities. Asian Pacific Journal ofTropical Biomedicine. 608-615.
44. Glazer AN (1994) Phycobiliproteins: a family of valuable, widely used fluorophores. J.Appl. Phycol. 6:105-112.
45. Rice-Evans CA, Miller NJ, Bolwell PG, Bramley PM, Pridham JB. The relativeantioxidant activities of plant-derived polyphenolic flavonoids. Free Radic Res 1995;22(4): 375-383.
46. Shalaby EA, Shanab SMM and Singh V (2010) Salt stress enhancement of antioxidantand antiviral efficiency of *Spirulina platensis*; J. Med. Plants Res. 4(24):.2622-2632.
47. Chinery R, Brockman JA, Peeler MO, Shyr Y, Beauchamp RD and Coffey R J (1997)Antioxidants enhance the cytotoxicity of chemotherapeutic agents in colorectal cancer-ap53-independent induction of p21(WAF1/CIP1) via c/ebp-beta. Nat. Med. 3:1233–1241.
48. Mazhar D, Ang R and Waxman J (2006) COX inhibitors and breast cancer. Br. J. Cancer.94: 346–350.

49. Singh SC, Sinha RP and Häder DP (2002) Role of lipids and fatty acids in stress tolerance in cyanobacteria. Acta protozoologica. 41:297-308.
50. Aboul-Enein AM, Shalaby EA, Abul-Ela F, Nasr-Allah AA, Mahmoud AM, El-ShemyHA, et al. (2011) Back to nature: Spotlight on cancer therapeutics. Vital Signs. 10: 8-9.
51. Wang H, Liu Y, Gao X, Carter CL and Liu ZR (2007) The recombinant beta subunit of Cphycocyanininhibits cell proliferation and induces apopotosis. Cancer Letters.247(1):150-158.
52. Mayer AMS and Hamann MT (2005) Marine pharmacology in 2001–2002: marinecompounds with antihelmintic, antibacterial,anticoagulant, antidiabetic, antifungal,anti-inflammatory, antimalarial, antiplatelet, antiprotozoal, antituberculosis, andantiviral activities; affecting the cardiovascular, immune and nervous systems andother miscellaneous mechanisms of action. Comparative Biochemistry and Phycology,Part C.140: 265-286.
53. Mendiola JA, Torres CF, Martín-Alvarez PJ, Santoyo S, Toré A, Arredondo BO,Señoráns FJ, Cifuentes A, Ibáñez E (2007) Use of supercritical CO2 to obtain extractswith antimicrobial activity from *Chaetoceros muelleri* microalga. A correlation with theirlipidic content. European Food Research and Technology. 224:505-510.
54. Herrero M, Ibañez E, Cifuentes A, Reglero G, Santoyo S (2006) *Dunaliella salina*microalga pressurized liquid extracts as potential antimicrobials. J. Food Protection.69:2471-2477.
55. Tramper J, Battershill C, Brandenburg W, Burgess G, Hill R, Luiten E, Müller W, OsingaR, Rorrer G, Tredici M, Uriz M,Wright P and Wijffels R (2003) What to do in marinebiotechnology? Biomolecular Engineering. 20:467-471.
56. Guedes AC, Barbosa CR, Amaro HM, Pereira CI, Malcata FX (2011) Microalgal andcyanobacterial cell extracts for use as natural antibacterial additives against foodpathogens. International J. Food Sci. and Technol. 46:862-870.
57. Sánchez F, Fernández JM., Acien FG, Rueda A, Perez-Parra J, Molina E (2008) Influenceof culture conditions on the productivity and lutein content of the new strainScenedesmus almeriensis. Process Biochemistry. 43:398-405.
58. Borowitzka MA (1995) Microalgae as sources of pharmaceuticals and other biologicallyactive compounds. J. Appl. Phycol.7:65-68.
59. Damonte EB, Pujol CA, Coto CE (2004) Prospects for the therapy and prevention ofDengue virus infections. Advances in Virus Research. 63:239-285.
60. Huleihe M, Ishanu V, Tal J and Arad S. (2001) Antiviral effect of red microalgapolysaccharides on Herpes simplex and Varicella zoster viruses. Journal of AppliedPhycology.13:127-134.
61. Geresh S and Arad (Malis) S (1991) The extracellular polysaccharides of the redmicroalgae: chemistry and rheology. Bioresouce Technology. 38:195-201.
62. Evans L, Callow M, Percival E, Fareed V (1974) Studies on the synthesis andcomposition of extracellular mucilage in the unicellular red alga *Rhodella*. J.Cell Science.16:1-21.
63. Ramus J and Robins DM (1975) The correlation of Golgi activity and polysaccharidesecretion in Porphyridium. J. Phycol. 11:70-74.
64. Tsekos I (1999) The sites of cellulose synthesis in algae: diversity and evolution ofcellulose-synthesizing enzyme complexes. Journal of Phycology. 35:635-655.
65. Keidan M, Friedlander M and Arad S (2009) Effect of brefeldin A on cell-wallpolysaccharide production in the red microalga Porphyridium sp. (Rhodophyta) J.Appl. Phycol. 21:707-717.
66. Ghannoum MA and Rice LB (1999) Antifungal agents: mode of action, mechanisms ofresistance, and correlation of these mechanisms with bacterial resistance. ClinicalMicrobiology Reviews.12:501-517.
67. Pratt R, Daniels TC, Eiler JB, Gunnison JB, Kumler WD et al (1944) Chlorellin, anantibacterial substance from *Chlorella*. Science. 99:351-352.
68. Ghasemi Y, Yazdi MT, Shafiee A, Amini M, Shokravi S, Zarrini G. Parsiguine (2004) anovel antimicrobial substance from Fischerella ambigua. Pharmaceutical Biology.42:318-322.
69. Ördög V, Stirk WA, Lenobel R, Bancírová M, Strand M and van Standen J (2004)Screening microalgae for some potentially useful agricultural and pharmaceuticalsecondary metabolites. J. Appl. Phycol. 16:309-314.

70. Ghasemi Y, Moradian A, Mohagheghzadeh A, Shokravi S, Morowvat MH (2007)Antifungal and antibacterial activity of the microalgae collected from paddy fields ofIran: characterization of antimicrobial activity of Chlorococcus dispersus. J. Biol. Sci.7:904-910.
71. Desbois AP, Mearns-Spragg A, Smith VJ (2009) A fatty acid from the diatomPhaeodactylum tricornutum is antibacterial against diverse bacteria including multiresistantStaphylococcus aureus (MRSA) Marine Biotechnology. 11:45-52.
72. Benkendorff K, Davis AR, Rogers CN and Bremner JB (2005) Free fatty acids and sterolsin the benthic spawn of aquatic molluscs, and their associated antimicrobial properties.J. Experimental Marine Biology and Ecology. 316:29-44.
73. Smith VJ, Desbois AP and Dyrynda EA (2010) Conventional and unconventionalantimicrobials from fish, marine invertebrates and micro-algae. Marine Drugs. 8:1213-1262.
74. Hewedy MA, Rahhal MMH and Ismail IA (2000) Pathological studies on soybeandamping-off disease. Egypt. J. Appl. Sci. 15:88-102.
75. Noaman NH, Khaleafa AF and Zaky SH (2004) Factors affecting antimicrobial activityof *Synechococcus leopoliensis*. Microbiol. Res. 159:395-402.
76. Kiviranta J, Abdel-Hamid A, Sivonen K, Niemelä SI and Carlberg G (2006) Toxicity ofcyanobacteria to mosquito larvae-screening of active compounds. Environ. Toxicol.Water Qual. 8: 63-71.
77. Hussien Manal Y, Abd El-All Azza AM and Mostafa Soha S M (2009) Bioactivity ofalgal extracellular byproducts on cercospora Leaf spot disease, growth performance and quality of sugar beet. The 4th Conference on Recent Technologies in Agriculture:Challenges of Agriculture Modernization, Nov. 3rd –5th , Special Edition of Bull. Fac.Agric., Cairo Univ. 1:119-129.
78. Oka Y, Koltai H, Bar-Eyal M, Mor M, Sharon E, Chet I and Spiegel Y (2000) Newstrategies for the control of plant parasitic nematodes. Pest Manage. Sci. 56:983-988.
79. Dhanam M, Kumar AC and Sowajanya AS (1994) Microcoleu vaginatus(Oscillatoriaceae), a blue-green alga (Cyanobacterium) parasitizing plant and soilnematodes. Indian Journal of Nematology. 24:125-132.
80. Shimizu Y (2003) Microalgal metabolites. Current Opinion in Microbiology. 6:236–242.
81. Shawky, Samaa M, Mostafa, Soha SM and Abd El-All, Azza AM (2009) Efficacy ofalgae, azolla and compost extract in controlling root knot nematode and its reflection oncucumber growth. Bull. Fac. Agric., Cairo Univ. 60:443-459.
82. Lardans, V and Dissous C (1998) Snail control strategies for reduction of schistosomiasistransmission. Parasitology Today 14:413-417.
83. Ahmad, R, Chu WL, Ismail Z, Lee HL and Phang SM (2004) Effect of ten chlorophyteson larval survival, development and adult body size of the mosquito Aedes aegypti.Southeast Asian J. Trop. Med. Public health. 35(1):79-87.
84. Mostafa Soha SM and Gawish Fathia A (2009) Towards to Control *Biomphalariaalexandrina* Snails and the Free Living Larval Stages of *Schistosoma mansoni* Using theMicroalga *Spirulina platensis*. Aust. J. Basic and Appl. Sci. 3(4):4112-4119.
85. Yamaguchi K (1997) Recent advances in microalgal bioscience in Japan, with specialreference to utilization of biomass and metabolites: a review. J. Appl. Phycol. 8:487-502.
86. Liang S, Xueming L, Chen F, and Chen Z (2004) Current microalgal health food R & Dactivities in China. Hydrobiologia. 512:45-48.
87. Soletto D, Binaghi L, Lodi A, Carvalho JCM, and Converti A (2005) Batch and fedbatchcultivations of *Spirulina platensis* using ammonium sulphate and urea as nitrogensources. Aquaculture. 243:217-224.
88. Rangel-Yagui, CO, Godoy Danesi ED, Carvalho J CM and Sato S (2004) Chlorophyllproduction from *Spirulina platensis*: cultivation with urea addition by fed-batchprocess.Bioresour. Technol. 92:133-14.
89. Metting FB (1996) Biodiversity and application of microalgae. J. Ind. Microbiol. 17:477-489.
90. Pugh N and Pasco DS (2001) Characterization of human monocyte activation by a watersoluble preparation of *Aphanizomenon flos-aquae*. Phytomedicine, 8:445-453.

91. Delisle H, Alladsoumgué M, Bégin F, Nandjingar K and Lasorsa C (1991) Householdfood consumption and nutritional adequacy in wadi zones of Chad, Central Africa.Ecol. Food Nutrit. 25:229-248.
92. Becker W (2004) Microalgae in human and animal nutrition. In: Richmond, A (ed.),microalgal culture. Handbook. Blackwell, Oxford. 312-351.
93. Muller-Feuga A (2004) Microalgae for aquaculture: the current global situation andfuture trends. In: Richmond A (ed) Handbook of microalgal culture. Blackwell Science.Oxford.pp 352–364.
94. Muller-Feuga A (2000) The role of microalgae in aquaculture: situation and trends. J.Appl. Phycol. 12: 527-534.
95. Horrigan LRS, Lawrence and Walker P (2002) How Sustainable Agriculture CanAddress the Environmental and Human Health Harms of Industrial Agriculture.Environ Health Perspect. 110:445-456.
96. Mahdi SS, Hassan GI, Samoon SA, Rather HA, Dar SA and Zehra B (2010) Bio-frtilizersin organic agriculture. Journal of Phytology. 2(10): 42-54.
97. Silva JA, Evensen CI, Bowen RL, Kirby R, Tsuji GY and Yost RS (2000) ManagingFertilizer Nutrients to Protect the Environment and Human Health. In: Plant NutrientManagement in Hawaii's Soils, Approaches for Tropical and Subtropical Agriculture. J.Silva and R. Uchida, eds. College of Tropical Agriculture and Human Resources,University of Hawaii at Manoa. Capter 1:7-22.
98. Earanna N and Govindan R (2002) Role of biofertilizers in mulberry production-Areview. Indian J. Seric. 41(2):92-99.
99. kleiner KT and Harper KT (1977) Soil properties in relation to cryptogamic groundcover in Canyon-lands National park. J. Range Managem. 30: 202-205.
100. Aly MHA, Abd El-All Azza AM and Mostafa Soha SM (2008) Enhancement of sugarbeet seed germination, plant growth, performance and biochemical compounds ascontributed by algal extracellular products. J. Agric. Sci., Mansoura Univ. 33(12):8429-8448.
101. Wake H, Akasata A, Umetsu H, Ozeki Y, Shimomura and Matsunaga (1992) T.Promotion of plantlet formation from the somatic embryos of carrot with a highmolecular weight extract of marine cyanobacterium. Plant cell Report.11:6265.
102. Bapat VA,Iyer RK and Rao PS (1996) Effect of cyanobacterial extract on somaticembryogenesis in tissue culture of sandalwood (Santalum album) J. Medicinal andAromatic Plant Science. 18(1):1014.
103. Storni DC, Zaccaro M, Cristina M, Ileana G, María SA and Gloria ZDC (2003)Enhancing rice callus regeneration by extracellular products of Tolypothrixtenuis(Cyanobacteria) World J. Microbiology and Biotechnology. 19(1):2934.
104. Zaccaro MC, Kato A, Zulpa G, Storni MM, Steyerthal N, Lobasso K and Stella AM(2006) Bioactivity of *Scytonema hofmanni* (Cyanobacteria) in *Lilium alexandrae* in vitropropagation. Electronic Journal of Biotechnology. 9(3):210214.
105. Selvarani V (1983) Studies on the influence on nitrogen fixing and non-nitrogen fixingblue green algae on the soil, growth and yield of paddy (*Oryza sativa*-IR 50) M. Sc.,Madurai Kamaraj University, Madurai.
106. Mahmoud AA, Mostafa Soha SM, Abd El-All, Azza AM and Hegazi AZ (2007) Effectof cyanobacterial inoculation in presence of organic and inorganic amendments oncarrot yield and sandy soil properties under drip irrigation regime. Egypt. J. of Appl.Sci. 22(12B): 716-733.
107. Ali Laila KM and Mostafa Soha SM (2009) Evaluation of potassium humate and*Spirulina platensis* as a bio-organic fertilizer for sesame plants grown under salinitystress. The 7th International Conference of Organic Agriculture. 13-15 December, Egypt.J. Res. 87(1):369-388.
108. Hegazi Amal Z, Mostafa Soha SM and Hamdino MIA (2010) Influence Of differentcyanobacterial application methods on growth and seed production of common Beanunder various levels of mineral nitrogen fertilization. Nature and Science. 8(11): 202-212.

109. Packer M (2009) Algal capture of carbon dioxide; biomass generation as a tool forgreenhouse gas mitigation with reference to New Zealand energy strategy and policy.Energy Policy. 37(9):3428-37.
110. Brennan L and Owende P (2010) Biofuels from microalgae- A review of technologiesfor production, processing, and extractions of biofuels and co-products. 14:557-77.
111. Li Y, Horsman M, Wu N, Lan CQ and Dubois-Calero N (2008) Biofuels frommicroalgae. Biotechnol Prog 24(4):815–820.
112. Haiduc AG et al (2009) SunCHem: an integrated process for the hydrothermalproduction of methane from microalgae and CO2 mitigation. J. Appl. Phycol. 21:529–541.
113. Mostafa SSM, Shalaby EA and Mahmoud GI (2012) Cultivating Microalgae inDomestic Wastewater for Biodiesel Production. Not Sci Biol. 4(1):56-65.
114. Aslan S and Kapdan IK (2006) Batch kinetics of nitrogen and phosphorus removalfromsynthetic wastewater by algae. Ecological Engineering. 28(1):64–70.
115. Gonzales LE, Canizares RO and Baena S (1997) Efficiency of ammonia andphosphorusremoval from a Colombian agroindustrial wastewater by the microalgae*Chlorealla vulgaris* and *Scenedesmus dimorphus*. Bioresource Technology. 60:259–62.
116. Lee K and Lee CG (2001) Effect of light/dark cycles on wastewater treatmentsbymicroalgae. Biotechnology and Bioprocess Engineering. 6:194–9.
117. Martı´nez ME, Sa´nchez S, Jime´nez JM, El Yousfi F, Mun˜oz L (2000) Nitrogenandphosphorus removal from urban wastewater by the microalgaScenedesmusobliquus. Bioresource Technology. 73(3): 263–72.
*118.* Olguı´n EJ, Galicia S, Mercado G and Perez T (2003) Annual productivity of *Spirulina*(Arthrospira) and nutrient removal in a pig wastewater recycle process under tropicalconditions. J. Appl. Phycol. 15:249–57.
119. Laliberte G, Lessard P, Noue J, Sylvestre S (1997) Effect of phosphorus additiononnutrient removal from wastewater with the cyanobacterium Phormidiumbohneri.Bioresource Technology. 59:227–33.
120. Dumas A, Laliberte G, Lessard P, De la Noue J (1998) Biotreatment of fish farmeffluents using the cyanobacterium *Phormidium bohneri*. Aquacultural Engineering.17(1):57–68.
121. Bilanovic D, Andargatchew A, Kroeger T and Shelef G (2009) Freshwater and marinemicroalgae sequestering of CO2 at different C and N concentrations—response surfacemethodology analysis. Energy Conversionand Management. 50:262–7.
122. Salih FM and Haase RA (2012) Potentials of microalgal biofuel production. Review, J.Petroleum Technol. and Alternative Fuels. 3(1):1-4.
123. Chisti Y (2007) Biodiesel from microalgae. Biotechnol Adv. 25(3):294-306.Afify AMR, Shalaby EA and Shanab SMM (2010) Enhancement of biodieselproduction from different species of algae. Grasas Y Aceites. 61(4):416-422.
124. Andersen RA (2005) Algal Culturing Techniques. San Diego, Elsevier Inc, p. 678.
125. Chen P Min M, Chen Y, Wang L, Li Y, Chen Q, Wang C, Wan Y, Wang X, Cheng Y,Deng S, Hennessy K, Lin X, Liu Y, Wang Y, Martinez B and Ruan R (2009) Review ofthe biological and engineering aspects of algae to fuels approach. Int. J. Agric. Biol. Eng.2(4):1-30.
126. Gadd GM and White C (1992) Removal of thorium from simulated acid processstreams by fungal biomass-potential for thorium desorption and reuse of biomass anddesorbent. J Chem Technol Biotechnol. 55:39–44.
127. Figueira MM, Volesky B, Ciminelli VST and Roddick FA (2000) Biosorption of metalsin brown seaweed biomass. Water Res. 34:196-204.
128. Sheng, Ping Xin, Lai, Heng Tan, Paul, Chen J and Yen- Peng, Ting (2005) Biosorptionperformance of two brown marine algae for removal of chromium and cadmium. J.Dispersion Sci. Technol. 25(5):679-686.
129. Lamaia C, Kruatrachuea M, Pokethitiyooka P, Upathamb ES and SoonthornsarathoolaV (2005) Toxicity and Accumulation of Lead and Cadmium in the Filamentous GreenAlga *Cladophora fracta* (O.F. Muller ex Vahl) Kutzing: A Laboratory Study, Science Asia.31, 121-127.

130. Shanab SMM and AM Essa (2007) Heavy metals tolerance, biosorption andbioaccumulation by some microalgae (Egyptian isolates) N. Egypt. J. Microbiol. 17:65-77.
131. Atici TO, Obali A, Altindag S, Ahiska and Aydin D (2010) The accumulation of heavymetals (Cd, Pb, Hg, Cr) and their state in phytoplanktonic algae and zooplanktonic organisms in Beysehir Lake and Mogan Lake, Turkey. African J. Biotechnol. (9):475-487.
132. Lefebvre DD, Kelly D and Budd K (2007) Biotransformation of HgII by cyanobacteria.Appl. Environ. Microbiol. 73:243-249.
133. Kelly DJA, Budd K and Lefebvre DD (2006) Mercury analysis of acid- and alkalinereducedbiological samples; identification of meta-cinnabar as the major biotransformedcompound in algae. Appl. Environ. Microbiol. 72: 361-367.
134. Lengke MF, Ravel B, Fleet ME, Wanger G, Gordon RA and Southham G (2006)Mechanisms of gold bioaccumulation by filamentous cyanobacteria from gold (III)-chloride complex. Environ. Sci. Technol. 40:6304-6309.
135. Essa AMM and Mostafa Soha SM (2011) Biomineralization of some heavy metals bycyanobacterial biogas. Egpt. J. Botany. In Press
136. Desmorieux H and Decaen N (2005) Convective drying of *Spirulina* in thin layer. J.Food Eng. 66: 497–503.
137. Jong-Yuh C and Mei-Fen S (2005) Potential hypoglycemic effects of *Chlorella* instreptozotocin-induced diabetic mice. Life Sci. 77:980–990.

# CHAPTER 10

# Recent advances of metabolomics in plant biotechnology

**Yozo Okazaki1, Kazuki Saito2**

*1 RIKEN Plant Science Center, Tsurumi-ku, Yokohama 230-0045, Japan*

*2 Graduate School of Pharmaceutical Sciences, Chiba University, Inage-ku, Chiba 263-8522, Japan*

## ABSTRACT

Biotechnology, including genetic modification, is a very important approach to regulate the production of particular metabolites in plants to improve their adaptation to environmental stress, to improve food quality, and to increase crop yield. Unfortunately, these approaches do not necessarily lead to the expected results due to the highly complex mechanisms underlying metabolic regulation in plants. In this context, metabolomics plays a key role in plant molecular biotechnology, where plant cells are modified by the expression of engineered genes, because we can obtain information on the metabolic status of cells via a snapshot of their metabolome. Although metabolome analysis could be used to evaluate the effect of foreign genes and understand the metabolic state of cells, there is no single analytical method for metabolomics because of the wide range of chemicals synthesized in plants. Here, we describe the basic analytical advancements in plant metabolomics and bioinformatics and the application of metabolomics to the biological study of plants.

**Keywords:** Metabolomics Mass, spectrometry, NMR, Functional genomics, GMO, Quantitative genetics, QTL

# 1. INTRODUCTION

The modification of metabolism by biotechnological techniques is often utilized for the optimal production of plant metabolites, which directly benefit human health and plant growth. For example, "golden rice" is a transgenic line of *Oryza sativa* that was genetically engineered to biosynthesize $\beta$-carotene, a pro-vitamin A, in the edible parts of rice (Ye et al. 2000; Yonekura-Sakakibara and Saito 2009). The introduction of transcription factors from the snapdragon to tomato increased the production of anthocyanins, which have health protective properties (Butelli et al. 2008). However, many similar approaches do not necessarily lead to the expected results, e.g., overexpression of foreign *S*-linalool synthase in transgenic petunia did not result in the expected accumulation of free linalool, but led to the accumulation of *S*-linalyl-$\beta$-D-glucoside (Lucker et al. 2001). These unexpected results suggest that highly complex regulatory systems control plant metabolism and also indicate the need for more precise information on plant metabolism. In this context, metabolomics plays a key role in the field of molecular biotechnology, where plant cells are modified by the expression of engineered genes. Metabolomic analysis provides us with in-depth information on cellular metabolism via a snapshot of the metabolome, often combined with data from other "omics" (Oksman-Caldentey and Saito 2005; Saito and Matsuda 2010).

Metabolomics is one of the omics approaches that can be used to acquire comprehensive information on metabolites. It aims to grasp the global state of metabolism in measured samples. Among the omics studies used in plant sciences, genomics was the first to emerge, and uncovered the genome sequences of several organisms, including *Arabidopsis* (Arabidopsis Genome Initiative 2000) and rice (Goff et al. 2002; Sasaki et al. 2002; Yu et al. 2002; International Rice Genome Sequencing Project 2005). There should be no doubt that the development of the automated DNA sequencer has led to the current progress of genomics. Other omics studies have also been developed as a result of technical innovations. Microarrays made high throughput analysis of mRNA expression feasible and led to the emergence of transcriptomics. Two-dimensional electrophoresis and mass spectrometry (MS) significantly contributed to the development of proteomics. Similarly, MS and nuclear magnetic resonance (NMR) spectroscopy have facilitated metabolomic studies. However, metabolomics is not as advanced as the other omics because there is a critical difference between metabolites and other molecules, i.e., DNA, RNA, and proteins are linear polymers consisting of a limited numbers of monomers, and the interpretation of RNA and protein sequences can be facilitated by genome information according to the central dogma of molecular biology, whereas metabolites comprise a more heterogeneous group than these polymers in terms of their physical and chemical properties, varying widely with respect to size, polarity, quantity, and stability. In addition, there are an estimated 200,000 plant metabolites (Fiehn 2002b; Dixon and Strack 2003), and many of these metabolites remain unknown. Thus, no single method has yet been developed for plant metabolomics, and researchers who want to acquire comprehensive metabolome information have to employ several methodologies according to the chemical properties of the metabolites. In spite of these difficulties, metabolomics, metabolic profiling, and metabolic fingerprinting have been employed in many biological studies. These techniques have been applied to the functional identification of unknown genes through the metabolic profiling of plants in which some genes are up- or down-regulated (Bino et al. 2004; Hirai et al. 2005; Oksman-Caldentey and Saito 2005; Tohge et al. 2005; Watanabe et al. 2008; Yonekura-Sakakibara et al. 2008; Okazaki et al. 2009; Matsuda et al. 2010), the discovery of biomarkers associated with disease phenotypes (Soga et al. 2006; Sreekumar et al. 2009), the safety assessment of genetically modified organisms (GMOs) (Baker et al. 2006; Beale et al. 2009; Kusano et al. 2011a), and the discovery of compounds involved in plant resistance to biotic and abiotic stresses (Leiss et al. 2009; Ward et al. 2010b; Kusano et al. 2011c). When combined with genomics, transcriptomics, and/or proteomics, metabolomics can also help to interpret and understand many complex biological processes; indeed, metabolomics is now widely recognized as a cornerstone of systems biology (Quackenbush 2002; Hall 2006; Saito and Matsuda 2010). In this review, we introduce the basic analytical protocols for plant metabolomics and bioinformatics and the practical application of metabolomics to the biological study of plants.

# 2. ANALYTICAL TECHNOLOGIES FOR THE PLANT METABOLOME

## 2.1 Mass spectrometry

MS is the most frequently used technique in metabolic studies. MS provides mass-to-charge ratio information, which enables the structure of metabolites to be determined. The main advantage of MS is its high sensitivity. In addition, the combination of chromatographic separation with MS increases the number of compounds that can be detected by reducing the complexity of the mass spectra and the matrix effect. There are several chromatographic techniques that can be combined with MS.

Gas chromatography (GC)–MS is used for metabolite profiling. Capillary GC uses a carrier gas to move analytes through a coated, fused silica capillary. GC–MS requires the analyte to be vaporized in order for its migration through the capillary; therefore, analytes must be volatile or amenable to chemical derivatization to render them volatile. Certain types of samples (terpenoids and essential oils) are particularly well suited for GC–MS analysis. With appropriate derivatization, more polar metabolites or metabolites with highly polar functional groups can also be analyzed, e.g., amino acids, sugars, organic acids, fatty acids, and amines. This indicates that many compounds associated with primary metabolism can be analyzed by GC–MS. Electron ionization (EI) is the most commonly used GC–MS ionization technique; it is robust, highly reproducible, and considered to be less affected by the matrix effect, e.g., ion suppression. In addition, EI generates informative and characteristic mass spectra due to the relatively high degree of fragmentation reactions, which are useful for compound identification. However, molecular ions, which effectively reduce the number of candidate structures or elemental compositions, are often undetected, suggesting that GC–EI–MS can be used for the targeted analysis of known primary metabolites. Mass spectral libraries of EI–MS are commercially and non-commercially available, and these can be routinely used for peak annotation. To date, GC–MS-based metabolic profiling has been frequently used in many metabolomic studies of plants, animals, and microorganisms (Fiehn et al. 2000; Roessner et al. 2001; Fernie et al. 2004; Shellie et al. 2005; Keurentjes et al. 2006; Schauer et al. 2006, 2008; Kusano et al. 2007, 2011a, c; Ralston-Hooper et al. 2008).

Liquid chromatography (LC)–MS is also used for metabolomic studies. The major ionization method for LC–MS is atmospheric ionization (API), which includes electrospray ionization (ESI) and atmospheric chemical ionization (APCI) (Codrea et al. 2007; Dunn 2008). These methods do not require the compound to be volatile, and are thus suited for compounds with heat-labile functional groups, chemically unstable substructures, high vapor points, and high molecular weights. In plant metabolomics, LC–MS is frequently used to profile secondary metabolites (e.g., phenylpropanoids, alkaloids, and saponins) and complex lipids (e.g., glycerolipids, and sphingolipids), often with a reversed-phase column (e.g., C18) for separation (De Vos et al. 2007; Naoumkina et al. 2007; Bottcher et al. 2008; Farag et al. 2008; Matsuda et al. 2009, 2010; Okazaki et al. 2011). Many of these compounds are not amenable to volatilization by GC. Secondary metabolites are not directly involved in the normal growth or development of plants, but they have important physiological roles, e.g., as toxic compounds for predators and attractors for pollinators. In addition, they have been historically used as pharmaceutical compounds and food additives for flavor, indicating the special importance of LC–MS-based analysis in plant metabolomics. In contrast to EI–MS, API–MS usually affords $[M+H]^+$ and $[M-H]^-$ as the main signals, which are useful for reducing the candidate structure of the detected compounds. By using collision-induced dissociation of tandem MS, i.e., MS/MS, more informative mass spectra with many fragments are obtained. Although various compounds can be analyzed by LC–MS, peak annotation is still troublesome due to the shortage of mass spectral libraries for API–MS and commercially available authentic compounds. Much effort is required for the more comprehensive identification of metabolites by the production of additional reference compounds (Nakabayashi et al. 2009, 2010) and the generation of more comprehensive databases (Kind and Fiehn 2006; Moco et al. 2006; Shinbo et al. 2006; Bocker and Rasche 2008; Horai et al. 2010). In addition to untargeted metabolomics, widely targeted metabolomic studies use a triple quadrupole mass spectrometer equipped with LC (Sawada et al. 2009; Albinsky et al. 2010b). In a targeted metabolomics strategy, predefined metabolite-specific signals (by selected reaction monitoring with tandem MS or selected ion monitoring) are often used to determine precisely and accurately the relative abundances and concentrations of a limited number of known and expected endogenous metabolites (Griffiths et al. 2010).

Capillary electrophoresis (CE)–MS is utilized to analyze a wide spectrum of ionic metabolites. Similar to LC–MS, API is the most suitable ionization method for CE–MS. In capillary zone electrophoresis, constituent ions migrate on the basis of their electrostatic force, which results from the charge and size of ions, in addition to the electro-osmotic flow derived from the capillary and the type of electrolyte used. Ionic compounds are separated at a high resolution in a narrow capillary. In metabolomic analyses using CE–MS, samples are often divided for cation and anion analyses. For cation analysis, Soga's method, using formic acid as an electrolyte, is experimentally convenient and provides excellent separation of metabolites with good reproducibility (Soga et al. 2006). In contrast, the routine analysis of anions at a high resolution is more difficult than for cations, although various approaches, including the use of coated capillaries, have been assessed (Soga et al. 2002; Harada et al. 2006). Recently, a platinum ESI electrospray needle has been developed that significantly improved the analysis of anions by CE–MS (Soga et al. 2009). Target ionic metabolites in CE–MS analyses include amino acids, organic acids, nucleotides, and sugar phosphates. The compounds detectable by CE–MS are rather similar to those detected by GC–MS, but CE–MS can analyze these compounds without derivatization. In addition, similar to LC–MS, molecular-related ions are also detectable with CE–MS. Since the metabolites suited for CE–MS analysis are physiologically important and common to all organisms, CE–MS has been used in a variety of metabolomic studies (Monton and Soga 2007; Oikawa et al. 2008; Sato et al. 2008; Ramautar et al. 2009; Urano et al. 2009; Ishikawa et al. 2010), including the identification of biomarkers for the progression of prostate cancer (Sreekumar et al. 2009), oxidative stress (Soga et al. 2006), the measurement of internal body time (Minami et al. 2009), and the functional analysis of unknown genes in *Arabidopsis* (Ohkama-Ohtsu et al. 2008, 2011; Watanabe et al. 2008).

As described above, many of the analytes in metabolomic studies are still unknown, and even the known compounds are difficult to obtain from commercial sources. Under such unfavorable circumstances, the accurate *m/z* values with ultra-high resolution generated by Fourier transform ion cyclotron resonance mass spectrometry (FT-ICR MS) are very important. FT-ICR MS is of value in metabolomic studies for the following two reasons. Firstly, different compounds with very similar molecular masses can be detected separately, even by their direct infusion without any chromatographic steps, and this can result in the rapid detection of a number of metabolites. Secondly, an accurate estimate of the chemical formula of the detected peaks can be acquired, which can facilitate annotation procedures for unknown compounds. MS/MS analysis coupled with FT-ICR MS is also helpful to estimate the chemical formula because fragment ions are also detected with high resolution and accuracy. In fact, several metabolites that accumulated in *Arabidopsis* following treatment with herbicides were identified by MS/MS analyses of FT-ICR MS (Oikawa et al. 2006). However, because of difficulties in hardware handling and the extremely large amount of data generated, the number of reports on metabolomic studies with FT-ICR MS is rather limited compared with other types of MS. The non-targeted metabolite analysis of strawberry fruit was the first published metabolomic study using FT-ICR MS (Aharoni et al. 2002). This technique has also been applied to *Arabidopsis* functional genomics (Hirai et al. 2004, 2005; Tohge et al. 2005), transgenic tobacco (Mungur et al. 2005), and the identification of metabolic biomarkers in Crohn's disease (Jansson et al. 2009). Hyphenation of FT-ICR MS and LC has increased the range of applications for FT-ICR MS in metabolomic studies of tomato (Iijima et al. 2008) and *Lotus japonicas* (Suzuki et al. 2008), although the real power of this MS approach still seems to be difficult to demonstrate due to the shortage of informatics technology. A recent report described a fine structural elucidation strategy with FT-ICR MS by using all of the isotopic peaks derived from $^{13}$C-, $^{15}$N-, $^{34}$S-, and $^{18}$O-substituted forms recorded under high magnetic field FT-ICR MS (Miura et al. 2010). In addition, the combination of FT-ICR MS and the labeling of metabolites with stable isotopes (e.g., $^{13}$C and $^{15}$N) has been introduced to elevate the quality of plant metabolite annotation (Hegeman et al. 2007; Giavalisco et al. 2008, 2009). According to these reports, the comparison of

MS data from $^{12}C/^{13}C$ labeling experiments enabled the removal of background noise peaks and the positive identification of compounds with a true biological origin, thus eliminating the ambiguity in chemical formula assignment and resulting in the clear association of one measured mass to one chemical formula.

Since the metabolome consists of a vast array of compounds, most current metabolomic analyses using a single analytical protocol can only detect a fraction of the metabolites in a complex biological sample; thus, the combination of multiple MS systems, considering the scope of each MS system, can reveal the overall metabolic status of a sample. For example, we established a multi-MS platform for metabolomic analysis that consisted of several MS techniques (GC–MS, CE–MS, and LC–MS). This platform has been applied to the functional identification of an unknown gene (Watanabe et al. 2008) and the evaluation of a genetically modified tomato (Kusano et al. 2011a), as described later. Of course, such a multi-MS system provides a large amount of metabolite information, but the analysis of this data depends on the availability of well-structured databases that describe the assayed metabolites. Chemical identifiers are notoriously incoherent and encompass a wide range of different referencing schemes with varying scope and coverage, since chemical databases use multiple types of identifiers in parallel, but lack a common primary key for reliable database consolidation. Thus, connecting identifiers of analytes found in experimental data with the identifiers of their parent metabolites in public databases can, therefore, be very important. Recently, a tool for consolidating metabolite identifiers was developed to enable contextual and multi-platform metabolomic data analysis (Redestig et al. 2010), and we utilized this tool in practice to prepare a single consensus dataset for our multi-MS platform.

## 2.2 Nuclear magnetic resonance spectroscopy

NMR spectroscopy is another technology that is frequently employed in metabolomic studies since it is able to determine the atomic state of compounds and enables the identification of metabolites that are otherwise unidentifiable by MS analysis (Ward et al. 2007, 2010a; Beale et al. 2009; Kim et al. 2011). NMR spectroscopy can yield detailed information on the quantities and identities of the metabolites present in extracts or in vivo (Kikuchi et al. 2004; Lindon et al. 2004; Wang et al. 2004; Krishnan et al. 2005; Clayton et al. 2006; Ratcliffe and Shachar-Hill 2006; Sekiyama and Kikuchi 2007; Tian et al. 2007; Hagel et al. 2008; Sekiyama et al. 2010). Metabolome analysis by NMR usually does not require any pretreatment including column chromatography and derivatization, which makes standardization straightforward and points to the great potential of NMR spectroscopy in metabolomics, although attentions should be paid to pH of samples because chemical shifts could be changed depending on pH. For metabolomic analysis by NMR, buffered solutions are frequently utilized to stabilize the pH of sample solutions. Among the advantages of NMR over MS-based methods are that it is non-destructive, non-biased, highly quantitative, and enables the identification of complex unidentified metabolites. Even compounds which are bound to insoluble polymers can also be analyzed by high-resolution solid state NMR, which can afford direct and intact structural information. The major disadvantage of NMR, relative to MS, is its low sensitivity. Signal overlap from many similar molecules in biological samples is a major problem that inhibits the accurate assignment of NMR signals. However, these disadvantages, such as its lack of sensitivity and resolution, are gradually being overcome by the development of cryogenic probes (Kovacs et al. 2005), higher-strength superconducting magnets, miniaturized radiofrequency coils (Van and Veenstra 2009), and multidimensional NMR techniques (Kikuchi et al. 2004; Sekiyama and Kikuchi 2007; Chikayama et al. 2010). Dynamic nuclear polarization (DNP) enhanced NMR technique, developed by a combination of NMR and electron paramagnetic resonance (EPR) (Frydman and Blazina 2007), may provide a new approach to acquire hyper-sensitive NMR spectra for metabolomics. Labeling the samples with stables isotopes, e.g., $^{13}C$ and $^{15}N$. is also a useful technique that selectively enhances the sensitivity of NMR and improves signal sharpness (Kikuchi and Hirayama 2007; Chikayama et al. 2010). Labeling of metabolites with isotopes receptive by NMR is also useful for metabolic flux analysis and fluxomics by tracking the selective signal enhancement of isotopologues (Mesnard and Ratcliffe 2005; Eisenreich and Bacher 2007; Sekiyama and Kikuchi 2007; Coquin et al. 2008; Bothwell and Griffin 2011; Fan and Lane 2011; Moseley et al. 2011; Ward et al. 2011). These suggest the importance of the complementary use of MS and NMR in metabolomics.

# 3. INFORMATICS FOR METABOLOMIC STUDIES

Extremely large amounts of data are generated by instrument analysis, particularly for the high-performance instruments frequently used for metabolome analysis that can detect tiny signals with high resolution. To handle the large amount of datasets generated and comprehend the metabolome data, automated software is needed that can identify peaks from raw MS or NMR data, align the peaks among samples, and identify and quantify each metabolite; therefore, informatics is an essential tool for processing large metabolomic datasets (Fukushima et al. 2009b; Tohge and Fernie 2009, 2010; Go 2010). For GC–MS data, an automated spectral deconvolution and identification system using the NIST mass spectral library is available (Ausloos et al. 1999). To process raw MS data acquired by LC–MS or CE–MS, several open source programs are available, including MetAlign (Vorst et al. 2005), MZmine2 (Katajamaa et al. 2006), XCMS (Smith et al. 2006; Benton et al. 2008), MET-IDEA (Broeckling et al. 2006), and Mass++ (http://groups.google.com/group/massplusplus). To analyze FT-ICR MS datasets, DrDmassPlus (Oikawa et al. 2006) can be used for peak picking, peak alignment, and some statistical analysis. These programs facilitate comprehensive data analysis for non-targeted metabolomic approaches.

As with any omics discipline, metabolite annotation and identification is highly dependent on the availability and quality of electronic databases. To annotate the peaks processed by the programs described above, several mass spectral databases can be used, including MassBank (Horai et al. 2010), METLIN (Smith et al. 2005), the Golm Metabolite Database (Kopka et al. 2005), FiehnLib (Kind et al. 2009), and Lipid Search (Taguchi et al. 2007). The Human Metabolome Database (Wishart et al. 2009) and the Madison Metabolomics Consortium Database (Markley et al. 2007; Cui et al. 2008) provide mass spectral and NMR data for human metabolome studies; they are also useful for annotating signals derived from primary metabolites in plants. Recently, "ReSpect," a depository of API–$MS^n$ data from phytochemicals drawn from the literature, has been developed (http://spectra.psc.riken.jp/) as one of the web applications of PRIMe, a platform of RIKEN Metabolomics (http://prime.psc.riken.jp/; (Akiyama et al. 2008)). Users can access $MS^n$ information at this website, and they can freely download the manually refined $MS^n$ data. Some of these databases are integrated with software for processing raw MS data, e.g., METLIN and XCMS, and MassBank and Mass++. In addition to these spectral databases, several compound databases, including KNApSAcK (Shinbo et al. 2006), Kyoto Encyclopedia of Genes and Genomes (KEGG) (Kanehisa et al. 2008), PubChem (Wheeler et al. 2008), LipidBank (Yasugi and Watanabe 2002), and LIPIDMAPS (Fahy et al. 2007), are readily accessible. The information on compound name, chemical formula, and molecular weight that is deposited in these databases is also helpful for peak identification when authentic compounds and standard mass and NMR spectra are not available. Several databases can also be used that focus mainly on the mass spectra or metabolite information from several plant species. In MS2T (Matsuda et al. 2009), LC–MS data acquired from various plant species, including *Arabidopsis*, rice, wheat, and several non-model plant species, e.g., pear, are deposited. KOMICS (Iijima et al. 2008) and MotoDB (Moco et al. 2006) provide information for the ions that are detected in tomato, while ARMeC (http://www.armec.org/MetaboliteLibrary/) is a database of metabolites from *Arabidopsis* and potato. These data can be used to validate metabolome data among various laboratory, and as a reference source for many researchers who have to set up metabolome analytical platforms.

To comprehend the biochemical process involved in metabolism from metabolome data, metabolic pathway databases are very useful. There are several pathway databases available for plant metabolomics, including PlantCyc (http://www.plantcyc.org/), MapMan (Thimm et al. 2004), KaPPA-View (Tokimatsu et al. 2005; Sakurai et al. 2011), Arabidopsis Reactome (http://www.arabidopsisreactome.org/about.html), and KEGG (Kanehisa et al. 2008). KEGG PLANT (http://www.kegg.jp/kegg/plant/) is a new interface of KEGG that focuses on phytochemicals, especially secondary metabolites. Some of these databases import the abundance information of metabolites (and transcriptome information), and display an integrated overview of the metabolic state by reflecting the abundance of each metabolite on the putative metabolic pathway in plants. These databases also include information on the enzymatic reactions underlying these metabolic pathways and information on the proteins or genes involved in each reaction step. Although these pathway databases are helpful, it is necessary to bear in mind that many metabolic pathways in plants are divided by subcellular and intercellular compartmentalization and are highly redundant when attempting to understand the metabolic state of plants from metabolome data.

The deposition of metabolomic data in publicly available databases is also an important issue that needs to be considered. These databases enable data sharing with researchers in different laboratories and facilitate the meta-analysis of the original data. These depositories include AtMetExpress (http://prime.psc.riken.jp/lcms/AtMetExpress/) (Matsuda et al. 2010), PlantMetabolomics.org (http://plantmetabolomics.vrac.iastate.edu/) (Bais et al. 2010), and MetabolomeExpress (https://www.metabolome-express.org/) (Carroll et al. 2010).

In metabolomic studies, statistical analysis is often employed to evaluate significant differences in the metabolites detected in different samples. The methodology typically employs multivariate analysis to statistically process the massive amount of analytical chemistry data generated by high-throughput and simultaneous metabolite analyses (Fiehn 2002a; Fukusaki and Kobayashi 2005; Hall 2006; Okada et al. 2010). Various statistical methods used in conventional genetic studies are applicable to metabolomic data by considering the amount of each metabolite as a trait value. Principal component analysis (PCA), a multivariate analysis method, is commonly used in metabolomic studies. The PCA model can provide an overview of all observations or samples in a data table by projecting and clustering each sample and highlighting any holistic differences in the complex metabolic state of each sample. Many reports have described the application of PCA to metabolomic data (Catchpole et al. 2005; Takahashi et al. 2005; Tarpley et al. 2005; Tohge et al. 2005; Baker et al. 2006; Dixon et al. 2006; Oikawa et al. 2006; Kim et al. 2007; Kusano et al. 2007; Moco et al. 2007; Ku et al. 2010b). In addition, several statistical analytical methods have been used to analyze metabolomic datasets, e.g., hierarchical cluster analysis (HCA) (Grata et al. 2007; Parveen et al. 2007), partial least squares discriminant analysis (PLS-DA) (Jonsson et al. 2004; Kusano et al. 2007; Ku et al. 2010a), and batch-learning self-organizing map analysis (BL-SOM) (Hirai et al. 2004, 2005; Matsuda et al. 2010). Depending on the objective of each study, the most appropriate statistical analytical method should be exploited to evaluate the available metabolomic data.

# 4. BIOTECHNOLOGICAL APPLICATIONS

There have been many biological applications of plant metabolomics. In this section, we introduce example case studies, some of which are summarized in Table 1.

## 4.1 Application of metabolomics to functional genomics

The functional characterization of gene function on the genome scale is one of the most important tasks in the post-genomic era. In modern *Arabidopsis* research, loss-of-function or gain-of-function mutant lines play an important role in deciphering gene function. A combination of the metabolomic approach with these bioresources has been demonstrated to be an effective strategy to uncover the role of genes with unknown functions. Using GC–MS and CE–MS, Watanabe et al. (2008) obtained a metabolome dataset from T-DNA insertion mutants of *Arabidopsis* with immature $\beta$-substituted alanine synthase (*bsas*) genes and its corresponding wild-type. Statistical analysis revealed that one unknown metabolite was found in the wild-type plant, but did not accumulate in one of the *bsas* mutants, *bsas3;1*. This compound was eventually identified as $\gamma$-glutamyl-$\beta$-cyanoalanine, and the *bsas3;1* gene was shown to be involved in the biosynthesis of the precursor of this dipeptide. Metabolomics is also effective in high-throughput mutant screening. High-throughput GC–MS analysis of *Arabidopsis* gain-of-function mutant lines overexpressing full-length cDNAs of *Arabidopsis* led to the finding of a new transcriptional factor regulating nitrogen metabolism (Albinsky et al. 2010a). The functional characterization of unknown genes using a metabolomics approach can also be highly accelerated by combinatory use of other omics, such as transcriptomics.

**Table 1** Application of metabolomics in plant biological studies

| Application | Species | Gene | Phenotype | Analytical technology | Range | Bioinformatics | Reference |
|---|---|---|---|---|---|---|---|
| Identification of metabolic genes | *A. thaliana* | $\beta$-substituted alanine synthase family genes | Accumulation of $\beta$-cyano-Ala derivatives in the KO line | GC–MS, CE–MS | Untargeted profiling | PCA, *t* test | Watanabe et al. (2008) |
| | *A. thaliana* | Flavonoid 3-*O*-glucosyltransferase: anthocyanin 5-*O*-glucosyltransferase | Decrease of flavonol 3-glucosides; deficiency in 5-glucosylation of anthocyanins | FT-ICR MS, LC–MS | Untargeted profiling, targeted profiling | PCA | Tohge et al. (2005) |
| | *A. thaliana* | Flavonol 3-*O*-arabinosyltransferase: UDP-rhamnose synthase | Decrease of flavonol arabinosides; decrease of flavonoid rhamnosylation | LC–MS | Untargeted profiling | PCA | Yonekura-Sakakibara et al. (2008) |
| | *A. thaliana* | Desulfo-glucosinoate sulfotransferase | – | FT-ICR MS | Untargeted profiling | BL-SOM | Hirai et al. (2005) |
| | *M. truncatula* | Peroxidases exhibiting aurone synthase activity | Accumulation of the aurone hispidol as a major response to yeast elicitor | LC–MS | Untargeted profiling | ANOVA | Farag et al. (2009) |
| | *A. thaliana* | Transporter for Met-derived glucosinolates | Decease of Met-derived glucosinolates and increase of Met | LC–MS | Wide targeted profiling | *t* test | Sawada et al. (2009) |
| | *A. thaliana* | UDP-glucose pyrophosphorylase | Deficiency of sulfolipid in the KO line | LC–MS | Untargeted profiling | *t* test, HCA | Okazaki et al. (2009) |
| Evaluation of GMOs | Tomato | Miraculin | >92% of detected metabolites had an acceptable range of variation considering the variations among several tomato cultivars | GC–MS, LC–MS, CE–MS | Untargeted profiling | PCA, ANOVA, OPLS-DA | Kusano et al. (2011a) |
| | Wheat | Wheat gluten coding gene | Stronger influence of site and year than genotype, although a slight increase of saccharides was observed in transgenic lines | $^{1}H$ NMR | Untargeted profiling, targeted profiling | PCA | Baker et al. (2006) |
| Metabolome QTL analysis | *A. thaliana* | – | Identification of QTLs for ~75% of the >2000 mass signals detected from recombinant inbred ecotype lines; discovery of mQTL hot spots | LC–MS | Untargeted profiling | ANOVA | Keurentjes et al. (2006) |
| | *A. thaliana* | – | Linkage of metabolic profile and biomass accumulation. Discovery of mQTL hot spots | GC–MS | Untargeted profiling | Pearson correlation | Lisec et al. (2008) |
| | Tomato | – | Association of >50% of the metabolic loci with QTLs that modify yield-associated traits | GC–MS | Untargeted profiling | ANOVA, Pearson correlation | Schauer et al. (2006) |

| | | | | | | | |
|---|---|---|---|---|---|---|---|
| Investigation of the stress response | *A. thaliana* | Chalcone synthase; sinapoylglucose:malate sinapoyltransferase | Reprogramming of primary and secondary metabolism against UV-B light | GC–MS, LC–MS | Untargeted profiling | Vector analysis, PCA | Kusano et al. (2011c) |
| | *A. thaliana* | 9-*cis*-Epoxycarotenoid dioxygenase | Global metabolite–metabolite correlations in dehydration-increased amino acids in wild-type, and reconstruction of strong correlations with raffinose in the KO line | GC–MS, CE–MS | Untargeted profiling | Pearson correlation | Urano et al. (2009) |
| | *A. thaliana* | – | Changes in phenolic and indolic compounds, and rapid alterations in amino acids and other nitrogenous compounds, specific classes of glucosinolates, disaccharides, and molecules that influence the prevalence of reactive oxygen species | Flow infusion ESI MS, GC–MS, NMR | Untargeted profiling | PCA, PLS-DA, *t* test | Ward et al. (2010b) |

*KO* knock out, *ANOVA* analysis of variance

Several groups have used transcriptome co-expression analysis with metabolome data to find new metabolic genes using a limited number of known genes on the basis of the assumption that genes involved in the same metabolic pathway are co-expressed by a shared regulatory mechanism (Saito et al. 2008). This strategy, using the "gene-to-metabolite" correlation, successfully revealed the function of genes involved in plant primary metabolism (Okazaki et al. 2009; Kusano et al. 2011b), secondary metabolism (Hirai et al. 2005; Tohge et al. 2005; Hirai et al. 2007; Yonekura-Sakakibara et al. 2007; Farag et al. 2008, 2009; Yonekura-Sakakibara et al. 2008; Kuzina et al. 2009; Matsuda et al. 2010), and the circadian clock (Fukushima et al. 2009a).

## 4.2 Application of metabolomics to quantitative genetics

Natural variation provides a valuable resource to study the genetic regulation of quantitative traits. Since metabolite levels in plant tissues (m-trait) is also a quantitative trait, quantitative trait loci (QTL) analysis of m-traits is useful to enhance our understanding of the genetic architecture underlying naturally variable phenotypes with respect to metabolism. For example, QTL analysis of the levels of vitamin E in *Arabidopsis* seeds revealed the several QTLs responsible for the control of levels of this compound (Gilliland et al. 2006). QTL analysis of m-traits of *Arabidopsis* also revealed several loci regulating the levels of secondary metabolites flavonoids and glucosinolates (Keurentjes et al. 2006; Chan et al. 2010). Recently, metabolome QTL (mQTL) analysis has made a comprehensive understanding of the genetic background of m-traits possible (Keurentjes et al. 2006; Schauer et al. 2006, 2008; Wentzell et al. 2007; Lisec et al. 2008; Rowe et al. 2008; Kliebenstein 2009; Chan et al. 2010). These reports revealed that the mQTLs are unevenly distributed throughout the genome, and there are several mQTL hot-spot regions (Keurentjes et al. 2006; Lisec et al. 2008; Rowe et al. 2008), suggesting the possibility that overall metabolic state of plant cells could be controlled by modification of small genomic regions. The relationship between m-traits and other important biological traits, such as yield, taste and biomass, has been focused on since these traits are likely to closely interact with plant metabolism (Schauer et al. 2006; Meyer et al. 2007; Lisec et al. 2008). The analysis of tomato fruit traits with metabolome data indicated that there are weak correlations among these traits (Schauer et al. 2006, 2008). Regression analysis of metabolome data for *Arabidopsis* biomass traits demonstrated that the growth rate of *Arabidopsis* seedlings is, to some extent, predictable from the metabolome signature (Meyer et al. 2007). In addition, combinatory analyses of mQTL and naturally variable loci that altered circadian clock output suggested possible links between clock function and metabolism in *Arabidopsis* (Kerwin et al. 2011). These pioneering studies suggest that the interaction between metabolite composition and other traits may be predictable through future advanced metabolome analysis. The use of natural variation and mapping populations are also useful to discover new metabolic genes because transformation is still highly difficult in many plant species, except *Arabidopsis* (Morreel et al. 2006; Schilmiller et al. 2010). The accumulation of metabolome data from these lines may be useful for crop improvement through metabolomics-assisted breeding (Fernie and Schauer 2009).

## 4.3 Evaluation of the substantial equivalence of genetically modified organisms

To improve crop properties at the genetic level, two different strategies can be employed: traditional breeding and genetic modification (GM); however, both strategies can generate "unintended phenotypic effects" in metabolome. The public often feel insecure about the use of these strategies, particularly with GMOs, although it is not that surprising considering that we currently cannot target transgenes to specific genomic sites of the host plant, and that little is known about the mechanisms that determine the sites at which integration occurs. In addition, we do not fully understand the function of the non-coding regions of the genome. There are an increasing number of GM crops, such as barley, maize, pea, rice, soybean, and wheat (Ricroch et al. 2011); thus, we have to provide a suitable protocol to evaluate the safety of these GMOs.

A major principle and guiding tool for the food safety assessment of GMOs is the concept of "substantial equivalence" (SE), according to the principles outlined in the Organization for Economic Cooperation and Development (OECD) consensus documents (OECD 2006), and further elaborated by the Food and Agriculture Organization of the United Nations/World Health Organization. The internationally accepted SE framework proposes that risk assessment of GMOs should begin by comparing them with traditional varieties. The goal of SE evaluation is not to draw a conclusion about the novel organism's safety status because that would require the testing of all compounds, which is impossible; instead, by examining a broad set of traits, it aims to obtain a picture of the magnitude of changes to use as a screen for possible problematic changes and a starting point for further investigations. Since metabolome analysis can provide a vast amount of metabolite information, several metabolomic studies have performed an SE evaluation of GM tomato (Le Gall et al. 2003), potato (Catchpole et al. 2005), and wheat (Baker et al. 2006). Recently, we analyzed and evaluated the SE of a transgenic tomato expressing a foreign gene coding for taste-modifying miraculin using a multiple MS-based metabolomics platform (Kusano et al. 2011a). The chosen multiple-MS-based metabolomics platform (GC–MS, CE–MS, and LC–MS) detected compounds that represent 86% of the estimated chemical diversity of the metabolites listed in the LycoCyc database, and showed that >92% had an acceptable range of variation, while simultaneously indicating a reproducible transformation-related metabolic signature.

## 4.4 Application of metabolomics to the investigation of adaptive responses against stresses

Plants possess defense mechanisms against various stresses so they can adapt to unfavorable surroundings. Since the synthesis of specific metabolites is one of the most well-known defense reactions, the investigation of defense reactions with a metabolomics approach can help the holistic understanding of the strategies of plants against stresses. Sunlight is indispensable to all plants, except for several parasitic species, but ultra-violet (UV) light is a harmful element causing the oxidation of DNA and cellular components by the production of free radicals. Metabolome analysis of the UV-B response of wild-type *Arabidopsis* and mutants for flavonoid or sinapoyl malate revealed that the early metabolic responses only occurred at the level of primary metabolites, suggesting that these effectively prime the cell to facilitate the later production of UV-B-absorbing secondary metabolites, such as flavonoid and sinapoyl malate (Kusano et al. 2011c). In addition to abiotic stress, biotic stress is also a serious problem for plants. Ward and coworkers investigated the metabolic changes in *Arabidopsis* associated with the establishment of disease, and demonstrated clear differences in the metabolome of *Arabidopsis* leaves infected with virulent *Pseudomonas syringae* compared to uninfected leaves (Ward et al. 2010b). In addition to the confirmation of changes in phenolic and indolic compounds, they identified rapid alterations in the abundance of amino acids and other nitrogenous compounds, specific classes of glucosinolates, disaccharides, and molecules that influence the prevalence of reactive oxygen species. These reports suggest that comprehensive reprogramming of metabolic pathways is involved in the interaction of plants with stresses. Recently, many studies have examined the overall metabolic changes associated with biotic or abiotic stresses using metabolome analysis (Urano et al. 2009; Widodo et al. 2009; Chaouch et al. 2010; Consonni et al. 2010; Mhamdi et al. 2010; Simon et al. 2010).

The discovery of biomarkers is also important to understand the defense reaction of plants, and it could help to determine the disease condition at the metabolite level, leading to the development of new drugs, early disease screening, and improved pest control. Biomarker discovery by metabolomics has been actively performed in the field of human metabolomics (Soga et al. 2006; Jansson et al. 2009; Koulman et al. 2009; Sreekumar et al. 2009). In such kind of studies, the unambiguous identification or annotation of most of the metabolites is not necessarily required if the discovered biomarkers are truly useful for early disease screening. Thus, the power of the technologies used in metabolomics may directly play an effective role in this field since we still expend considerable time on processing the information derived from many unknown metabolites obtained by metabolome analysis.

# REFERENCES


1. Aharoni A, Ric de Vos CH, Verhoeven HA, Maliepaard CA, Kruppa G, Bino R, Goodenowe DB (2002) Nontargeted metabolome analysis by use of Fourier transform ion cyclotron mass spectrometry. OMICS 6:217–234. doi:10.1089/15362310260256882
2. Akiyama K, Chikayama E, Yuasa H, Shimada Y, Tohge T, Shinozaki K, Hirai MY, Sakurai T, Kikuchi J, Saito K (2008) PRIMe: a Web site that assembles tools for metabolomics and transcriptomics. In Silico Biol 8:339–345
3. Albinsky D, Kusano M, Higuchi M, Hayashi N, Kobayashi M, Fukushima A, Mori M, Ichikawa T, Matsui K, Kuroda H, Horii Y, Tsumoto Y, Sakakibara H, Hirochika H, Matsui M, Saito K (2010a) Metabolomic screening applied to rice FOX Arabidopsis lines leads to the identification of a gene-changing nitrogen metabolism. Mol Plant 3:125–142. doi:10.1093/mp/ssp069
4. Albinsky D, Sawada Y, Kuwahara A, Nagano M, Hirai A, Saito K, Hirai MY (2010b) Widely targeted metabolomics and coexpression analysis as tools to identify genes involved in the side-chain elongation steps of aliphatic glucosinolate biosynthesis. Amino Acids 39:1067–1075. doi:10.1007/s00726-010-0681-5
5. Arabidopsis Genome Initiative (2000) Analysis of the genome sequence of the flowering plant Arabidopsis thaliana. Nature 408:796–815. doi:10.1038/35048692
6. Ausloos P, Clifton CL, Lias SG, Mikaya AI, Stein SE, Tchekhovskoi DV, Sparkman OD, Zaikin V, ZhuD(1999) The critical evaluation of a comprehensive mass spectral library. J Am Soc Mass Spectrom 10:287–299. doi:10.1016/S1044-0305(98)00159-7
7. Bais P, Moon SM, He K, Leitao R, Dreher K, Walk T, Sucaet Y, Barkan L, Wohlgemuth G, Roth MR, Wurtele ES, Dixon P, Fiehn O, Lange BM, Shulaev V, Sumner LW, Welti R, Nikolau BJ, Rhee SY, Dickerson JA (2010) PlantMetabolomics.org: a web portal for plant metabolomics experiments. Plant Physiol 152:1807–1816. doi:10.1104/pp.109.151027
8. Baker JM, Hawkins ND, Ward JL, Lovegrove A, Napier JA, Shewry PR, Beale MH (2006) A metabolomic study of substantial equivalence of field-grown genetically modified wheat. Plant Biotechnol J 4:381–392. doi:10.1111/j.1467-7652.2006.00197.x
9. Beale MH, Ward JL, Baker JM (2009) Establishing substantial equivalence: metabolomics. Methods Mol Biol 478:289–303. doi:10.1007/978-1-59745-379-0_17
10. Benton HP, Wong DM, Trauger SA, Siuzdak G (2008) XCMS2: processing tandem mass spectrometry data for metabolite identification and structural characterization. Anal Chem 80: 6382–6389. doi:10.1021/ac800795f
11. Bino RJ, Hall RD, Fiehn O, Kopka J, Saito K, Draper J, Nikolau BJ, Mendes P, Roessner-Tunali U, Beale MH, Trethewey RN, Lange BM, Wurtele ES, Sumner LW (2004) Potential of metabolomics as a functional genomics tool. Trends Plant Sci 9:418–425. doi: 10.1016/j.tplants.2004.07.004
12. Bocker S, Rasche F (2008) Towards de novo identification of metabolites by analyzing tandem mass spectra. Bioinformatics 24:i49–i55. doi:10.1093/bioinformatics/btn270

13. Bothwell JH, Griffin JL (2011) An introduction to biological nuclear magnetic resonance spectroscopy. Biol Rev Camb Philos Soc 86:493–510. doi:10.1111/j.1469-185X.2010.00157.x
14. Bottcher C, von Roepenack-Lahaye E, Schmidt J, Schmotz C, Neumann S, Scheel D, Clemens S (2008) Metabolome analysis of biosynthetic mutants reveals a diversity of metabolic changes and allows identification of a large number of new compounds in Arabidopsis. Plant Physiol 147:2107–2120. doi:10.1104/pp.108.117754
15. Broeckling CD, Reddy IR, Duran AL, Zhao X, Sumner LW (2006) MET-IDEA: data extraction tool for mass spectrometry-based metabolomics. Anal Chem 78:4334–4341. doi:10.1021/ac0521596
16. Butelli E, Titta L, Giorgio M, Mock HP, Matros A, Peterek S, Schijlen EG, Hall RD, Bovy AG, Luo J, Martin C (2008) Enrichment of tomato fruit with health-promoting anthocyanins by expression of select transcription factors. Nat Biotechnol 26:1301–1308. doi:10.1038/nbt.1506
17. Carroll AJ, Badger MR, Harvey Millar A (2010) The Metabolome- Express Project: enabling web-based processing, analysis and transparent dissemination of GC/MS metabolomics datasets. BMC Bioinform 11:376. doi:10.1186/1471-2105-11-376
18. Catchpole GS, Beckmann M, Enot DP, Mondhe M, Zywicki B, Taylor J, Hardy N, Smith A, King RD, Kell DB, Fiehn O, Draper J (2005) Hierarchical metabolomics demonstrates substantial compositional similarity between genetically modified and conventional potato crops. Proc Natl Acad Sci USA 102:14458–14462. doi:10.1073/pnas.0503955102
19. Chan EK, Rowe HC, Kliebenstein DJ (2010) Understanding the evolution of defense metabolites in Arabidopsis thaliana using genome-wide association mapping. Genetics 185:991–1007. doi: 10.1534/genetics.109.108522
20. Chaouch S, Queval G, Vanderauwera S, Mhamdi A, Vandorpe M, Langlois-Meurinne M, Van Breusegem F, Saindrenan P, Noctor G (2010) Peroxisomal hydrogen peroxide is coupled to biotic defense responses by ISOCHORISMATE SYNTHASE1 in a daylength-related manner. Plant Physiol 153:1692–1705. doi: 10.1104/pp.110.153957
21. Chikayama E, Sekiyama Y, Okamoto M, Nakanishi Y, Tsuboi Y, Akiyama K, Saito K, Shinozaki K, Kikuchi J (2010) Statistical indices for simultaneous large-scale metabolite detections for a single NMR spectrum. Anal Chem 82:1653–1658. doi:10.1021/ac9022023
22. Clayton TA, Lindon JC, Cloarec O, Antti H, Charuel C, Hanton G, Provost JP, Le Net JL, Baker D, Walley RJ, Everett JR, Nicholson JK (2006) Pharmaco-metabonomic phenotyping and personalized drug treatment. Nature 440:1073–1077. doi: 10.1038/nature04648
23. Codrea MC, Jimenez CR, Heringa J, Marchiori E (2007) Tools for computational processing of LC–MS datasets: a user's perspective. Comput Methods Programs Biomed 86:281–290. doi: 10.1016/j.cmpb.2007.03.001
24. Consonni C, Bednarek P, Humphry M, Francocci F, Ferrari S, Harzen A, Ver Loren van Themaat E, Panstruga R (2010) Tryptophanderived metabolites are required for antifungal defense in the Arabidopsis mlo2 mutant. Plant Physiol 152:1544–1561. doi: 10.1104/pp.109.147660
25. Coquin L, Feala JD, McCulloch AD, Paternostro G (2008) Metabolomic and flux-balance analysis of age-related decline of hypoxia tolerance in Drosophila muscle tissue. Mol Syst Biol 4:233. doi:10.1038/msb.2008.71
26. Cui Q, Lewis IA, Hegeman AD, Anderson ME, Li J, Schulte CF, Westler WM, Eghbalnia HR, Sussman MR, Markley JL (2008) Metabolite identification via the Madison Metabolomics Consortium Database. Nat Biotechnol 26:162–164. doi:10.1038/nbt0 208-162
27. De Vos RC, Moco S, Lommen A, Keurentjes JJ, Bino RJ, Hall RD (2007) Untargeted large-scale plant metabolomics using liquid chromatography coupled to mass spectrometry. Nat Protoc 2: 778–791. doi:10.1038/nprot.2007.95
28. Dixon RA, Strack D (2003) Phytochemistry meets genome analysis, and beyond. Phytochemistry 62:815–816. doi:10.1016/S0031-9422(02)00712-4
29. Dixon RA, Gang DR, Charlton AJ, Fiehn O, Kuiper HA, Reynolds TL, Tjeerdema RS, Jeffery EH, German JB, Ridley WP, Seiber JN (2006) Applications of metabolomics in agriculture. J Agric Food Chem 54:8984–8994. doi:10.1021/jf061218t

30. Dunn WB (2008) Current trends and future requirements for the mass spectrometric investigation of microbial, mammalian and plant metabolomes. Phys Biol 5:11001. doi:10.1088/1478-3975/5/1/011001
31. Eisenreich W, Bacher A (2007) Advances of high-resolution NMR techniques in the structural and metabolic analysis of plant biochemistry. Phytochemistry 68:2799–2815. doi:10.1016/j.phytochem.2007.09.028
32. Fahy E, Sud M, Cotter D, Subramaniam S (2007) LIPID MAPS online tools for lipid research. Nucleic Acids Res 35:W606–W612. doi:10.1093/nar/gkm324
33. Fan TW, Lane AN (2011) NMR-based stable isotope resolved metabolomics in systems biochemistry. J Biomol NMR 49:267–280. doi:10.1007/s10858-011-9484-6
34. Farag MA, Huhman DV, Dixon RA, Sumner LW (2008) Metabolomics reveals novel pathways and differential mechanistic and elicitor-specific responses in phenylpropanoid and isoflavonoid biosynthesis in Medicago truncatula cell cultures. Plant Physiol 146:387–402. doi:10.1104/pp.107.108431
35. Farag MA, Deavours BE, de Fatima A, Naoumkina M, Dixon RA, Sumner LW (2009) Integrated metabolite and transcript profiling identify a biosynthetic mechanism for hispidol in Medicago truncatula cell cultures. Plant Physiol 151:1096–1113. doi:10.1104/pp.109.141481
36. Fernie AR, Schauer N (2009) Metabolomics-assisted breeding: a viable option for crop improvement? Trends Genet 25:39–48. doi:10.1016/j.tig.2008.10.010
37. Fernie AR, Trethewey RN, Krotzky AJ, Willmitzer L (2004) Metabolite profiling: from diagnostics to systems biology. Nat Rev Mol Cell Biol 5:763–769. doi:10.1038/nrm1451
38. Fiehn O (2002) Metabolomics—the link between genotypes and phenotypes. Plant Mol Biol 48:155–171. doi:10.1023/A:1013713905833
39. Fiehn O, Kopka J, Do¨rmann P, Altmann T, Trethewey RN, Willmitzer L (2000) Metabolite profiling for plant functional genomics. Nat Biotechnol 18:1157–1161. doi:10. 1038/81137
40. Frydman L, Blazina D (2007) Ultrafast two-dimensional nuclear magnetic resonance spectroscopy of hyperpolarized solutions. Nat Phys 3:415–419. doi:10.1038/nphys597
41. Fukusaki E, Kobayashi A (2005) Plant metabolomics: potential for practical operation. J Biosci Bioeng 100:347–354. doi:10.1263/jbb.100.347
42. Fukushima A, Kusano M, Nakamichi N, Kobayashi M, Hayashi N, Sakakibara H, Mizuno T, Saito K (2009a) Impact of clockassociated Arabidopsis pseudo-response regulators in metabolic coordination. Proc Natl Acad Sci USA 106:7251–7256. doi:10.1073/ pnas.0900952106
43. Fukushima A, Kusano M, Redestig H, Arita M, Saito K (2009b) Integrated omics approaches in plant systems biology. Curr Opin Chem Biol 13:532–538. doi:10.1016/j.cbpa.2009.09.022
44. Giavalisco P, Hummel J, Lisec J, Inostroza AC, Catchpole G, Willmitzer L (2008) High-resolution direct infusion-based mass spectrometry in combination with whole 13C metabolome isotope labeling allows unambiguous assignment of chemical sum formulas. Anal Chem 80:9417–9425. doi:10.1021/ac8014627
45. Giavalisco P, Kohl K, Hummel J, Seiwert B, Willmitzer L (2009) 13C isotope-labeled metabolomes allowing for improved compound annotation and relative quantification in liquid chromatography–mass spectrometry-based metabolomic research. Anal Chem 81:6546–6551. doi:10.1021/ac900979e
46. Gilliland LU, Magallanes-Lundback M, Hemming C, Supplee A, Koornneef M, Bentsink L, Dellapenna D (2006) Genetic basis for natural variation in seed vitamin E levels in Arabidopsis thaliana. Proc Natl Acad Sci USA 103:18834–18841. doi:10.1073/ pnas.0606221103
47. Go EP (2010) Database resources in metabolomics: an overview. J Neuroimmune Pharmacol 5:18–30. doi:10.1007/s11481-009-9157-3
48. Goff SA, Ricke D, Lan TH, Presting G, Wang R, Dunn M, Glazebrook J, Sessions A, Oeller P, Varma H, Hadley D, Hutchison D, Martin C, Katagiri F, Lange BM, Moughamer T, Xia Y, Budworth P, Zhong J, Miguel T, Paszkowski U, Zhang S, Colbert M, Sun WL, Chen L, Cooper B, Park S, Wood TC, Mao L, Quail P, Wing R, Dean R, Yu Y, Zharkikh A, Shen R,

Sahasrabudhe S, Thomas A, Cannings R, Gutin A, Pruss D, Reid J, Tavtigian S, Mitchell J, Eldredge G, Scholl T, Miller RM, Bhatnagar S, Adey N, Rubano T, Tusneem N, Robinson R, Feldhaus J, Macalma T, Oliphant A, Briggs S (2002) A draft sequence of the rice genome (Oryza sativa L. ssp. japonica). Science 296:92–100. doi:10.1126/science.1068275

49. Grata E, Boccard J, Glauser G, Carrupt PA, Farmer EE, Wolfender JL, Rudaz S (2007) Development of a two-step screening ESI–TOF–MS method for rapid determination of significant stressinduced metabolome modifications in plant leaf extracts: the wound response in Arabidopsis thaliana as a case study. J Sep Sci 30:2268–2278. doi:10.1002/jssc.200700143
50. Griffiths WJ, Koal T, Wang Y, Kohl M, Enot DP, Deigner HP (2010) Targeted metabolomics for biomarker discovery. Angew Chem 49:5426–5445. doi:10.1002/anie.200905579
51. Hagel JM, Weljie AM, Vogel HJ, Facchini PJ (2008) Quantitative 1H nuclear magnetic resonance metabolite profiling as a functional genomics platform to investigate alkaloid biosynthesis in opium poppy. Plant Physiol 147:1805–1821. doi:10.1104/pp.108.120493
52. Hall RD (2006) Plant metabolomics: from holistic hope, to hype, to hot topic. New Phytol 169:453–468. doi:10.1111/j.1469-8137.2005.01632.x
53. Harada K, Fukusaki E, Kobayashi A (2006) Pressure-assisted capillary electrophoresis mass spectrometry using combination of polarity reversion and electroosmotic flow for metabolomics anion analysis. J Biosci Bioeng 101:403–409. doi:10.1263/jbb.101.403
54. Hegeman AD, Schulte CF, Cui Q, Lewis IA, Huttlin EL, Eghbalnia H, Harms AC, Ulrich EL, Markley JL, Sussman MR (2007) Stable isotope assisted assignment of elemental compositions for metabolomics. Anal Chem 79:6912–6921. doi:10.1021/ac070346t
55. Hirai MY, Yano M, Goodenowe DB, Kanaya S, Kimura T, Awazuhara M, Arita M, Fujiwara T, Saito K (2004) Integration of transcriptomics and metabolomics for understanding of global responses to nutritional stresses in Arabidopsis thaliana. Proc Natl Acad Sci USA 101:10205–10210. doi:10.1073/pnas.0611629104
56. Hirai MY, Klein M, Fujikawa Y, Yano M, Goodenowe DB, Yamazaki Y, Kanaya S, Nakamura Y, Kitayama M, Suzuki H, Sakurai N, Shibata D, Tokuhisa J, Reichelt M, Gershenzon J, Papenbrock J, Saito K (2005) Elucidation of gene-to-gene and metabolite-togene networks in Arabidopsis by integration of metabolomics and transcriptomics. J Biol Chem 280:25590–25595. doi:10.1074/jbc.M502332200
57. Hirai MY, Sugiyama K, Sawada Y, Tohge T, Obayashi T, Suzuki A, Araki R, Sakurai N, Suzuki H, Aoki K, Goda H, Nishizawa OI, Shibata D, Saito K (2007) Omics-based identification of Arabidopsis Myb transcription factors regulating aliphatic glucosinolate biosynthesis. Proc Natl Acad Sci USA 104:6478–6483. doi:10.1073/pnas.0611629104
58. Horai H, Arita M, Kanaya S, Nihei Y, Ikeda T, Suwa K, Ojima Y, Tanaka K, Tanaka S, Aoshima K, Oda Y, Kakazu Y, Kusano M, Tohge T, Matsuda F, Sawada Y, Hirai MY, Nakanishi H, Ikeda K, Akimoto N, Maoka T, Takahashi H, Ara T, Sakurai N, Suzuki H, Shibata D, Neumann S, Iida T, Funatsu K, Matsuura F, Soga T, Taguchi R, Saito K, Nishioka T (2010) MassBank: a public repository for sharing mass spectral data for life sciences. J Mass Spectrom 45:703–714. doi:10.1002/jms.1777
59. Iijima Y, Nakamura Y, Ogata Y, Tanaka K, Sakurai N, Suda K, Suzuki T, Suzuki H, Okazaki K, Kitayama M, Kanaya S, Aoki K, Shibata D (2008) Metabolite annotations based on the integration of mass spectral information. Plant J 54:949–962. doi:10.1111/j.1365-313X.2008.03434.x
60. International Rice Genome Sequencing Project (2005) The map-based sequence of the rice genome. Nature 436:793–800. doi: 10.1038/nature03895
61. Ishikawa T, Takahara K, Hirabayashi T, Matsumura H, Fujisawa S, Terauchi R, Uchimiya H, Kawai-Yamada M (2010) Metabolome analysis of response to oxidative stress in rice suspension cells overexpressing cell death suppressor Bax inhibitor-1. Plant Cell Physiol 51:9–20. doi:10.1093/pcp/pcp162
62. Jansson J, Willing B, Lucio M, Fekete A, Dicksved J, Halfvarson J, Tysk C, Schmitt-Kopplin P (2009) Metabolomics reveals metabolic biomarkers of Crohn's disease. PLoS ONE 4:e6386. doi:10.1371/journal.pone.0006386

63. Jonsson P, Gullberg J, Nordstrom A, Kusano M, Kowalczyk M, Sjostrom M, Moritz T (2004) A strategy for identifying differences in large series of metabolomic samples analyzed by GC/MS. Anal Chem 76:1738–1745. doi:10.1021/ac0352427
64. Kanehisa M, Araki M, Goto S, Hattori M, Hirakawa M, Itoh M, Katayama T, Kawashima S, Okuda S, Tokimatsu T, Yamanishi Y (2008) KEGG for linking genomes to life and the environment. Nucleic Acids Res 36:D480–D484. doi:10.1093/nar/gkm882
65. Katajamaa M, Miettinen J, Oresic M (2006) MZmine: toolbox for processing and visualization of mass spectrometry based molecular profile data. Bioinformatics 22:634–636. doi:10. 1093/bioinformatics/btk039
66. Kerwin RE, Jimenez-Gomez JM, Fulop D, Harmer SL, Maloof JN, Kliebenstein DJ (2011) Network quantitative trait loci mapping of circadian clock outputs identifies metabolic pathway-to-clock linkages in Arabidopsis. Plant Cell 23:471–485. doi: 10.1105/tpc.110.082065
67. Keurentjes JJ, Fu J, de Vos CH, Lommen A, Hall RD, Bino RJ, van der Plas LH, Jansen RC, Vreugdenhil D, Koornneef M (2006) The genetics of plant metabolism. Nat Genet 38:842–849. doi: 10.1038/ng1815
68. Kikuchi J, Hirayama T (2007) Practical aspects of uniform stable isotope labeling of higher plants for heteronuclear NMR-based metabolomics. Methods Mol Biol 358:273–286
69. Kikuchi J, Shinozaki K, Hirayama T (2004) Stable isotope labeling of Arabidopsis thaliana for an NMR-based metabolomics approach. Plant Cell Physiol 45:1099–1104. doi:10.1093/pcp/pch117
70. Kim JK, Bamba T, Harada K, Fukusaki E, Kobayashi A (2007) Timecourse metabolic profiling in Arabidopsis thaliana cell cultures after salt stress treatment. J Exp Bot 58:415–424. doi:10. 1093/jxb/erl216
71. Kim HK, Choi YH, Verpoorte R (2011) NMR-based plant metabolomics: where do we stand, where do we go? Trends Biotechnol. doi:10.1016/j.tibtech.2011.02.001
72. Kind T, Fiehn O (2006)Metabolomic database annotations via query of elemental compositions: mass accuracy is insufficient even at less than 1 ppm. BMC Bioinform 7:234. doi:10.1186/1471-2105-7-234
73. Kind T, Wohlgemuth G, Lee do Y, Lu Y, Palazoglu M, Shahbaz S, Fiehn O (2009) FiehnLib: mass spectral and retention index libraries for metabolomics based on quadrupole and time-offlight gas chromatography/mass spectrometry. Anal Chem 81:10038–10048. doi:10.1021/ac9019522
74. Kliebenstein D (2009) Advancing genetic theory and application by metabolic quantitative trait loci analysis. Plant Cell 21:1637–1646. doi:10.1105/tpc.109.067611
75. Kopka J, Schauer N, Krueger S, Birkemeyer C, Usadel B, Bergmuller E, Do¨rmann P, Weckwerth W, Gibon Y, Stitt M, Willmitzer L, Fernie AR, Steinhauser D (2005) GMD@CSB.DB: the Golm Metabolome Database. Bioinformatics 21:1635–1638. doi: 10.1093/bioinformatics/bti236
76. Koulman A, Lane GA, Harrison SJ, Volmer DA (2009) From differentiating metabolites to biomarkers. Anal Bioanal Chem 394:663–670. doi:10.1007/s00216-009-2690-3
77. Kovacs H, Moskau D, Spraul M (2005) Cryogenically cooled probes—a leap in NMR technology. Prog Nucl Magn Reson Spectrosc 46:131–155. doi:10.1016/ j.pnmrs. 2005.03.001
78. Krishnan P, Kruger NJ, Ratcliffe RG (2005) Metabolite fingerprinting and profiling in plants using NMR. J Exp Bot 56:255–265. doi: 10.1093/jxb/eri010
79. Ku KM, Choi JN, Kim J, Kim JK, Yoo LG, Lee SJ, Hong YS, Lee CH (2010a) Metabolomics analysis reveals the compositional differences of shade grown tea (Camellia sinensis L.). J Agric Food Chem 58:418–426. doi:10.1021/jf902929h
80. Ku KM, Kim J, Park HJ, Liu KH, Lee CH (2010b) Application of metabolomics in the analysis of manufacturing type of pu-erh tea and composition changes with different postfermentation year. J Agric Food Chem 58:345–352. doi:10.1021/jf902818c
81. Kusano M, Fukushima A, Kobayashi M, Hayashi N, Jonsson P, Moritz T, Ebana K, Saito K (2007) Application of a metabolomic method combining one-dimensional and two-dimensional gas chromatography-time-of-flight/mass spectrometry to metabolic

phenotyping of natural variants in rice. J Chromatogr B Analyt Technol Biomed Life Sci 855:71–79. doi:10.1016/j.jchromb.2007.05.002

82. Kusano M, Redestig H, Hirai T, Oikawa A, Matsuda F, Fukushima A, Arita M, Watanabe S, Yano M, Hiwasa-Tanase K, Ezura H, Saito K (2011a) Covering chemical diversity of geneticallymodified tomatoes using metabolomics for objective substantial equivalence assessment. PLoS ONE 6:e16989. doi:10.1371/journal.pone.0016989
83. Kusano M, Tabuchi M, Fukushima A, Funayama K, Diaz C, Kobayashi M, Hayashi N, Tsuchiya YN, Takahashi H, Kamata A, Yamaya T, Saito K (2011b) Metabolomics data reveal a crucial role of cytosolic glutamine synthetase 1;1 in coordinating metabolic balance in rice. Plant J 66:456–466. doi:10.1111/j.1365-313X.2011.04506.x
84. Kusano M, Tohge T, Fukushima A, Kobayashi M, Hayashi N, Otsuki H, Kondou Y, Goto H, Kawashima M, Matsuda F, Niida R, Matsui M, Saito K, Fernie AR (2011c) Metabolomics reveals comprehensive reprogramming involving two independent metabolic responses of Arabidopsis to ultraviolet-B light. Plant J. doi:10.1111/j.1365-313X.2011.04599.x
85. Kuzina V, Ekstrom CT, Andersen SB, Nielsen JK, Olsen CE, Bak S (2009) Identification of defense compounds in Barbarea vulgaris against the herbivore Phyllotreta nemorum by an ecometabolomic approach. Plant Physiol 151:1977–1990. doi:10.1104/pp.109.136952
86. Le Gall G, Colquhoun IJ, Davis AL, Collins GJ, Verhoeyen ME (2003) Metabolite profiling of tomato (Lycopersicon esculentum) using 1H NMR spectroscopy as a tool to detect potential unintended effects following a genetic modification. J Agric Food Chem 51:2447–2456. doi:10.1021/jf0259967
87. Leiss KA, Maltese F, Choi YH, Verpoorte R, Klinkhamer PG (2009) Identification of chlorogenic acid as a resistance factor for thrips in Chrysanthemum. Plant Physiol 150:1567–1575. doi:10.1104/pp.109.138131
88. Lindon JC, Holmes E, Nicholson JK (2004) Metabonomics: systems biology in pharmaceutical research and development. Curr Opin Mol Ther 6:265–272
89. Lisec J, Meyer RC, Steinfath M, Redestig H, Becher M, Witucka- Wall H, Fiehn O, Torjek O, Selbig J, Altmann T, Willmitzer L (2008) Identification of metabolic and biomass QTL in Arabidopsis thaliana in a parallel analysis of RIL and IL populations. Plant J 53:960–972. doi:10.1111/j.1365-313X.2007.03383.x
90. Lucker J, Bouwmeester HJ, Schwab W, Blaas J, van der Plas LH, Verhoeven HA (2001) Expression of Clarkia S-linalool synthase in transgenic petunia plants results in the accumulation of Slinalyl- beta-D-glucopyranoside. Plant J 27:315–324. doi:10.1046/j.1365-313x.2001.01097.x
91. Markley JL, Anderson ME, Cui Q, Eghbalnia HR, Lewis IA, Hegeman AD, Li J, Schulte CF, Sussman MR, Westler WM, Ulrich EL, Zolnai Z (2007) New bioinformatics resources for metabolomics. Pac Symp Biocomput 12:157–168
92. Matsuda F, Yonekura-Sakakibara K, Niida R, Kuromori T, Shinozaki K, Saito K (2009) MS/MS spectral tag-based annotation of nontargeted profile of plant secondary metabolites. Plant J57:555–577. doi:10.1111/j.1365-313X.2008.03705.x
93. Matsuda F, Hirai MY, Sasaki E, Akiyama K, Yonekura-Sakakibara K, Provart NJ, Sakurai T, Shimada Y, Saito K (2010) AtMetExpress development: a phytochemical atlas of Arabidopsis development. Plant Physiol 152:566–578. doi:10.1104/pp.109.148031
94. Mesnard F, Ratcliffe RG (2005) NMR analysis of plant nitrogen metabolism. Photosynth Res 83:163–180. doi:10.1007/s11120-004-2081-8
95. Meyer RC, Steinfath M, Lisec J, Becher M, Witucka-Wall H, Torjek O, Fiehn O, Eckardt A, Willmitzer L, Selbig J, Altmann T (2007) The metabolic signature related to high plant growth rate in Arabidopsis thaliana. Proc Natl Acad Sci USA 104:4759–4764. doi:10.1073/pnas.0609709104
96. Mhamdi A, Mauve C, Gouia H, Saindrenan P, Hodges M, Noctor G (2010) Cytosolic NADP-dependent isocitrate dehydrogenase contributes to redox homeostasis and the regulation of pathogen responses in Arabidopsis leaves. Plant Cell Environ 33:1112–1123. doi:10.1111/j.1365-3040.2010.02133.x
97. Minami Y, Kasukawa T, Kakazu Y, Iigo M, Sugimoto M, Ikeda S, Yasui A, van der Horst GT, Soga T, Ueda HR (2009) Measurement of internal body time by blood metabolomics. Proc Natl Acad Sci USA 106:9890–9895. doi:10.1073/pnas.0900617106

98. Miura D, Tsuji Y, Takahashi K, Wariishi H, Saito K (2010) A strategy for the determination of the elemental composition by Fourier transform ion cyclotron resonance mass spectrometry based on isotopic peak ratios. Anal Chem 82:5887–5891. doi:10.1021/ac902931x
99. Moco S, Bino RJ, Vorst O, Verhoeven HA, de Groot J, van Beek TA, Vervoort J, de Vos CH (2006) A liquid chromatography-mass spectrometry-based metabolome database for tomato. Plant Physiol 141:1205–1218. doi:10.1104/pp.106.078428
100. Moco S, Capanoglu E, Tikunov Y, Bino RJ, Boyacioglu D, Hall RD, Vervoort J, De Vos RC (2007) Tissue specialization at the metabolite level is perceived during the development of tomato fruit. J Exp Bot 58:4131–4146. doi:10.1093/jxb/erm271
101. Monton MR, Soga T (2007) Metabolome analysis by capillary electrophoresis-mass spectrometry. J Chromatogr A 1168:237–246. doi:10.1016/j.chroma.2007.02.065
102. Morreel K, Goeminne G, Storme V, Sterck L, Ralph J, Coppieters W, Breyne P, Steenackers M, Georges M, Messens E, Boerjan W (2006) Genetical metabolomics of flavonoid biosynthesis in Populus: a case study. Plant J 47:224–237. doi:10.1111/j.1365-313X.2006.02786.x
103. Moseley HN, Lane AN, Belshoff AC, Higashi RM, Fan TW (2011) A novel deconvolution method for modeling UDP-GlcNAc biosynthetic pathways based on 13C mass isotopologue profiles under non steady-state conditions. BMC Biol 9:37. doi:10.1186/1741-7007-9-37
104. Mungur R, Glass AD, Goodenow DB, Lightfoot DA (2005) Metabolite fingerprinting in transgenic Nicotiana tabacum altered by the Escherichia coli glutamate dehydrogenase gene. J Biomed Biotechnol 2005:198–214. doi:10.1155/JBB.2005.198
105. Nakabayashi R, Kusano M, Kobayashi M, Tohge T, Yonekura- Sakakibara K, Kogure N, Yamazaki M, Kitajima M, Saito K, Takayama H (2009) Metabolomics-oriented isolation and structure elucidation of 37 compounds including two anthocyanins from Arabidopsis thaliana. Phytochemistry 70:1017–1029. doi: 10.1016/j.phytochem.2009.03.021
106. Nakabayashi R, Yamazaki M, Saito K (2010) A polyhedral approach for understanding flavonoid biosynthesis in Arabidopsis. N Biotechnol 27:829–836. doi:10.1016/j.nbt.2010.03.004
107. Naoumkina M, Farag MA, Sumner LW, Tang Y, Liu CJ, Dixon RA (2007) Different mechanisms for phytoalexin induction by pathogen and wound signals in Medicago truncatula. Proc Natl Acad Sci USA 104:17909–17915. doi:10.1073/pnas.0708697104
108. OECD (2006) An Introduction to the Food/Feed Safety Consensus Documents of the Task Force. Series on the Safety of Novel Foods and Feeds, No 14. Organization for Economic Cooperation and Development, Paris, pp 7–9
109. Ohkama-Ohtsu N, Oikawa A, Zhao P, Xiang C, Saito K, Oliver DJ (2008) A gamma-glutamyl transpeptidase-independent pathway of glutathione catabolism to glutamate via 5-oxoproline in Arabidopsis. Plant Physiol 148:1603–1613. doi:10.1104/pp.108.125716
110. Ohkama-Ohtsu N, Sasaki-Sekimoto Y, Oikawa A, Jikumaru Y, Shinoda S, Inoue E, Kamide Y, Yokoyama T, Hirai MY, Shirasu K, Kamiya Y, Oliver DJ, Saito K (2011) 12-oxo-phytodienoic acid–glutathione conjugate is transported into the vacuole in Arabidopsis. Plant Cell Physiol 52:205–209. doi:10.1093/pcp/pcq181
111. Oikawa A, Nakamura Y, Ogura T, Kimura A, Suzuki H, Sakurai N, Shinbo Y, Shibata D, Kanaya S, Ohta D (2006) Clarification of pathway-specific inhibition by Fourier transform ion cyclotron resonance/mass spectrometry-based metabolic phenotyping studies. Plant Physiol 142:398–413. doi:10.1104/pp.106.080317
112. Oikawa A, Matsuda F, Kusano M, Okazaki Y, Saito K (2008) Rice metabolomics. Rice 1:63–71. doi:10.1073/pnas.0708697104
113. Okada T, Afendi FM, Altaf-Ul-Amin M, Takahashi H, Nakamura K, Kanaya S (2010) Metabolomics of medicinal plants: the importance of multivariate analysis of analytical chemistry data. Curr Comput Aided Drug Des 6:179–196
114. Okazaki Y, Shimojima M, Sawada Y, Toyooka K, Narisawa T, Mochida K, Tanaka H, Matsuda F, Hirai A, Hirai MY, Ohta H, Saito K (2009) A chloroplastic UDP-glucose pyrophosphorylase from Arabidopsis is the committed enzyme for the first step of sulfolipid biosynthesis. Plant Cell 21:892–909. doi:10.1105/tpc.108.063925

115. Okazaki Y, Kamide Y, Hirai MY, Saito K (2011) Plant lipidomics based on hydrophilic interaction chromatography coupled to ion trap time-of-flight mass spectrometry. Metabolomics (in press). doi:10.1007/s11306-011-0318-z
116. Oksman-Caldentey KM, Saito K (2005) Integrating genomics and metabolomics for engineering plant metabolic pathways. Curr Opin Biotechnol 16:174–179. doi:10.1016/j.copbio.2005.02.007
117. Parveen I, Moorby JM, Fraser MD, Allison GG, Kopka J (2007) Application of gas chromatography–mass spectrometry metabolite profiling techniques to the analysis of heathland plant diets of sheep. J Agric Food Chem 55:1129–1138. doi:10.1021/jf062995w
118. Quackenbush J (2002) Extracting biology from high-dimensional biological data. J Exp Biol 48:1507–1517. doi:10.1242/jeb.004432
119. Ralston-Hooper K, Hopf A, Oh C, Zhang X, Adamec J, Sepulveda MS (2008) Development of GC9GC/TOF–MS metabolomics for use in ecotoxicological studies with invertebrates. Aquat Toxicol 88:48–52. doi:10.1016/j.aquatox.2008.03.002
120. Ramautar R, Somsen GW, de Jong GJ (2009) CE-MS in metabolomics. Electrophoresis 30:276–291. doi:10.1002/elps.200800512
121. Ratcliffe RG, Shachar-Hill Y (2006) Measuring multiple fluxes through plant metabolic networks. Plant J 45:490–511. doi: 10.1111/j.1365-313X.2005.02649.x
122. Redestig H, Kusano M, Fukushima A, Matsuda F, Saito K, Arita M (2010) Consolidating metabolite identifiers to enable contextual and multi-platform metabolomics data analysis. BMC Bioinform 11:214. doi:10.1186/1471-2105-11-214
123. Ricroch AE, Berge JB, Kuntz M (2011) Evaluation of genetically engineered crops using transcriptomic, proteomic, and metabolomic profiling techniques. Plant Physiol 155:1752–1761. doi: 10.1104/pp.111.173609
124. Roessner U, Luedemann A, Brust D, Fiehn O, Linke T, Willmitzer L, Fernie A (2001) Metabolic profiling allows comprehensive phenotyping of genetically or environmentally modified plant systems. Plant Cell 13:11–29. doi:10.1105/tpc.13.1.11
125. Rowe HC, Hansen BG, Halkier BA, Kliebenstein DJ (2008) Biochemical networks and epistasis shape the Arabidopsis thaliana metabolome. Plant Cell 20:1199–1216. doi:10.1105/tpc.108.058131
126. Saito K, Matsuda F (2010) Metabolomics for functional genomics, systems biology, and biotechnology. Annu Rev Plant Biol 61:463–489. doi:10.1146/annurev.arplant.043008.092035
127. Saito K, Hirai MY, Yonekura-Sakakibara K (2008) Decoding genes with coexpression networks and metabolomics—'majority report by precogs'. Trends Plant Sci 13:36–43. doi:10.1016/j.tplants.2007.10.006
128. Sakurai N, Ara T, Ogata Y, Sano R, Ohno T, Sugiyama K, Hiruta A, Yamazaki K, Yano K, Aoki K, Aharoni A, Hamada K, Yokoyama K, Kawamura S, Otsuka H, Tokimatsu T, Kanehisa M, Suzuki H, Saito K, Shibata D (2011) KaPPA-View4: a metabolic pathway database for representation and analysis of correlation networks of gene co-expression and metabolite coaccumulation and omics data. Nucleic Acids Res 39:D677–D684. doi:10.1093/nar/gkq989
129. Sasaki T, Matsumoto T, Yamamoto K, Sakata K, Baba T, Katayose Y, Wu J, Niimura Y, Cheng Z, Nagamura Y, Antonio BA, Kanamori H, Hosokawa S, Masukawa M, Arikawa K, Chiden Y, Hayashi M, Okamoto M, Ando T, Aoki H, Arita K, Hamada M, Harada C, Hijishita S, Honda M, Ichikawa Y, Idonuma A, Iijima M, Ikeda M, Ikeno M, Ito S, Ito T, Ito Y, Iwabuchi A, Kamiya K, Karasawa W, Katagiri S, Kikuta A, Kobayashi N, Kono I, Machita K, Maehara T, Mizuno H, Mizubayashi T, Mukai Y, Nagasaki H, Nakashima M, Nakama Y, Nakamichi Y, Nakamura M, Namiki N, Negishi M, Ohta I, Ono N, Saji S, Sakai K, Shibata M, Shimokawa T, Shomura A, Song J, Takazaki Y, Terasawa K, Tsuji K, Waki K, Yamagata H, Yamane H, Yoshiki S, Yoshihara R, Yukawa K, Zhong H, Iwama H, Endo T, Ito H, Hahn JH, Kim HI, Eun MY, Yano M, Jiang J, Gojobori T (2002) The genome sequence and structure of rice chromosome 1. Nature 420:312–316. doi:10.1038/nature01184
130. Sato S, Arita M, Soga T, Nishioka T, Tomita M (2008) Time-resolved metabolomics reveals metabolic modulation in rice foliage. BMC Syst Biol 2:51. doi:10.1186/1752-0509-2-51

131. Sawada Y, Akiyama K, Sakata A, Kuwahara A, Otsuki H, Sakurai T, Saito K, Hirai MY (2009) Widely targeted metabolomics based on large-scale MS/MS data for elucidating metabolite accumulation patterns in plants. Plant Cell Physiol 50:37–47. doi: 10.1093/pcp/pcn183
132. Schauer N, Semel Y, Roessner U, Gur A, Balbo I, Carrari F, Pleban T, Perez-Melis A, Bruedigam C, Kopka J, Willmitzer L, Zamir D, Fernie AR (2006) Comprehensive metabolic profiling and phenotyping of interspecific introgression lines for tomato improvement. Nat Biotechnol 24:447–454. doi:10.1038/nbt1192
133. Schauer N, Semel Y, Balbo I, Steinfath M, Repsilber D, Selbig J, Pleban T, Zamir D, Fernie AR (2008) Mode of inheritance of primary metabolic traits in tomato. Plant Cell 20:509–523. doi: 10.1105/tpc.107.056523
134. Schilmiller A, Shi F, Kim J, Charbonneau AL, Holmes D, Daniel Jones A, Last RL (2010) Mass spectrometry screening reveals widespread diversity in trichome specialized metabolites of tomato chromosomal substitution lines. Plant J 62:391–403. doi: 10.1111/j.1365-313X.2010.04154.x
135. Sekiyama Y, Kikuchi J (2007) Towards dynamic metabolic network measurements by multi-dimensional NMR-based fluxomics. Phytochemistry 68:2320–2329. doi:10.1016/j.phytochem.2007.04.011
136. Sekiyama Y, Chikayama E, Kikuchi J (2010) Profiling polar and semipolar plant metabolites throughout extraction processes using a combined solution-state and high-resolution magic angle spinning NMR approach. Anal Chem 82:1643–1652. doi:10.1021/ac9019076
137. Shellie RA, Welthagen W, Zrostlikova J, Spranger J, Ristow M, Fiehn O, Zimmermann R (2005) Statistical methods for comparing comprehensive two-dimensional gas chromatography-time-offlight mass spectrometry results: metabolomic analysis of mouse tissue extracts. J Chromatogr A 1086:83–90. doi:10.1016/j.c hroma.2005.05.088
138. Shinbo Y, Nakamura Y, Altaf-Ul-Amin M, Asahi H, Kurokawa K, Arita M, Saito K, Ohta D, Shibata D, Kanaya S (2006) KNApSAcK: a comprehensive species-metabolite relationship database. In: Saito K, Dixon R, Willmitzer L (eds) Biotechnology in agriculture and forestry. Springer, Heidelberg, pp 165–184
139. Simon C, Langlois-Meurinne M, Bellvert F, Garmier M, Didierlaurent L, Massoud K, Chaouch S, Marie A, Bodo B, Kauffmann S, Noctor G, Saindrenan P (2010) The differential spatial distribution of secondary metabolites in Arabidopsis leaves reacting hypersensitively to Pseudomonas syringae pv tomato is dependent on the oxidative burst. J Exp Bot 61:3355–3370. doi: 10.1093/jxb/erq157
140. Smith CA, O'Maille G, Want EJ, Qin C, Trauger SA, Brandon TR, Custodio DE, Abagyan R, Siuzdak G (2005) METLIN: a metabolite mass spectral database. Ther Drug Monit 27:747–751
141. Smith CA, Want EJ, O'Maille G, Abagyan R, Siuzdak G (2006) XCMS: processing mass spectrometry data for metabolite profiling using nonlinear peak alignment, matching, and identification. Anal Chem 78:779–787. doi:10.1021/ac051437y
142. Soga T, Ueno Y, Naraoka H, Matsuda K, Tomita M, Nishioka T (2002) Pressure-assisted capillary electrophoresis electrospray ionization mass spectrometry for analysis of multivalent anions. Anal Chem 74:6224–6229. doi:10.1021/ac0202684
143. Soga T, Baran R, Suematsu M, Ueno Y, Ikeda S, Sakurakawa T, Kakazu Y, Ishikawa T, Robert M, Nishioka T, Tomita M (2006) Differential metabolomics reveals ophthalmic acid as an oxidative stress biomarker indicating hepatic glutathione consumption. J Biol Chem 281:16768–16776. doi:10.1074/jbc.M601876200
144. Soga T, Igarashi K, Ito C, Mizobuchi K, Zimmermann HP, Tomita M (2009) Metabolomic profiling of anionic metabolites by capillary electrophoresis mass spectrometry. Anal Chem 81:6165–6174. doi:10.1021/ac900675k
145. Sreekumar A, Poisson LM, Rajendiran TM, Khan AP, Cao Q, Yu J, Laxman B, Mehra R, Lonigro RJ, Li Y, Nyati MK, Ahsan A, Kalyana-Sundaram S, Han B, Cao X, Byun J, Omenn GS, Ghosh D, Pennathur S, Alexander DC, Berger A, Shuster JR, Wei JT, Varambally S, Beecher C, Chinnaiyan AM (2009) Metabolomic profiles delineate potential role for sarcosine in prostate cancer progression. Nature 457:910–914. doi:10.1038/nature07762

146. Suzuki H, Sasaki R, Ogata Y, Nakamura Y, Sakurai N, Kitajima M, Takayama H, Kanaya S, Aoki K, Shibata D, Saito K (2008)\ Metabolic profiling of flavonoids in Lotus japonicus using liquid chromatography Fourier transform ion cyclotron resonance mass spectrometry. Phytochemistry 69:99–111. doi:10.1016/j.phytochem.2007.06.017
147. Taguchi R, Nishijima M, Shimizu T (2007) Basic analytical systems for lipidomics by mass spectrometry in Japan. Methods Enzymol 432:185–211. doi:10.1016/S0076-6879(07)32008-9
148. Takahashi H, Hotta Y, Hayashi M, Kawai-Yamada M, Komatsu S, Uchiyama H (2005) High throughput metabolome and proteome analysis of transgenic rice plants (Oryza sativa L.). Plant Biotechnol 22:47–50. doi:10.5511/plantbiotechnology.22.47
149. Tarpley L, Duran AL, Kebrom TH, Sumner LW (2005) Biomarker metabolites capturing the metabolite variance present in a rice plant developmental period. BMC Plant Biol 5:8. doi:10.1186/1471-2229-5-8
150. Thimm O, Blasing O, Gibon Y, Nagel A, Meyer S, Kruger P, Selbig J, Muller LA, Rhee SY, Stitt M (2004) MAPMAN: a user-driven tool to display genomics data sets onto diagrams of metabolic pathways and other biological processes. Plant J 37:914–939. doi:10.1111/j.1365-313X.2004.02016.x
151. Tian C, Chikayama E, Tsuboi Y, Kuromori T, Shinozaki K, Kikuchi J, Hirayama T (2007) Top-down phenomics of Arabidopsis thaliana: metabolic profiling by one- and two-dimensional nuclear magnetic resonance spectroscopy and transcriptome analysis of albino mutants. J Biol Chem 282:18532–18541. doi: 10.1074/jbc.M700549200
152. Tohge T, Fernie AR (2009) Web-based resources for mass-spectrometry- based metabolomics: a user's guide. Phytochemistry 70:450–456. doi:10.1016/j.phytochem.2009.02.004
153. Tohge T, Fernie AR (2010) Combining genetic diversity, informatics and metabolomics to facilitate annotation of plant gene function. Nat Protoc 5:1210–1227. doi:10.1038/nprot.2010.82
154. Tohge T, Nishiyama Y, Hirai MY, Yano M, Nakajima J, Awazuhara M, Inoue E, Takahashi H, Goodenowe DB, Kitayama M, Noji M, Yamazaki M, Saito K (2005) Functional genomics by integrated analysis of metabolome and transcriptome of Arabidopsis plants over-expressing an MYB transcription factor. Plant J 42:218–235. doi:10.1111/j.1365-313X.2005.02371.x
155. Tokimatsu T, Sakurai N, Suzuki H, Ohta H, Nishitani K, Koyama T, Umezawa T, Misawa N, Saito K, Shibata D (2005) KaPPA-view: a web-based analysis tool for integration of transcript and metabolite data on plant metabolic pathway maps. Plant Physiol 138:1289–1300. doi:10.1104/pp.105.060525
156. Urano K, Maruyama K, Ogata Y, Morishita Y, Takeda M, Sakurai N, Suzuki H, Saito K, Shibata D, Kobayashi M, Yamaguchi- Shinozaki K, Shinozaki K (2009) Characterization of the ABAregulated global responses to dehydration in Arabidopsis by metabolomics. Plant J 57:1065–1078. doi:10.1111/j.1365-313X.2008.03748.x
157. Van QN, Veenstra TD (2009) How close is the bench to the bedside? Metabolic profiling in cancer research. Genome Med 1:5. doi:10.1186/gm5
158. Vorst O, de Vos CH, Lommen A, Staps RV, Visser RG, Bino RJ, Hall RD (2005) A non-directed approach to the differential analysis of multiple LC–MS-derived metabolic profiles. Metabolomics 1:169–180. doi:10.1007/s11306-005-4432-7
159. Wang Y, Holmes E, Nicholson JK, Cloarec O, Chollet J, Tanner M, Singer BH, Utzinger J (2004) Metabonomic investigations in mice infected with Schistosoma mansoni: an approach for biomarker identification. Proc Natl Acad Sci USA 101:12676–12681. doi: 10.1073/pnas.0404878101
160. Ward JL, Baker JM, Beale MH (2007) Recent applications of NMR
161. spectroscopy in plant metabolomics. FEBS J 274:1126–1131.
162. doi:10.1111/j.1742-4658.2007.05675.x
163. Ward JL, Baker JM, Miller SJ, Deborde C, Maucourt M, Biais B, Rolin D, Moing A, Moco S, Vervoort J, Lommen A, Schafer H, Humpfer E, Beale MH (2010a) An inter-laboratory comparison demonstrates that [1H]-NMR metabolite fingerprinting is a robust technique

for collaborative plant metabolomic data collection. Metabolomics 6:263–273. doi:10.1007/s11306-010-0200-4
164. Ward JL, Forcat S, Beckmann M, Bennett M, Miller SJ, Baker JM, Hawkins ND, Vermeer CP, Lu C, Lin W, Truman WM, Beale MH, Draper J, Mansfield JW, Grant M (2010b) The metabolic transition during disease following infection of Arabidopsis thaliana by Pseudomonas syringae pv. tomato. Plant J. doi: 10.1111/j.1365-313X.2010.04254.x
165. Ward JL, Baker JM, Llewellyn AM, Hawkins ND, Beale MH (2011) Metabolomic analysis of Arabidopsis reveals hemiterpenoid glycosides as products of a nitrate ion-regulated, carbon flux overflow. Proc Natl Acad Sci USA. doi:10.1073/pnas.1018875108
166. Watanabe M, Kusano M, Oikawa A, Fukushima A, Noji M, Saito K (2008) Physiological roles of the beta-substituted alanine synthase gene family in Arabidopsis. Plant Physiol
167. 146:310–320. doi:10.1104/pp.107.106831
168. Wentzell AM, Rowe HC, Hansen BG, Ticconi C, Halkier BA, Kliebenstein DJ (2007) Linking metabolic QTLs with network and cis-eQTLs controlling biosynthetic pathways. PLoS Genet 3:1687–1701. doi:10.1371/journal.pgen.0030162
169. Wheeler DL, Barrett T, Benson DA, Bryant SH, Canese K, Chetvernin V, Church DM, Dicuccio M, Edgar R, Federhen S, Feolo M, Geer LY, Helmberg W, Kapustin Y, Khovayko O, Landsman D, Lipman DJ, Madden TL, Maglott DR, Miller V, Ostell J, Pruitt KD, Schuler GD, Shumway M, Sequeira E, Sherry ST, Sirotkin K, Souvorov A, Starchenko G, Tatusov RL, Tatusova TA, Wagner L, Yaschenko E (2008) Database resources of the National Center for Biotechnology Information. Nucleic Acids Res 36:D13–D21. doi:10.1093/nar/gkm1000
170. Widodo, Patterson JH, Newbigin E, Tester M, Bacic A, Roessner U (2009) Metabolic responses to salt stress of barley (Hordeum vulgare L.) cultivars, Sahara and Clipper, which differ in salinity tolerance. J Exp Bot 60:4089–4103. doi:10.1093/jxb/erp243
171. Wishart DS, Knox C, Guo AC, Eisner R, Young N, Gautam B, Hau DD, Psychogios N, Dong E, Bouatra S, Mandal R, Sinelnikov I, Xia J, Jia L, Cruz JA, Lim E, Sobsey CA, Shrivastava S, Huang P, Liu P, Fang L, Peng J, Fradette R, Cheng D, Tzur D, Clements M, Lewis A, De Souza A, Zuniga A, Dawe M, Xiong Y, Clive D, Greiner R, Nazyrova A, Shaykhutdinov R, Li L, Vogel HJ, Forsythe I (2009) HMDB: a knowledgebase for the human metabolome. Nucleic Acids Res 37:D603–D610. doi:10.1093/nar/gkn810
172. Yasugi E, Watanabe K (2002) LIPIDBANK for Web, the newly developed lipid database. Tanpakushitsu Kakusan Koso 47: 837–841
173. Ye X, Al-Babili S, Kloti A, Zhang J, Lucca P, Beyer P, Potrykus I (2000) Engineering the provitamin A (b-carotene) biosynthetic pathway into (carotenoid-free) rice endosperm. Science 287:303–305. doi:10.1126/science.287.5451.303
174. Yonekura-Sakakibara K, Saito K (2009) Functional genomics for plant natural product biosynthesis. Nat Prod Rep 26:1466–1487. doi:10.1039/B817077K
175. Yonekura-Sakakibara K, Tohge T, Niida R, Saito K (2007) Identification of a flavonol 7-O-rhamnosyltransferase gene determining flavonoid pattern in Arabidopsis by transcriptome coexpression analysis and reverse genetics. J Biol Chem 282:14932–14941. doi:10.1074/jbc.M611498200
176. Yonekura-Sakakibara K, Tohge T, Matsuda F, Nakabayashi R, Takayama H, Niida R, Watanabe-Takahashi A, Inoue E, Saito K (2008) Comprehensive flavonol profiling and transcriptome coexpression analysis leading to decoding gene-metabolite correlations in Arabidopsis. Plant Cell 20:2160–2176. doi: 10.1105/tpc.108.058040
177. Yu J, Hu S, Wang J, Wong GK, Li S, Liu B, Deng Y, Dai L, Zhou Y, Zhang X, Cao M, Liu J, Sun J, Tang J, Chen Y, Huang X, Lin W, Ye C, Tong W, Cong L, Geng J, Han Y, Li L, Li W, Hu G, Li J, Liu Z, Qi Q, Li T, Wang X, Lu H, Wu T, Zhu M, Ni P, Han H, Dong W, Ren X, Feng X, Cui P, Li X, Wang H, Xu X, Zhai W, Xu Z, Zhang J, He S, Xu J, Zhang K, Zheng X, Dong J, Zeng W, Tao L, Ye J, Tan J, Chen X, He J, Liu D, Tian W, Tian C, Xia H, Bao Q, Li G, Gao H, Cao T, Zhao W, Li P, Chen W, Zhang Y, Hu J, Liu S, Yang J, Zhang G, Xiong Y, Li Z, Mao L, Zhou C, Zhu Z, Chen R, Hao B, Zheng W, Chen S, Guo W, Tao M, Zhu L, Yuan L, Yang H (2002) A draft sequence of the rice genome (Oryza sativa L. ssp. indica). Science 296:79–92. doi: 10.1126/science.1068037

# CHAPTER 11

# Biotechnology can Improve a Traditional Product as Table Olives

**Maria Tufariello, Giovanni Mita and Gianluca Bleve***

## ABSTRACT

Table olives are fermented vegetables very popular in the world and especially in the Mediterranean countries. Five main styles (Spanish or Sevillian, Castelvetrano, Siciliano, Californian, and Greek) are diffused to produce commercial products, beside several traditional styles. Although the main preparation methods of table olives are known for a long time, they are not yet optimized systems, and each of them is characterized by advantages and disadvantages. The use of NaOH for green olive debittering is responsible for the elimination of many aroma compounds and nutritionally important molecules. High volumes of heavily contaminated wastewaters are produced during olive processing. Spontaneous fermentation processes used to ferment black or green olives are difficult either to monitor or control. Microbial starters, selected for specific bio/technological and safety traits, can be useful to (i) improve the table olives organoleptic characteristics, (ii) control the fermentation process and significantly reduce the time to obtain a final product, (iii) monitor the correct evolution of the process, (iv) ensure the maintenance and/or improvement of nutritional and healthy features of the product, (v) protect table olives from undesired spoilage and pathogenic microorganisms, (vi) produce table olives as a carrier of microorganisms with probiotics characters, and (vii) enhance product stability and shelf life.

**Keywords**: table olives, starters, organoleptic traits, nutritional characteristics, probiotics

# 1. INTRODUCTION

Table olives are one of the most important and popular fermented vegetables in Western world and in particular in Southern European countries. Table olives world production was estimated to be 2,742,500 tons in 2015–2016 season (Figure 1).

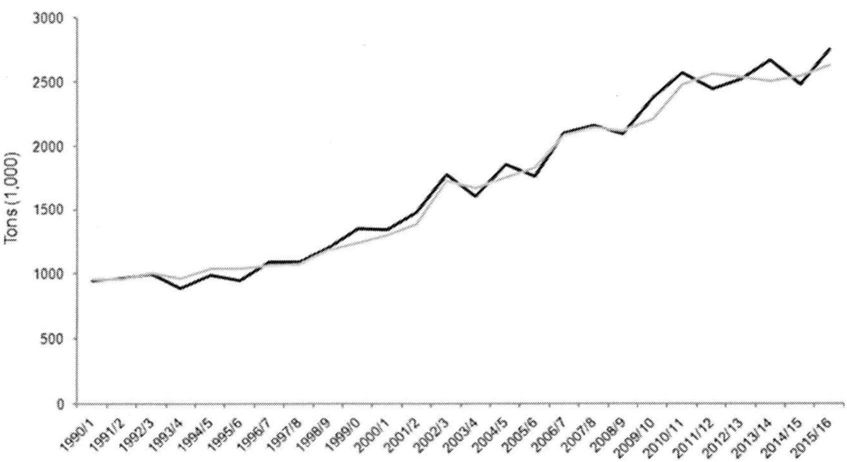


**Figure 1.** Table olives world production ( ) and consumptio (——— ). Adapted from data reported in [1].

The 29% of this production (796,000 tons) is located in the European Union (EU). Spain has a leading position in table olive production with 514,000 tons, followed by Greece (210,000 tons), Italy (50,000 tons), and Portugal (17,500 tons) [1]. Among the countries of the Mediterranean Basin, Egypt, Turkey, Algeria, and Morocco are the main producers and consumers (Figure 2).

According to International Olive Council Standard, the term "table olive" means the product prepared from the sound fruits of varieties of the cultivated olive trees that are chosen for their production of olives whose volume, shape, flesh-to-stone ratio, fine flesh, taste, firmness, and ease of detachment from the stone make them particularly suitable for processing; treated to remove its bitterness and preserved by natural fermentation, or by heat treatment with or without the addition of preservatives; packed with or without covering liquid.

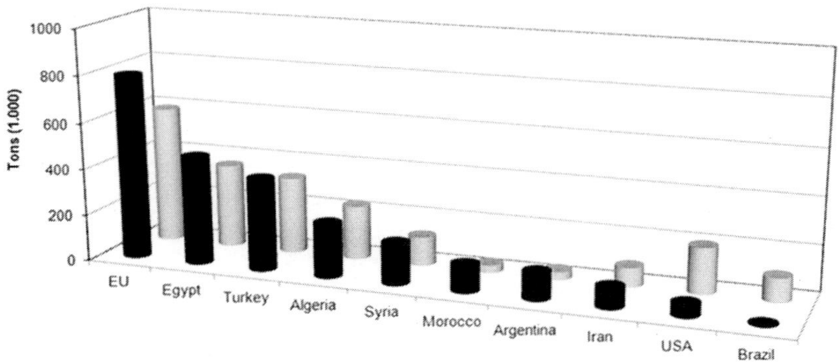


**Figure 2.** Table olives production and imports in main producing and importer countries. Adapted from data reported in [1].

Table olives are classified according to the degree of ripeness of the drupes (green olives, olives turning color, and black olives), trade preparations (treated olives, natural olives, dehydrated and/or shriveled olives, olives darkened by oxidation, and specialties), and styles (whole, pitted, stuffed, salad, and other). They are produced by processing raw olives with the objective of eliminating their natural bitterness, which is mainly due to oleuropein and other phenolics [2]. The main commercial types of table olives are processed according to five styles: Spanish (or Sevillian), Castelvetrano, Siciliano, Californian, and Greek [3], although several other traditional styles also exist for the preparation of treated and natural table olives [4].

The two main commercial table olives preparations, lye-treated olives (Spanish and Castelvetrano styles) and brine-soaked olives (Greek style) are industrially produced by spontaneous fermentation, but, currently, these processes are difficult to be monitored and controlled [5]. The spontaneous process cannot ensure either the correct evolution of the process or the good quality and safety standards of the final product. Controls of the presence of biogenic amines and toxins in table olives commercial preparations need to be increased [6].

In green olives productions, the NaOH is used as chemical debittering system. This treatment is economic, characterized by a simple implementation, and an easy standardization. However, simultaneously to the debittering effect, it causes the elimination of many aroma compounds together with nutritional and health important molecules. The process produces also high volumes of heavily contaminated wastewaters. Besides, the use of NaOH for debittering organic table olives is prohibited in many countries [7].

The employment of starter cultures of *Lactobacillus plantarum* and *L. pentosus* can be used as an alternative to NaOH for debittering. This strategy has the advantage to control the fermentation process and to improve the quality of the final product [8]. Lactic acid fermentation is considered the key step in spontaneous fermentation processes. It promotes (i) debittering of the olives through oleuropein hydrolysis, (ii) lowering of brine pH, which prevents the growth of spoilage and pathogenic microorganisms, and (iii) the enhancement of a correct flavor and texture profile in the final product [9, 10].

It has also been demonstrated that yeasts, producing desirable metabolites and volatile compounds, are able to improve the organoleptic properties. Yeasts can also enhance the growth of lactic acid bacteria (LAB) and degrade phenolic compounds. A role of yeasts as starters has been recently proposed for production of table olive [11–15].

There is an increasing interest in lowering NaCl concentration (now 8–10%) and in shortening the fermentation time (8–12 months) in order to obtain a healthier product suitable to reach the market very soon.

Recently, Bleve et al. described a novel method based on the sequential use of autochthonous yeast and bacterial strains to shorten the time of fermentation, to standardize the process, and to improve organoleptic and nutritional properties of olives [16].

The future challenges will be to investigate some strains for their probiotic characteristics, in order to produce functional olives. Indeed, several studies demonstrated that the use of LAB as starter for table olive production can produce beneficial effects on human health [17, 18]. Also yeast strains have been evaluated for their probiotic properties [13, 15]. The nutritional and health-related compounds associated to fermented table olives (or derivatives) could be assessed by *in vitro* and *in vivo* analyses. The results of these assays can produce precise information on the importance of these compounds for the prevention and/or treatment of several human and animal diseases (i.e., gastrointestinal, cardiovascular, neurodegenerative diseases, and tumors).

Table olives are considered by many food scientists as the "food of the future" owing to the healthy bioactive compounds they contain. In fact, table olives, together with olive oil, represent an important food of the Mediterranean diet and are perceived to have positive nutritional and therapeutic effects. Monounsaturated fatty acids, as found in olives are known to be healthier than polyunsaturated and saturated fats. In addition, epidemiological studies indicate that olive biophenols have a role in lowering incidence of several chronic and heart diseases [19, 20].

## 2. PRODUCTION METHODS

Table olives, directly harvested from trees, need to be processed in order to reduce or eliminate their bitter taste and to obtain a product ready to be consumed. Different commercial preparations of table olives are produced using procedures inherited and opportunely modified from traditional methods. As previously extensively described by Boskou et al. [21], the main methods including fermentation steps to obtain the final product are water-cured olives produced by soaking olives in water over a week or more and then placing them in brine where a fermentation process can occur; Greek-style or "natural" olives and Sicilian-style green olives spontaneously fermented in brines; lye-treated olives (Spanish or Sevillian style, Castelvetrano method) produced by a first treatment with alkali (NaOH) and, after washing olives with water to remove NaOH, by a second step in brine to obtain a partial or complete fermentation of the drupes.

Other methods for black olives not involving fermentation are known as Californian and Spanish styles. The drupes are debittered by lye and soaking in brine. They are also aerated insufflating air to oxidize the pigments and immersed in ferrous gluconate or ferrous lactate solution in order to stabilize a uniform black color. Table olives can also be produced by traditional methods diffused in Mediterranean Basin using lime (CaO) and olive wood ash, or they can be dried and debittered without chemicals by using salt or heat treatment. Some cultivars of olives resulted naturally debittered also by parasite fungi directly on the tree without necessity of further treatment. They can also undergo a natural sweetening during ripening on the tree, although genetic and biochemical mechanisms involved in this last phenomenon are until now unknown.

## 3. BIOTECHNOLOGICAL APPROACHES TO PRODUCE TABLE OLIVES

The fermentation process, generally performed by indigenous microorganisms, is one of the best and oldest procedures of treating food products to transform and preserve them. However, as already demonstrated in several food products (wine, beer, bread, yogurt, cheese, sake, chocolate, etc.), spontaneous fermentations are uncontrolled and not predictable. These spontaneous processes are inefficient since they do not ensure the expected quality and safety characteristics of the final product, the sensorial and structure features, the limitation, or absence of growth of harmful or undesired spoilage organisms [22].

In order to obtain a more controlled process and to improve the quality and safety levels of table olives, the selection and use of starter cultures is diffusing. In fact, several studies demonstrated the usefulness and the benefits of starters in table olives production [5, 9, 15, 23].

## 3.1. Starter selection

The microbiota associated to olives can be different among the different cultivars. Microorganisms detected in table olives and brines belong to members of bacteria (lactic acid bacteria, Enterobacteriaceae, *Pseudomonas, Staphylococcus, Clostridium*, etc.), yeasts, and moulds. Enterobacteriaceae, *Clostridium*, and *Pseudomonas* are generally associated to raw olives and to the beginning of fermentation. They are completely eliminated at the end of the process, especially due to the low pH [24–26] (Figure 3).

The presence of hazardous pathogens such as *C. botulinum* has to be adduced to incorrect processing, heat treatment, packaging, and transportation [27].

The most studied group of bacteria is lactic acid bacteria (homo and hetero fermentative), since they are responsible for the sugar conversion to organic acids and in particular to lactic acid. The different table olives production methods can influence microbial population present in raw olives and their evolution during fermentation (Figure 3). *Lactobacillus coryniformis, L. plantarum, L. pentosus*, and *Leuconostoc mesenteroides* have been detected and isolated in Spanish style green olives across the process. Also *Enterococcus* spp., *Pseudomonas* spp., and *Staphylococcus* spp. are associated to olives produced by this method. Bacterial biodiversity associated to natural green and natural cracked green olives is richer than that present in Spanish style olives, treated with NaOH [28] (Figure 3). In black olives, bacteria belonging to Enterobacteriaceae, *Kocuria, Swaminathania, Acetobacter*, and *Pseudomonas* were detected only at the initial stage of fermentation, except for *Swaminathania* that has been found, in some cases, also at the end of fermentation. Bleve et al. [29, 30] reported the presence of LAB (*Lactobacillus* sp., *L. plantarum, L. pentosus, Leuconostoc mesenteroides*) associated to the final stage of fermentation (120–180 days) in Leccino and Kalamàta cultivars (Figure 3).

Yeasts found in olives belong to the genera *Candida, Debaryomyces, Hanseniaspora, Issatchenkia, Kluyveromyces, Pichia, Rhodotorula,*

*Saccharomyces, Torulaspora, Wickerhamomyces, Zygosaccharomyces, Zygotorulaspora*, with some differences between green and black olives (Figure 4). Yeasts are detectable throughout the fermentation process in all table olive cultivars.

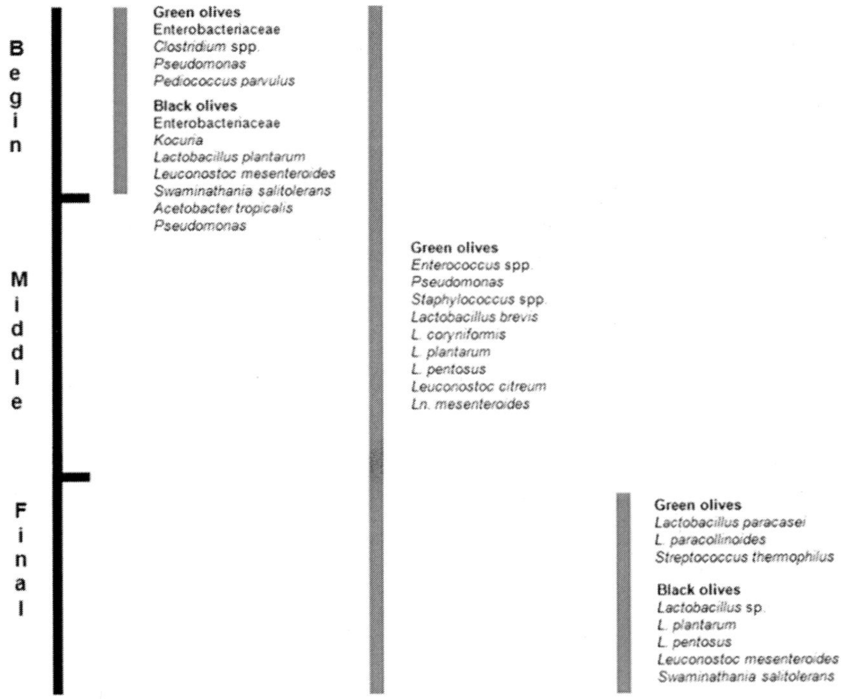


**Figure 3.** Main genera and species of bacteria associated to different production stages of green and black table olives. Adapted from Heperkan et al. [32]; Bleve et al. [29], [30].

Also the mould genera *Aureobasidium* and *Geotrichum* have been isolated from green and black naturally fermented olives, whereas isolates belonging to the genus *Penicillium* and *Aspergillus* were isolated by naturally fermented black olives [21, 31, 32].

The introduction of LAB and yeasts starter cultures in table olives production can also be motivated by the difficulty to monitor and control spontaneous fermentation in the industrially production of black as well as several cultivars of green olives [33–35]. Starter cultures are preparations of microorganisms, live, or resting, generally present in high cell number, which can be added to enhance, accelerate, and improve a fermentation process by their metabolic activities.

LAB have been considered very important since they are able to debitter olives, low brine pH, limit the spoilage and the presence of pathogens, and develop a correct flavor and texture in the final product. Several studies proposed the use of *Lactobacillus plantarum* and/or of *L. pentosus* as starter cultures among the possible available technological approaches [9, 10, 15, 33, 36–38].

The use of yeasts as starters cultures has been recently proposed for production of table olive [11, 13, 14, 35, 39], since they can improve the organoleptic properties [11, 40], enhance the growth of LAB [12, 41], and biodegrade phenolic compounds [42].

Moreover, the possibility to use simultaneous or sequential inocula of yeasts and LAB in green and black olives has been proposed [12, 15, 37, 41, 43]. The presence of yeasts together with LAB can produce a significant improvement of the sensorial quality of olives. They can also favor LAB growth rate, help in Enterobacteriaceae reduction, sensitively shorten the time needed to obtain the final product.

Moulds can be responsible of undesirable effects on table olives quality. They can alter olive taste and appearance and can be responsible of mycotoxins production [32]. These microorganisms need to be deeply studied in order to evaluate their possible positive role in table olive processing.

LAB and yeast strains to be used as starter cultures can be selected among microorganisms associated to a spontaneous fermentation. In a first step, they can be selected on the basis of their characteristics and abilities: (i) to lower pH (by homo or hetero-fermentative metabolism for LAB, by fermentative metabolism for yeasts); (ii) to survive and to grow in the presence of different constraints (poor nutrient substrate like olives in brine, low pH, high salt level, presence of phenols, wide range of temperatures), (iii) to produce lactic acid and other organic acids; (iv) to metabolize phenols and, in particular, to degrade oleuropein, which is the main compound responsible for the bitter taste in olives; (v) to develop desired flavors (volatile compounds); (vi) to produce no biogenic amines, which represent an emerging concern in table olives, wine and other fermented products [44–46]; (vii) to possess esterase and lipase activities that have a role in improving the aromatic profile of fermented olives (by increasing their free fatty acid content); (viii) to have no proteolytic and pectolytic activities, which could have a negative impact on olive quality since they are related to olive softening; (ix) to have functional (probiotics and health-promoting) properties. Several laboratory tests have been developed to select yeasts and LAB for all of these features. In a laboratory-scale, the most promising isolates can be tested for their ability to dominate the indigenous microbiota by predominant growth or by production of antagonistic substances during table olives fermentation.

**Green olives: cracked and directly brined olives, Spanish style green**

| | |
|---|---|
| *Candida apicola* | *Issatchenkia occidentalis* |
| *C. boidinii* | *K. lactis* |
| *C. diddensiae* | *K. marxianus* |
| *C. krusei* | *P. anomala* |
| *C. oleophila* | *P. guillermondii* |
| *C. parapsilosis* | *P. kluyveri* |
| *C. quercitrusa* | *P. membranifaciens* |
| *C. sorbosa* | *R. glutinis* |
| *C. tropicalis* | *R. minuta* |
| *C. rugosa* | *Rhodotorula mucilaginosa* |
| *Citeromyces matritensis* | *Saccharomyces cerevisiae* |
| *D. hansenii* | *T. delbrueckii* |
| | *Zygotorulaspora mrakii* |

**Black olives**

| | |
|---|---|
| *Candida* sp. | *Pichia anomala* |
| *C. boidinii* | *P. membranifaciens* |
| *C. olivae* | *Pichia* sp. |
| *C. saitoana* | *R. glutinis* |
| *C. tartarivorans* | *S. cerevisiae* |
| *Cryptococcus laurentis* | *Torulaspora delbrueckii* |
| *Debarymyces carsonii* | *Wickerhamomyces anomalus* |
| *Debaryomyces etchelsii* | *Zygosaccharomyces mrakii* |
| *Debaryomyces hansenii* | *Zygosaccharomyces* sp. |
| *Debaryomyces* sp. | *W. saturnus var. mrakii* |
| *Hanseniaspora guillermondii* | |

**Figure 4.** Main species of yeasts associated to green and black table olives. Adapted from Heperkan et al. [32]; Bleve et al. [29], [30].

The selected LAB and yeast isolates can be then tested in a pilot-scale fermentation (200 kg) in order to mimic the industrial conditions of fermentation. The best performing isolates can be proposed for industrial-scale fermentation in tanks of 3–8 tons.

## 3.2. Influence of starter cultures on table olives chemical and aromatic profile

The distribution and structure of the chemical constituents of olive fruit is complex and depend on variety, cultivation practices, geographical origin, and the level of maturation. Olive fruit's average composition is water (50%), protein (1.6%), fat (22%), carbohydrate (19.1%), cellulose (5.8%), inorganic substances (1.5%) and phenolic compounds (1–3%).

Both in green and black olive fermentations, lactic, citric, tartaric, and acetic acids were found to be the major metabolic products in drupes and in the brines [9, 47, 48], responsible for a decrease in pH value (about 4.0), satisfactory for naturally black olive fermentation [29, 30].

Although in the literature there are many data about aroma compounds in olive oil, very little is known about the quali-quantitative composition of volatile compounds in table olives. Among table olives, more attention has been placed on the characterization of the volatile fraction of the fermented black olives. Little is known about volatile fraction of green olives, probably because their volatile profiles are less rich, due to the NaOH treatment; the latter affects many precursors of the volatile compounds.

The formation of flavor compounds in table olives is a dynamic process mainly occurring during fermentation carried out by LAB and yeasts, along with a variety of contaminating microorganisms, which produce a variety of volatile compounds [49]. Volatile and semivolatile organic compounds are responsible for the olive complex flavor that in turn can influence the consumer's preference. The "green odor" of unripe olives was associated to the presence of C5 and C6 volatile compounds (alcohols and aldehydes) originating from the activity of lypoxygenase metabolic pathway [50]. Hexanol and 2-hexenal are the major contributors to the characteristic green odor of olives and of many fruit and vegetable fermented foods. In spontaneous fermentation of black olives, the main product is ethanol that derives from the metabolic activity of different yeasts and hetero-fermentative LAB and is very important for the organoleptic properties of the final product [51].

C6 alcohols such as 1-hexanol and *cis*-3-hexen-1-ol, characterized by a "vegetal" and "herbaceous" aromas, seem to be linked to the different yeast strain used [52]. As already observed in wines, the relevant presence of the ethyl-acetate ester at the end of fermentation adds complexity to the aroma of the final product [53]. The high level of isoamyl alcohols indicates the role of yeasts in driving the process. In particular, 2 + 3 methyl-1-butanol (isoamyl alcohol, fruity-winey notes), hexanol (fruity-green notes) and *cis*-3-hexen-1-ol (green notes) are very important both in olives and brines. Other higher alcohols (1-propanol and 2-methyl-1 propanol) derive from the reduction process of aldehydes, but can also be linked to the microbial deamination process of amino acids [54]. Hexanal, (*Z*)-hex-3-enol, hexanol, (*Z*)-hex-3-enol acetate and hexyl acetate, detectable at various concentrations were reported to be related to the lipoxygenase activity [29, 30, 49].

Fatty acids, formed enzymatically during fermentations constitute an important group of aroma compounds that can contribute to the aroma complexity of table olives [29, 30]. Terpenes production is closely linked to cultivars, geographical area, climatic conditions and proliferation of specific pests and microorganisms characteristic of a given production area [55]. The presence of styrene can increase during fermentation [15]. This compound could be linked to an environmental contaminants and/or produced by L-phenylalanine deamination and decarboxylation of *trans*-cinammic acid [60] or by the dehydration of 2-phenylethanol.

The use of selected starter cultures has been proposed for Spanish- and Greek-type, green and black, olives to improve fermentation performance. Starters can accelerate and control the process, reduce undesired off-flavors and enhance quality of the final product by the development of typical and peculiar sensorial and taste characteristics.

In the evolution of volatile compounds during spontaneous fermentation of different black olive cultivars (Leccino, Cellina di Nardò, Conservolea and Kalamàta), Tufariello et al. [15] identified three main temporary steps characterized by the presence of chemical descriptors: aldehydes at the first stage (30 days), higher alcohols and styrene in the middle (90 days), and ethyl esters and fatty acids at the end of fermentation (third fermentation stage, 180 days). These descriptors could help in monitoring the fermentation process of other black olive cultivars as well as of naturally fermented green olives.

In starter-driven fermentations carried out by sequential inoculum of yeast LAB strain, three main stages have been described. The first stage (30 days) is characterized by high aldehydes content, compounds responsible of herbaceous flavors in fruits and vegetables. The second stage (60 days) is characterized by the presence of higher alcohols, styrene [56] and terpenes, compounds correlated with the metabolic activities of inoculated yeast starter strains. The third final fermentation stage (90 days), mainly characterized by the presence of acetate esters (isoamyl acetate, ethyl acetate), esters (ethyl hexanoate and ethyl octanoate), and acids, probably due to the different pathways undertaken by LAB enzymes.

The use of starter microorganisms significantly reduced the time of fermentation process from 180 to 90 days. The first stage of fermentation shifted from 90 (spontaneous fermentation) to 60 days (starter-driven fermentation) and the second step shifted from 180 (spontaneous fermentation) to 90 days (starter-driven fermentation). The use of sequential inoculation strategy of selected yeast and LAB starters produced a volatile profile richer in compounds that can be associated to attributes such as fruity, winey-sweet and herbaceous. A significant reduction of volatile phenols and hydrocarbons was observed.

For Moresca and Kalamàta table olives inoculated with selected starter cultures of *L. plantarum*, a shift from herbal notes to fruity, sweet, and floral profile has been reported. In inoculated samples, a significant increase of higher alcohols (isoamylalcohols, 1-propanol, 2-methyl-1-propanol, phenylethylalcohol), esters (ethyl butanoate), and acetate esters (isoamyl acetate, ethyl acetate) was also observed [57].

Grounta et al. [58] demonstrated that the use of *L. pentosus* B281 produced table olives with good physical and chemical features and sensory properties highly appreciated by expert panelists. The coinoculation of *L. pentosus* B281 and *P. membranifaciens* M3A in brines of Conservolea olives developed a proper fermentation process, producing a final product with good sensory attributes and a milder acid gustatory sensation. This product could be suitable for consumers who do not appreciate the acid taste of natural black fermented olives and prefer milder tastes. No off-odors associated to abnormal fermentation (i.e., butyric, putrid fermentation, or zapateria spoilage) were detected by the panelists [58].

In green olives, most studies have been carried out for the selection of starter cultures able to control fermentation in brines after lye treatment. This process generally needs 3–7 months to be completed. It is mainly driven by LAB belonging to lactobacilli, *Leuconostoc* and *Pediococcus* spp. There are few studies on the volatile fraction and its evolution during the fermentation process. Panagou and Tassou [59] studied the evolution of the volatiles in green table olives (Conservolea cv.) treated with NaOH and then inoculated with *L. plantarum* or *L. pentosus*. The use of starters produced an acceleration of fermentation process. The final products contained increased concentrations of lactic and acetic acid as well as volatile molecules such as ethanol, methanol, acetate esters and isobutyric, isovaleric, and propionic acids. The sensorial characteristics ascribable to typical lactic fermentation were obtained in the final product also inoculating the strains *L. pentosus* B281 and *L. plantarum* B282, as single or combined cultures to ferment Spanish-style produced Halkidiki green olives [8].

In order to improve the fermentation of directly brined green olives, the application of the "pied de cuve" technology has been proposed [38]. Partially fermented brines deriving from a previous spontaneous fermentation were used to produce green olives with improved aroma and taste complexity. In comparison with brines deriving from previous fermentations performed with the starter *L. pentosus* OM13, undesired off-odors and off-flavors were not detected and a good control of microorganism spoilage was obtained. The use of selected strains of lactobacilli and yeasts accelerated the fermentation process of directly brined Bella di Cerignola green olives [37], a cultivar traditionally debittered in the Spanish style. In inoculated samples, major compounds that significantly increased were ethanol, acetic acid and ethyl acetate. There was also an increase in the level of esters (fruity nuances), alcohols (fruity, floral and sweet notes), and acids except for propanoic acid. A decrease of aldehydes content was also observed [37].

Volatile compounds confer peculiar sensorial characteristics and contribute to the aroma "fingerprint" of single table olives variety. Then, it could be important to promote the use of new descriptors other than those linked to taste (crispness, sourness, bitterness, and astringency), appearance (brightness, intense green color, etc.) [2]. New descriptors should be able to describe floral, fruity, green, winey as well other similar notes (Figure 5).

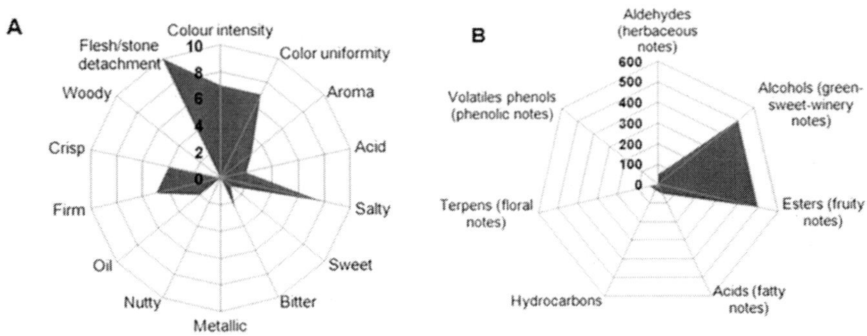


**Figure 5.** Spider plot showing the main organoleptic attribute intensities identified by trained panel (A) and (B) radar plot of all volatiles classes associated to black olives fermented by yeast and LAB starters.

To link chemical data to sensory data, it is necessary to evaluate the perception thresholds of the volatiles, defined as the lowest concentration capable of producing a sensation. The contribution of each volatile compound to odor profile can be quantified by its odor activity value (OAV). OAV is the ratio of the compound concentration to its odor threshold. In the table olives sector, these thresholds are not available, so it is not still possible to establish the role of each volatile compound as odorant in the multiplicity of olives aroma components.

## 3.3. Influence of microbial fermentation on table olives nutritional profile

In the past decades, olive oil and table olives have been attracting interest, mostly due to their beneficial effects on health. Table olives contain several nutritional components that largely depend on the olive variety, the cultivation conditions, the maturation stage of the olive fruit, and the processing method. The consumption of table olives thus allows the dietary introduction of bioactive components, such as triterpenic acids, α-tocopherol, biophenols, and fatty acids. These compounds are known to be responsible for a variety of health benefits. More specifically, olive fruits are remarkably rich in maslinic and oleanolic acids [60]. These triterpenic acids are located in the epicarp of the olive fruit and they constitute the main substances of the surface waxes [61, 62]. Some studies indicate that maslinic and oleanolic acids possess health beneficial activities such as anti-inflammatory, antioxidant [63–65], antimicrobial [66], antiviral [67], cardioprotective [68], antihypertensive [69], antihyperlipidemic [70, 71], antidiabetic [72, 73], and even antitumor [73–77]. It is worthwhile noting that the content of these bioactive compounds in table olives is significantly higher than in olive oil [78].

NaOH treated green and black olives contain low levels of these compounds in comparison with naturally fermented olives [78]. Indeed, the NaOH treatment leads to the solubilization of maslinic and oleanolic acids into the alkaline and washing solutions. The resulting final product contained significantly reduced levels of these compounds.

In Greek-style preparations, the fermentation of black table olives driven by selected starter cultures can preserve the triterpenic acid content. The amount of these molecules was around 1000–2000 mg/kg olive flesh, much higher than the values observed in extra virgin olive oils [Bleve G., unpublished]. These observations confirm that table olives can be considered a dietary natural source of triterpenic acids.

The health benefits of olive oil and table olives are also attributed to their high content in monounsaturated fatty acid (MUFA). Commonly recognized as a high-fat food (about 80–85% of the calories in olives come from fat), olives provide a high content of oleic acid. Linoleic acid and α-linolenic acid are present in small amounts. Owing to the content of MUFA, the consumption of table olives can prevent and reduce the risk of cardiovascular diseases, regulate cholesterol levels, stimulate transcription of LDL-cholesterol receptor mRNA, and reduce breast cancer risks [79–81]. In Spanish-, Californian-, and Greek-style processes, triglylglycerols composition remains unaffected, although fatty acid composition of both green and black olives, shows differences depending on the ripeness degree. The concentration of oleic acid, the most abundant fatty acid in green and black olives, showed differences depending on the stage of maturity the producing methods [82]. When considering the PUFA/SFA ratio, green and directly brined table olives showed a value >0.4. This is a value recommended by the nutritional guidelines [83]. In particular, a significantly high PUFA/SFA ratio was found in directly brined olives [84, 85]. The use of selected starter cultures for black olives fermentation ensured a PUFA/SFA ratio >0.4 in the final product.

The olive fruit is also highly valuable for the presence of α-tocopherol (TC), β-carotene (BC) [80], and biophenols. TC acts as the major radical scavenging antioxidant and efficiently interrupts the propagation of lipid oxidation chain [86]. Several works described a protective action of TC on human health against different pathologies. It contributes to reduce the effects of inflammations and it defends the body against the negative effects of free radicals [87]. When cultivars of black table olives were fermented using selected starter cultures, Vitamin E and carotenes levels were found more constant.

Among biophenols, tyrosol, hydroxytyrosol, luteolin, and oleuropein are the main species found in olives [88–93]. The latter compound is mainly responsible for the bitter taste of unprocessed olives. Other phenolic compounds are verbascoside, 3,4-dihydroxyphenylglycol [94], anthocyanins, flavonoids, and phenolic acids [88]. Mechanisms postulated for chemical and microbial oleuropein degradation and the effects on valuable phenols caused by different table olives processing methods are described in details by Boskou et al. [21].

In Greek-style fermented olives (driven by natural microbiota or by starter cultures), a higher content of total phenols was detected than that observed in lye treated olives [15, 29, 30, 37, 95, 96]. During the process, a complete hydrolysis of oleuropein and its aglycone takes place in olive flesh by yeasts and LAB β-glycosidase and esterase activities. In fact, high levels of hydroxytyrosol and tyrosol, together with verbascoside, caffeic acid, vanillic acid, and hydrocaffeic acid were detected in the final products. Table olives are a very good source of hydroxytyrosol which is known to possess a high antioxidant and free radical scavenging activity [95, 97]. Table olives together with virgin olive oil are the only edible source of hydroxytyrosol; in olive oil, however, the bound forms of this compound prevail.

Changes in the profiles of bioactive compounds caused by metabolic activities of microbial starters can produce variations in the bioaccessibility and/or bioavailability of these metabolites. The role of microorganisms and the effects of their activities on these bioactive compounds need to be further elucidated. Table olives can be used to ensure a "positive" or "optimal" dietary intake of these compounds. They represent a source of phytochemicals useful for the prevention of several diseases and the promotion of human health (Figure 6).

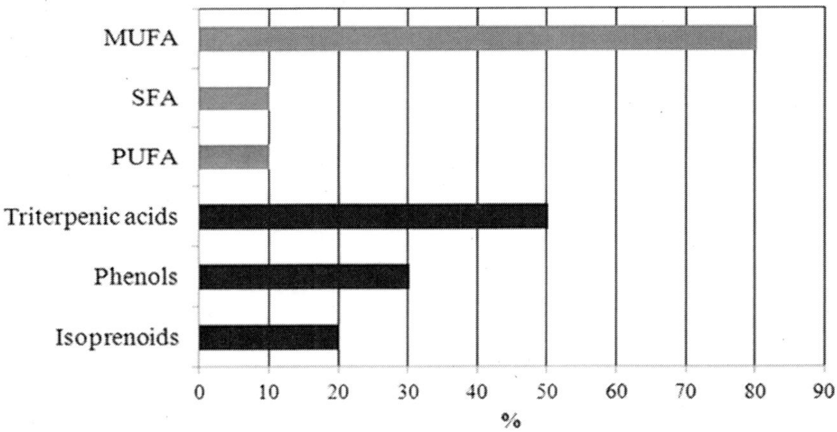


**Figure 6.** Profile of the main nutritional traits (% on drupe fresh weight) associated to black table olives fermented by yeast and LAB starters.

## 3.4. Use of starter cultures as probiotics

Another important character for the selection of potential starter cultures is referred to probiotic traits that beneficially influence intestinal microflora and health [98]. LAB, mainly the strains belonging to genera *Lactobacillus, Bifidobacterium*, and *Streptococcus*, are microorganisms generally considered for probiotics preparations. The term "probiotic" is defined by a United Nations and World Health Organization Expert Panel as "live microorganisms which when administered in adequate amounts confer a health benefit on the host" [99]. Probiotics are live microorganisms that have a beneficial effect on the host by influencing the composition and or metabolic activity of the flora of the gastrointestinal (GI) tract. Selected LAB probiotics strains can have beneficial properties. They can enhance the immune system responses, improve resistance to infection, protect against certain types of cancer, lower serum cholesterol levels, and reduce the incidence of coronary heart disease. They are also involved in the prevention or treatment of peptic ulcer disease, treatment of intractable diarrhea during antibiotic therapy, reduction of allergic inflammation, production of antimicrobial substances, reduction of symptoms of lactose intolerance, and the enhancement of the nutrients bioavailability [100–102].

Probiotic characteristics are strongly required for microorganisms to be used for food production. These attributes can be conferred to a food by microbial preparations different or by the same microorganisms used as starter cultures. In the first case, the interaction between probiotics strains and traditional starter cultures must be considered, since some probiotics strains may have effects on organoleptic properties of the food product or can influence negatively the starter culture bacteria [103]. In the latter case, strains to be proposed as starters need to be selected also for probiotics traits.

Recent studies have focused on the use of table olives as a carrier of LAB probiotic strains as well as on the evaluation of the fermentation performances of these probiotics [39, 101, 104, 105]. *L. paracasei* strain (IMPC2.1) was able to successfully colonize both the olive surface [102] and human gut [104], also driving the fermentation [106]. Moreover, LAB strains directly isolated from fermented olives have been proposed as probiotics starters instead of bacteria from human and animal sources. By this approach, some strain of *L. pentosus, L. plantarum*, and *L. paracasei* ssp. *paracasei* isolated from fermented olives showed desirable *in vitro* probiotic properties as well as good aptitude to be employed as starter cultures [8, 18].

During the past few years, some researchers have identified yeast species with potential probiotic properties, such as *C. boidinii, C. oleophila, D. hansenii*, and *P. membranifaciens* [35, 107, 108].

As proposed by different authors [14, 23, 109], for these microorganisms, probiotics traits are: (i) the resistance and/or survival to gastric pH conditions and to bile salts, (ii) the capacity to adhere to intestinal mucosa, and (iii) the antimicrobial activities against intestinal and food-borne pathogens. In addition, several other health promoting factors can be considered in order to promote yeast starters as probiotics. They are the production of B vitamins and the reduction of the intestinal proinflammatory response as an antagonistic effect of yeasts or probiotic bacteria toward pathogen; the ability to reduce cholesterol serum levels [110]; the ability to biodegrade phytate complexes, responsible for sequestering nutritional divalent minerals; the capability to synthesize natural folates, essential cofactors in the biosynthesis of nucleotides and crucial for cellular replication and growth [111]; the ability to produce a number of bioactive compounds. Anyway, in order to obtain probiotics and healthy olives, it is necessary that yeasts adhere to olive skin and survive during storage/packaging. The copresence of yeasts and LAB in the biofilm associated to epidermis of natural black olives (Greek-style fermented Conservolea) and Spanish-style olives (Gordal and Manzanilla) indicate that the coinoculation of yeasts and LAB as multifunctional starter is a good strategy for carrying probiotics by table olives [35, 58].

## 3.5. Influence of starters in bioremediation of table olive processing wastewaters (topw)

During table olives processing, clean water is used and a large quantity of wastewater is produced depending on cultivar, maturity, and type of treatment from 0.5 l/kg to 6 l/kg. In 2013/2014, 2.7 million tons of table olives were produced in the world. The production of 1.2–14 million tons of wastewaters can be estimated. The volumes of wastewaters depends on the different table olives processing methods: for Spanish style 2–3.5 l/kg of olives; for California green ripe olives 1.5–3.5 l/kg of olives, for California black ripe olives 2–6.5 l/kg of olives, for naturally black olives (Greek style) 1 l/kg of olives [112]. The availability of the water as a resource and the environmental impact deriving from its use are very important matters for many table olives producer countries. The main problems associated to TOPW are their high chemical oxygen demand (COD) up to 35 g/l, biological oxygen demand (BOD) ranging from 0.6 to 38.3 g/l, different pH values (alkaline, up to 9–13, for waters deriving by lye-treatment and acidic, 3.6–4.4, for fermentation brines), the presence of several water-soluble phenols and polyphenols, the high salt levels (56–77 g/l) in Greek-style olives.

The different composition of TOPW (produced by different table olives processing methods) requires the development of diverse approaches for their management and treatment. The contemporary presence of high organic matter content (reducing sugars, organic acids), suitable to be used by microorganisms, and of compounds that affect microbial growth and metabolism (phenols), renders biological approaches for their remediation very difficult.

There are different methods based on the use of aerobic and anaerobic processes of TOPW [113, 114]. There are also strategies that combine chemical and biological processes, using a pretreatment with *Aspergillus* sp. in order to degrade phenols, very toxic, and able to limit the activities of anaerobic digestion [115, 116]. Several studies demonstrated that the use of aerobic and anaerobic biodegradation of wastewaters can significantly reduce organic load expressed by COD (aerobic treatment between 50 and 70% and anaerobic between 81 and 94%) [116–119].

The use of starter microorganisms in table olive processing can represent a useful system to mitigate the presence of chemical pollutants in TOPW. They are phenols (such as oleuropein and derivates, anthocyanins), NaOH in the lye and sodium content in brine. Selected microorganisms, by their metabolic activities can reduce the presence of phenols. They can also efficiently dominate the spontaneous microflora and facilitate reduction of NaCl concentration in brines. Concerning NaOH in lye, a future challenge can be to develop new systems able to reduce or eliminate the use of NaOH in lye. These strategies can allow to perform a more natural transformation process by the combination of technological and biotechnological approaches.

## REFERENCES


1. http://www.internationaloliveoil.org/estaticos/view/132-world-table-olive-figures [Accessed: 04/03/2016].
2. International Olive oil Council (2004). Trade Standard Applying to Table Olives. Resolution No. RES-2/91-IV/04.
3. Garrido-Fernandez A, Fernandez Diez MJ, Adams MR. Olives and table olives. In: Garrido A, Fernandez MJ, Fernandez D, Adams MR, editors. Table Olives: Production and Processing. London: Chapman and Hall; 1997. p. 10–22, 289–367.
4. Lanza B. Nutritional and sensory quality of table olives. In: Muzzalupo I, editor. Olive Germplasm – The Olive Cultivation, Table Olive and Olive Oil Industry in Italy, 2012. InTech. ISBN 978-953-51-0883-2. DOI: 10.5772/51723.
5. Corsetti A, Perpetuini G, Schirone M, Tofalo R, Suzzi G. Application of starter cultures to table olive fermentation: an overview on the experimental studies. Frontiers in Microbiology. 2012; 3: 248–253. DOI: 10.3389/fmicb.2012.00248.
6. Medina-Pradas E, Arroyo-López FN. Presence of toxic microbial metabolites in table olives. Frontiers in Microbiology. 2015; 6: 1–6. DOI: 10.3389/fmicb.2015.00873.
7. García A, Romero C, Medina E, García P, De Castro A, Brenes M. Debittering of olives by polyphenol oxidation. Journal of Agricultural and Food Chemistry. 2008; 56: 11862–11867. DOI: 10.1021/jf802967y.
8. Blana VA, Grounta A, Tassou CC, Nychas GJ, Panagou EZ. Inoculated fermentation of green olives with potential probiotic Lactobacillus pentosus and Lactobacillus plantarum starter cultures isolated from industrially fermented olives. Food Microbiology. 2014; 38: 208–218. DOI: 10.1016/j.fm.2013. 09.007.
9. Panagou EZ, Schillinger U, Franz CMAP, Nychasa GJE. Microbiological and biochemical profile of cv. Conservolea naturally black olives during controlled fermentation with selected strains of lactic acid bacteria. Food Microbiology. 2008; 25: 348–358. DOI: 10.1016/j.fm.2007.10.005.
10. Servili M, Settanni L, Veneziani G, Esposto S, Massitti O, Taticchi A, Urbani S, Montedoro GF, Corsetti A. The Use of Lactobacillus pentosus 1MO to shorten the debittering process time of black table olives (Cv. Itrana and Leccino): a pilot-scale application. Journal of Agricultural and Food Chemistry. 2006; 54: 3869–3875. DOI: 10.1021/jf053206y.
11. Arroyo-López FN, Querol A, Bautista-Gallego J, Garrido-Fernández A. Role of yeasts in table olive production. International Journal of Food Microbiology. 2008; 128:189–196. DOI: 10.1016/j.ijfoodmicro.2008.08.018.
12. Segovia Bravo KA, Arroyo Lopez FN, García García P, Duran Quintana MC, Garrido Fernandez A. Treatment of green table olive solutions with ozone. Effect on their polyphenol content and on Lactobacillus pentosus and Saccharomyces cerevisiae growth. International Journal of Food Microbiology. 2007; 114: 60–68. DOI: 10.1016/j.ijfoodmicro.2006.09.032.
13. Bevilacqua A, Beneduce L, Sinigaglia M, Corbo MR. Selection of yeasts as starter cultures for table olives. Journal of Food Science. 2013; 78: M742–M751. DOI: 10.1111/1750-3841.12117.
14. Bonatsou S, Benítez A, Rodríguez-Gómez F, Panagou EZ, Arroyo-López FN. Selection of yeasts with multi functional features for application as starters in natural black table olive processing. Food Microbiology. 2015; 46: 66–73. DOI: 10.1016/j.fm.2014.07.011.
15. Tufariello M, Durante M, Ramires FA, Grieco F, Tommasi L, Perbellini E, Falco V, Tasioula-Margari M, Logrieco AF, Mita G, Bleve G. New process for production of fermented black

table olives using selected autochthonous microbial resources. Frontiers in Microbiology. 2015; 6: 1007–1022. DOI: 10.3389/fmicb.2015.01007.

16. Bleve G, Tufariello M, Durante M, Perbellini E, Mita G, Ramires AF, Grieco F, Logrieco AF. Metodo per la Produzione di Olive da Tavola Fermentate. 2015. European Patent Application No. 14197402.2. Munich: European Patent Office (EPO).
17. Lavermicocca P, Valerio F, Lonigro SL, De Angelis M, Morelli L, Callegari ML, Rizzello GC, Visconti A. Study of the adhesion and survival of lactobacilli and bifidobacteria on table olives with the aim of formulating a new probiotics food. Applied Environmental Microbiology. 2005; 71: 4233–4240. DOI: 10.1128/AEM.71.8.4233-4240.2005.
18. Argyri A, Zoumpopoulou G, Karatzas KA, Tsakalidou E, Nychas GJE, Panagou EZ, Tassou CC. Selection of potential probiotic lactic acid bacteria from fermented olives by in vitro tests. Food Microbiology. 2013; 33: 282–291. DOI: 10.1016/j.fm.2012.10.005.
19. Barbaro B, Toietta G, Maggio R, Arciello M, Tarocchi M, Galli A, Balsano C. Effects of the Olive-Derived Polyphenol Oleuropein on Human Health. International Journal of Molecular Sciences. 2014; 15: 18508–18524. DOI: 10.3390/ijms151018508.
20. Charoenprasert S, Mitchell A. Factors influencing phenolic compounds in table olives (Olea europaea). Journal of Agricultural and Food Chemistry. 2012; 60: 7081–7095. DOI: 10.1021/jf3017699.
21. Boskou D, Camposeo S, Clodoveo ML. Table olives as sources of bioactive compounds. In: Boskou D, editor. Olives and Olive Oil Bioactive Constituents,; 2015.Urbana, IL, USA: AOCS Press. p 179–216, 217—260. ISBN: 978-1-630670-41-2.
22. Steensels J, Verstrepen KJ. Taming wild yeast: potential of conventional and nonconventional yeasts in industrial fermentations. Annual Review of Microbiology. 2014; 68: 61–80. DOI: 10.1146/annurev-micro-091213-113025.
23. Heperkan D. Microbiota of table olive fermentations and criteria of selection for their use as starters. Frontiers in Microbiology. 2013; 4: 1–11. DOI: 10.3389/fmicb.2013.00143.
24. Tassou CC, Panagou EZ, Nychas GJE. Microbial colonization of naturally fermented olives. In: Preedy VR, Watson RR, editors. Olives and Olive Oil in Health and Disease. Singapore: Academic Press is an imprint of Elsevier; 2010. p. 397–406.
25. Randazzo CL, Ribbera A, Pitino I, Romeo FV, Caggia C. Diversity of bacterial population of table olives assessed by PCR-DGGE analysis. Food Microbiology. 2012; 32: 87–96. DOI: 10.1016/j.fm.2012.04.013.
26. Alves M, Gonçalves T, Quintas T. Microbial quality and yeast population dynamics in cracked green table olives fermentations. Food Control. 2012; 23: 363–368. DOI: 10.1016/ j.foodcont.2011.07.033.
27. Jalava K, Selby K, Pihlajasaari A, Kolho E, Dahlsten E, Forss N, Bäcklund T, Korkeala H, Honkanen-Buzalski T, Hulkko T, Derman Y, Järvinen A, Kotilainen H, Kultanen L, Ruutu P, Lyytikaïnen O, Lindström M. Two cases of food-borne botulism in Finland caused by conserved olives. Eurosurveillance. 2011; 16: 1–3.
28. Cocolin L, Alessandria V, Botta C, Gorra R, De Filippis F, Ercolini D, Rantsiou K. NaOH-debittering induces changes in bacterial ecology during table olives fermentation. PLoS One. 2013; 8: 1–11. DOI: 10.1371/journal.pone.0069074.
29. Bleve G, Tufariello M, Durante M, Perbellini E, Ramires FA, Grieco F, Cappello MS, De Domenico S, Mita G, Tasioula-Margari M, Logrieco AF. Physico-chemical and microbiological characterization of spontaneous fermentation of Cellina di Nardò and Leccino table olives. Frontiers in Microbiology. 2014; 5: 570, 1–18. DOI: 10.3389/fmicb.2014.00570.
30. Bleve G, Tufariello M, Durante M, Grieco F, Ramires FA, Mita G, Tasioula-Margari M, Logrieco AF. Physico-chemical and microbiological characterization of natural fermentation process of Conservolea and Kalamata table olives and developement of a protocol for the pre-selection of fermentation starters. Food Microbiology. 2015; 46: 368–382. DOI: 10.1016/j.fm.2014.08.021.
31. Nisiotou AA, Chorianopoulos N, Nychas GJ, Panagou EZ. 2010. Yeast heterogeneity during spontaneous fermentation of black conservolea olives in different brine solutions. Journal of Applied Microbiology. 2010; 108: 396–405. DOI: 10.1111/j.1365-2672.2009.04424.

32. Heperkan D, Erol-Meriç B, Sismanoglu G, Dalkiliç G, Güler FK. Mint: mycobiota, mycotoxigenic fungi and citrinin production in black olives. Advances in Experimental Medicine and Biology. 2006; 571: 203–210. DOI: 10.1007/0-387-28391-9_13.
33. De Castro A, Montano A, Casado FJ, Sanchez AH, Rejano L. Utilization of Enterococcus casselifavus and Lactobacillus pentosus as starter cultures for Spanish-style green olive fermentation. Food Microbiology. 2002. 19: 637–644. DOI: 10.1006/fmic.2002.0466.
34. Abriouel H, Benomar N, Lucas R, Gálvez A. Culture-independent study of the diversity of microbial populations in brines during fermentation of naturally fermented Aloreña green table olives. International Journal of Food Microbiology. 2011; 144: 487–496. DOI: 10.1016/j. ijfoodmicro.2010.11.006.
35. Arroyo-López FN, Romero-Gil V, Bautista-Gallego J, Rodriguez-Gómez F, Jiménez-Díaz R, García-García P, Querol A, Garrido-Fernandez A. Yeasts in table olive processing: desirable or spoilage microorganisms? International Journal of Food Microbiology. 2012; 160: 42–49. DOI: 10.1016/j.ijfoodmicro.2012.08.003.
36. Leal-Sánchez MV, Ruiz-Barba JL, Sánchez AH, Rejano L, Jiménez-Díaz R, Garrido A. Fermentation profile and optimization of green olive fermentation using Lactobacillus plantarum LPCO10 as a starter culture. Food Microbiology. 2003; 20: 421–430. DOI: 10.1016/S0740-0020(02)00147-8.
37. De Angelis M, Campanella D, Cosmai L, Summo C, Rizzello CG, Caponio F. Microbiota and metabolome of un-started and started Greek-type fermentation of Bella di Cerignola table olives. Food Microbiology. 2015; 52: 18-30. DOI: 0.1016/j.fm.2015.06.002.
38. Martorana A, Alfonzo A, Settanni L, Corona O, La Croce F, Caruso T, Moschetti G, Francesca N. An innovative method to produce green table olives based on "pied de cuve" technology. Food Microbiology. 2015; 50: 126–140. DOI: 10.1016/j.fm.2015.03.008.
39. Arroyo-López FN, Romero-Gil V, Bautista-Gallego J, Rodriguez-Gómez F, Jiménez-Díaz R, García-García P, Querol A, Garrido-Fernández A. Potential benefits of the application of yeast starters in table olive processing. Frontiers in Microbiology. 2012; 3: 161–164. DOI: 10.3389/fmicb.2012.00161.
40. Garrido Fernández A, Garcia P, Brenes M. Olive fermentations. In: Remand HJ, Reed G, editors. Biotechnology. Weinheim: VCH Press; 1995. p. 539–627.
41. Tsapatsaris S, Kotzekidou P. Application of central composite design and response surface methodology to the fermentation of olive juice by Lactobacillus plantarum and Debaryomyces hansenii. International Journal of Food Microbiology. 2004; 95: 157–163. DOI: 10.1016/ j.ijfoodmicro.2004.02.011.
42. Ettayebi K, Errachidi F, Jamai L, Tahri-Jouti MA, Sendile K, Ettayebi M. Biodegradation of polyphenols with immobilized Candida tropicalis under metabolic induction. FEMS Microbiology Letters. 2003; 223: 215–219. DOI: 10.1016/S0378-1097(03)00380-X.
43. Hurtado A, Ben Othman N, Chammem N, Hamdi M, Ferrer S, Reguant C, Bordons A, Rozés N. Characterization of Lactobacillus isolates from fermented olives and their bacteriocin gene profiles. Food Microbiology. 2011; 28: 1514–1518. DOI: 10.1016/j.fm.2011.07.010.
44. Garcia PG, Barranco CR, Quintana MC, Fernandez AG. Biogenic amine formation and "zapatera" spoilage of fermented green olives: effect of storage temperature and debittering process. Journal of Food Protection. 2004; 67: 117–123. DOI: 10.1016/j.fm.2007.10.005.
45. Spano G, Russo P, Lonvaud-Funel A, Lucas P, Alexandre H, Grandvalet C, Coton E, Coton M, Barnavon L, Bach B, Rattray F, Bunte A, Magni C, Ladero V, Alvarez M, Fernández M, Lopez P, de Palencia PF, Corbi A, Trip H, Lolkema JS. Biogenic amines in fermented foods. European Journal of Clinical Nutritional. 2010; 64: 95–100. DOI: 10.1038/ejcn.2010.218.
46. Tristezza M, Vetrano C, Bleve G, Spano G, Capozzi V, Logrieco A, Mita G, Grieco F. Biodiversity and safety aspects of yeast strains characterized from vineyards and spontaneous fermentations in the Apulia Region, Italy. Food microbiology. 2013; 36: 335–342. DOI: 10.1016/J.FM.2013.07.001.
47. Nychas GJN, Panagou EZ, Parker ML, Waldron KW, Tassou CC. Microbial colonization of naturally black olives during fermentation and associated biochemical activities in the cover brine. Letters in Applied Microbiology. 2002; 34: 173–177. DOI: 10.1046/j.1472-765x.2002.01077.x.

48. Chorianopoulos NG, Boziaris IS, Stamatiou A, Nuchas GJE. Microbial association and acidity development of unheated and pasteurised green table olives fermented using glucose or sucrose supplements at various levels. Food Microbiology. 2005; 22: 117–124. DOI: 10.1016/j.fm.2004. 04.010.
49. Sabatini N, Marsilio V. Volatile compounds in table olives (Olea europaea L., Nocellara del Belice cultivar). Food Chemistry. 2008; 107: 1522–1528. DOI: 10.1016/j.food chem.2007.10.008.
50. Angerosa F, Servili M, Selvaggini R, Taticchi A, Esposto S, Montedoro G. Volatile compounds in virgin olive oil: occurrence and their relationship with the quality. Journal of Chromatography A. 2004; 1054: 17–31.
51. Fleming HP, Etchells JL, Bell TA. Vapor analysis of fermented Spanish-type green olives by gas chromatography. Journal of Food Science. 1969; 34: 419–422. DOI: 10.1111/j.1365-2621.1969.tb12794.x.
52. Torrens J, Urpi P, Montserrat RA, Vichi S, López-Tamames E, Buxaderas S. Different commercial yeast strains affecting the volatile and sensory profile of cava base wine. International Journal of Food Microbiology. 2008; 124: 48-57. DOI: 10.1016/j.ijfoodmicro.2008.02.023.
53. Mallouchos A, Skandamis P, Loukatos P, Komaitis M, Koutinas A, Kanellaki M. Volatile compounds of wines produced by cells immobilized on grape skins. Journal of Agricultural and Food Chemistry. 2003; 51: 3060–3066. DOI: 10.1021/jf026177p.
54. McFeeters RF. Fermentation microorganisms and flavor changes in fermented foods. Journal of Food Science. 2004; 69: 35–37. DOI: 10.1111/j.1365-2621.2004.tb17876.x.
55. Damascelli A, Palmisano F. 2013 Sesquiterpene fingerprinting by headspace SPME–GC–MS: preliminary study for a simple and powerful analytical tool for traceability of olive oils. Food Analytical Methods. 2012; 6: 900–905. DOI: 10.1007/s12161-012-9500-9.
56. Gilbert-López B, Robles-Molina J, García-Reyes JF, Molina-Díaz A. Rapid determination of BTEXS in olives and olive oil by headspace-gas chromatography/mass spectrometry (HS-GC-MS). Talanta. 2010; 83: 391–399. DOI: 10.1016/j.talanta.2010.09.052.
57. Sabatini N, Mucciarella M, Marsilio V. Volatile compounds in uninoculated and inoculated table olives with Lactobacillus plantarum (Olea europaea L., cv. Morescaand Kalamata). Food Science and Technology. 2008; 41: 2017–2022. DOI: 10.1016/j.lwt.2007.12.002.
58. Grounta A, Doulgeraki AI, Nychas GJE, Panagou EZ. Biofilm formation on Conservolea natural black olives during single and combined inoculation with a functional Lactobacillus pentosus starter culture. Food Microbiology. 2016; 56: 35–44. DOI: 10.1016/j.fm.2015.12.002.
59. Panagou EZ, Tassou CC. Changes in volatile compounds and related biochemical profile during controlled fermentation of cv. Conservolea green olives. Food Microbiology. 2006; 23: 738–746. DOI: 0.1016/j.fm.2006.02.005.
60. Stiti N, Triki S, Hartmann MA. Formation of triterpenoids throughout Olea europaea fruit ontogeny. Lipids. 2007; 42: 55–67.
61. Bianchi G. Lipids and phenols in table olives. European Journal of Lipid Science and Technology. 2003; 105: 229–242. DOI: 10.1002/ejlt.200390046.
62. Guinda Á, Rada M, Delgado T, Gutiérrez-Adánez P, Castellano JM. Pentacyclic triterpenoids from olive fruit and leaf. Journal of Agriculture of Food Chemistry. 2010; 58: 9685–9691. DOI: 10.1021/jf102039t.
63. Ismaili H, Milella L, Fkih-Tetouani S, Ilidrissi A, Camporese A, Sosa S, Altinier G, Della Loggia R, Aquino R. In vivo topical anti-inflammatory and in vitro antioxidant activities of two extracts of Thymus satureioides leaves. Journal of Ethnopharmacology. 2004; 91: 31–36. DOI: 10.1016/j.jep.2003.11.013.
64. Liu J. Pharmacology of oleanolic acid and ursolic acid. Journal of Ethnopharmacology. 1995; 49: 57–68. DOI: 10.1016/0378-8741(95)90032-2.
65. Tsai SJ, Yin MC. Antioxidative and anti-inflammatory protection of oleanolic acid and ursolic acid in PC12 cells. Journal of Food Science. 2008; 73: 174–178. DOI: 10.1111/j.1750-3841.2008.00864.x.

66. Horiuchi K, Shiota S, Hatano T, Yoshida T, Kuroda T, Tsuchiya T. Antimicrobial activity of oleanolic acid from Salvia officinalis and related compounds on vancomycin-resistant Enterococci (VRE). Biological and Pharmaceutical Bulletin. 2007; 30: 1147–1149.
67. Parra A, Rivas F, Lopez PE, Garcia-Granados A, Martinez A, Albericio F, Marquez N, Muñoz E. Solution- and solid-phase synthesis and anti-HIV activity of maslinic acid derivatives containing amino acids and peptides. Bioorganic & Medicinal Chemistry. 2009; 17: 1139–1145.
68. Allouche Y, Beltran G, Gaforio JJ, Uceda M, Mesa MD. Antioxidant and antiatherogenic activities of pentacyclic triterpenic diols and acids. Food and Chemical Toxicology. 2010; 48: 2885–2890. DOI: 10.1016/j.fct.2010.07.022.
69. Rodriguez-Rodriguez R, Perona JS, Herrera MD, Ruiz-Gutierrez V. Triterpenic compounds from "orujo" olive oil elicit vasorelaxation in aorta from spontaneously hypertensive rats. Journal of Agricultural and Food Chemistry. 2006; 54: 2096–2102. DOI: 10.1021/jf0528512.
70. Liu J, Sun H, Duan W, Mu D, Zhang L. Maslinic acid reduces blood glucose in KK-Ay mice. Biological and Pharmaceutical Bulletin. 2007; 30: 2075–2078. DOI: 10.1248/bpb.30.2075.
71. Liu J, Sun H, Wang X, Mu D, Liao H, Zhang L. Effects of oleanolic acid and maslinic acid on hyperlipidemia. Drug Development Research. 2007; 68: 261–266. DOI: 10.1002/ddr.20187.
72. Sato H, Genet C, Strehle A, Thomas C, Lobstein A, Wagner A, Mioskowski C, Auwerx J, Saladin R. Anti-hyperglycemic activity of a TGR5 agonist isolated from Olea europaea. Biochemical and Biophysical Research Communications. 2007; 362: 793–798.
73. Hsum YW, Yew WT, Hong PLV, Soo KK, Hoon LS, Chieng YC, Mooi LY. Cancer chemopreventive activity of maslinic acid: suppression of COX-2 expression and inhibition of NF-κB and AP-1 activation in Raji cells. Planta Medica. 2011; 77: 152–157. DOI: 10.1055/s-0030-1250203.
74. Juan ME, Planas JM. Effects of pentacyclic triterpenes from olives on colon cancer. In: Watson RR, Preedy VR, editors. Bioactive Foods and Extracts: Cancer Treatment and Prevention. New York: CRC Press; 2010. p. 403–413.
75. Li C, Yang Z, Zhai C, Qiu W, Li D, Yi Z, Wang L, Tang J, Qian M, Luo J, Liu M. Maslinic acid potentiates the anti-tumor activity of tumor necrosis factor a by inhibiting NF-κB signaling pathway. Molecular Cancer. 2010; 9: 1–13. DOI: 10.1186/1476-4598-9-73.
76. Reyes FJ, Centelles JJ, Lupianez JA, Cascante M. (2Alpha, 3beta)-2,3-dihydroxyolean-12-en-28-oic acid, a new natural triterpene from Olea europea, induces caspase dependent apoptosis selectively in colon adenocarcinoma cells. FEBS Letters. 2006; 580: 6302–6310. DOI: 10.1017/S0007114508882979.
77. Reyes-Zurita FJ, Rufino-Palomares EE, Lupianez JA, Cascante M. Maslinic acid, a natural triterpene from Olea europaea L., induces apoptosis in HT29 human colon cancer cells via the mitochondrial apoptotic pathway. Cancer Letters. 2009; 273: 44–54. DOI: 10.1016/j.canlet.2008.07.033.
78. Romero C, García A, Medina E, Ruíz-Méndez Ma V, de Castro A, Brenes M. Triterpenic acids in table olives. Food Chemistry. 2010; 118: 670–674. DOI: 10.1016/j.foodchem.2009.05.037.
79. Owen RW, Haubner R, Würtele G, Hull WE, Spiegelhalder B, Bartsch H. Olives and olive oil in cancer prevention. European Journal of Cancer Prevention. 2004; 13: 319–326. DOI: 10.1097/01.cej.0000130221.19480.7e.
80. Sakouhi F, Harrabi S, Absalon C, Sbei K, Boukhchina Kallel H. α-Tocopherol and fatty acids contents of some Tunisian table olives (Olea europea L.): changes in their composition during ripening and processing. Food Chemistry. 2008; 108: 833–839. DOI: 10.1016/j.foodchem.2007.11.043.
81. Kastorini CM, Milionis HJ, Goudevenos JA, Panagiotakos DB. Mediterranean diet and coronary heart disease: is obesity a link—a systematic review. Nutrition, Metabolism and Cardiovascular Diseases. 2010; 20: 536–551. DOI: 10.1016/j.numecd.2010.04.006.

82. Ünal K, Nergiz C. The effect of table olive preparing methods and storage on the composition and nutritive value of olives. Grasas Aceites. 2003; 54: 71–76. DOI: 10.3989/gya.2003.v54.i1.280.
83. Wood JD, Richardson RI, Nute GR, Fisher AV, Campo MM, Kasapidou E, Sheard PR, Enser M. Effects of fatty acids on meat quality: a review. Meat Science. 2003; 66: 21–32. DOI: 10.1016/S0309-1740(03)00022-6.
84. Lopez A, Montano A, Garcia P, Garrido A. Fatty acid profile of table olives and its multivariate characterization using unsupervised (PCA) and supervised (DA) chemometrics. Journal of Agricultural and Food Chemistry. 2006; 54: 6747–6753. DOI: 10.1021/jf0612474.
85. Uylaser V, Yildiz G. Fatty acid profile and mineral content of commercial table olives from Turkey. Notulae Botanicae Horti Agrobotanici Cluj-Napoca. 2013; 41: 518–523.
86. Kalman A, Mujahid C, Mottier P, Heudi O. Determination of α-tocopherol in infant foods by liquid chromatography combined with atmospheric pressure chemical ionisation mass spectrometry. Rapid Communications in Mass Spectrometry. 2003; 17: 723–727. DOI: 10.1002/rcm.970.
87. Bogani P, Galli C, Villa M, Visioli F. Postprandial anti-inflammatory and antioxidant effects of extra virgin olive oil. Atherosclerosis. 2007; 190: 181–186. DOI: 10.1016/j.atherosclerosis. 2006.01.011.
88. Macheix, JJ, Fleuriet A, Billot J. Fruit phenolics, vol. 92.Boca Raton, FL: CRC Press; 1990. p.111–112.
89. Ryan D, Robards K. Phenolic compounds in olives. Analyst. 1998; 123: 31R–44R.
90. Borzillo A, Iannotta N, Uccella N. Oinotria table olives: quality evaluation during ripening and processing by biomolecular components. European Food Research and Technology. 2000; 212: 113–121. DOI: 10.1007/s002170000178.
91. Uccella N. Olive biophenols: novel ethnic and technological approach. Trends Food Science and Technology. 2001; 11: 328–339. DOI: 10.3989/gya.095811.
92. Saija A, Ucella N. Olive biophenols; functional effects on human wellbeing. Trends Food Science and Technology. 2001; 11: 357–363.
93. Panizzi L, Scarpati ML, Oriente G. The constitution of oleuropein, a bitter glucoside of the olive with hypotensive action. Gazzetta Chimica Italiana. 1960; 90: 1449–1485.
94. Bianchi G, Pozzi N. 3,4-Dihydroxyphenylglycol, a major C6–C2 phenolic in Olea europaea fruits. Phytochemistry. 1994; 35: 1335–1337.
95. Blekas G, Psomiadou E, Tsimidou M, Boskou D. On the importance of total polar phenols to monitor the stability of Greek virgin olive oil. European Journal of Lipid Science and Technology. 2002; 140: 340–346. DOI: 10.1002/1438-9312(200206)104:6<340.
96. Pistarino E, Aliakbarian B, Casazza AA, Paini M, Cosulich ME, Perego P. Combined effect of starter culture and temperature on phenolic compounds during fermentation of Taggiasca black olives. Food Chemistry. 2013; 138: 2043–2049. DOI: 10.1016/j.foodchem.2012.11.021.
97. Leenen R, Roodenburg AJC, Vissers MN, Schuurbiers JAE, van Putte KPAM, Wiseman SA, van de Put FHMM. Supplementation of plasma with olive oil phenols and extracts: influence on LDL oxidation. Journal of Agricultural and Food Chemistry. 2002; 50: 1290–1297. DOI: 10.1021/jf010968u.
98. Doulgeraki AI, Hondrodimou O, Iliopoulos V, Panagou EZ. Lactic acid bacteria and yeast heterogeneity during aerobic and modified atmosphere packaging storage of natural black Conservolea olives in polyethylene pouches. Food Control. 2012; 26: 49–57. DOI: 10.1016/j.foodcont.2012.01.006.
99. FAO/WHO. Health and nutritional properties of probiotics in food including powder milk with live lactic acid bacteria.Cordoba (Argentina): American Córdoba Park Hotel; 2001.
100. Parvez S, Kang M, Chung HS, Cho C, Hong M-C, Shin M-K, Bae H. Survey and mechanism of skin depigmenting and lightening agents. Review. Phytotherapy Research. 2006; 20: 921–934. DOI: 10.1002/ptr.1954.
101. De Bellis P, Valerio F, Sisto A, Lonigro SL, Lavermicocca P. Probiotic table olives: microbial populations adhering on olive surface in fermentation sets inoculated with the probiotic

strain Lactobacillus paracasei IMPC2.1 in an industrial plant. International Journal of Food Microbiology. 2010; 140: 6–13. DOI: 10.1016/j.ijfoodmicro.2010.02.024.
102. Mena B, Aryana KJ. Influence of ethanol on probiotic and culture bacteria Lactobacillus bulgaricus and Streptococcus thermophilus within a therapeutic product. Journal of Medical Microbiology. 2012; 2: 70–76. DOI: 10.4236/ojmm.2012.23010.
103. Tamime AY, Saarela M, Korslund Sondergaard A, Mistry VV, Shah NP. Production and maintenance of viability of probiotic micro-organisms in dairy products. In: Tamime AY, editor. Probiotic Dairy Products. London: Blackwell Publishing Ltd; 2005. p. 39–72.
104. Lavermicocca P, Valerio F, Lisa Lonigro S, De Angelis M, Morelli L, Callegari ML, Rizzello CG, Visconti A. Study of adhesion and survival of Lactobacilli and Bifidobacteria on table olives with the aim of formulating a new probiotic food. Applied Environmental Microbiology. 2005; 71: 4233–4240. DOI: 10.1128/AEM.71.8.4233-4240.2005.
105. Saravanos E, Kagli D, Zoumpopoulou G, Panagou EZ, Tassou CC. Use of probiotic lactic acid bacteria as starter cultures in Spanish-style green olive fermentation and determination of their survival using PFGE. Food Microbiology. 2008; 1–4 September, Aberdeen, UK.
106. Valerio A, D'Antona G, Nisoli E. Branched-chain amino acids, mitochondrial biogenesis, and healthspan: an evolutionary perspective. Aging. 2011; 3: 464–478.
107. Psani M, Kotzekidou P. Technological characteristics of yeast strains and their potential as starter adjuncts in Greek-style black olive fermentation. World Journal of Microbiology and Biotechnology. 2006; 22: 1329–1336. DOI: 10.1007/s11274-006-9180-y.
108. Silva T, Reto M, Sol M, Peito A, Peres CM, Peres C, Malcata XF. Characterization of yeasts from Portuguese brined olives, with a focus on their potentially probiotic behavior. LWT Food Science and Technology. 2011; 44: 1349–1354. DOI: 10.1016/j.lwt.2011.01.029.
109. Bevilacqua A, Corbo MR, Sinigaglia M. Selection of yeasts as starter cultures for table olives: a step-by-step procedure. Frontiers in Microbiology. 2012; 3: 194–202. DOI: 10.3389/ fmicb.2012.00194.
110. Kourelis A, Kotzamanidis C, Litopoulou-Tzanetaki E, Scouras ZG, Tzanetakis N, Yiangou M. Preliminary probiotic selection of dairy and human 227 yeast strains. Journal of Biological Research Thessaloniki. 2010; 13: 93–104.
111. Moslehi-Jenabian S, Lindegaard Pedersen L, Jespersen L. Beneficial effects of probiotic and food borne yeasts on human health. Nutrients. 2010; 2: 449–473. DOI: 10.3390/nu2040449.
112. Cappelletti GM, Nicoletti GM, Russo C. Wastewater from table olive industries. In: Fernando Sebastian Garcia Einschlag, editor. Waste Water – Evaluation and Management. InTech; 2011. ISBN: 978-953-307-233-3. Available from: http://www.intechopen.com/books/waste-water-evaluation-and-management/wastewater-from-table-olive-industries.
113. Aggelis GG, Gavala HN, Lyberatos G. Combined and separate aerobic and anaerobic biotreatment of green olive debittering wastewater. Journal of Agricultural Engineering Research. 2001; 80: 283–92.
114. Rivas FJ, Beltran FJ, Alvarez P, Frades J, Gimeno O. Joint aerobic biodegradation of wastewater from table olive manufacturing industries and urban wastewater. Bioprocess Engineering. 2000; 23: 283–286.
115. Borja R, Martin A, Alonso V, Garcia I, Banks CJ. Influence of different aerobic pretreatments on the kinetics of anaerobic digestion of olive-mill wastewater. Water Research. 1995; 29: 489–495.
116. Kyriacou A, Lasaridi KE, Kotsou M, Balis C, Pilidis G. Combined bioremediation and advanced oxidation of green table olive processing wastewater. Process Biochemistry. 2005; 40: 1401–1408.
117. Beltran J, Gonzalez T, Garcia J. Kinetics of the biodegradation of green table olive wastewaters by aerobic and anaerobic treatments. Journal of Hazardous Materials. 2008; 154: 839–845.
118. Brenes M, Garcìa P, Romero C, Garrido A. Treatment of green table olive wastewaters by an activated-sludge process. Journal of Chemical Technology and Biotechnology. 2000; 75: 459–463.

119. Kotsou M, Kyriacou A, Lasaridi K, Pilidis G. Integrated aerobic biological treatment and chemical oxidation with Fenton's reagent for the processing of green table olive wastewater. Process Biochemistry. 2004; 39: 1653–1660.

# CHAPTER 12

# Anthocyanin accumulation enhanced in *Lc*-transgenic cotton under light and increased resistance to bollworm

**Xiaoping Fan[1], Bohong Fan[1], Yuxiang Wang[1], Weicai Yang[2]**

[1] *Institute of Cotton Research, Academy of ShanXi Agricultural Science, Yuncheng City, Shanxi Province 044000, China*

[2] *Key Laboratory of Molecular and Developmental Biology and National Centre for Plant Gene Research (Beijing), Institute of Genetics and Developmental Biology, Chinese Academy of Sciences, Datun Road, Beijing 100101, China*

## ABSTRACT

Breeding of naturally colored cotton fiber has been hampered by the limited germplasm, an alternative way is to use transgenic approach to create more germplasm for breeding. Here, we report our effort to engineer anthocyanin production in cotton. The maize *Lc* gene, under the control of the constitutive 35S promoter, was introduced into cotton through genetic transformation. Our data showed that the expression of the *Lc* gene alone is sufficient to trigger the accumulation of anthocyanin in a variety of cell types including fiber cells in cotton. However, the accumulation of colored anthocyanin in cotton fibers requires the participation of light signaling. These data indicate that it is feasible to engineer colored fibers through transgenic approach in cotton. Furthermore, we showed that the *Lc*-transgenic cotton plants are resistant to cotton bollworm. These transgenic plants are, therefore, potentially useful for cotton breeding against cotton bollworm.

**Keywords:** *Lc* gene, *Gossypium hirsutum,* Anthocyanin, Cotton fiber

# 1. INTRODUCTION

Naturally colored cotton clothing and textile products become more and more popular in people's life. They not only have the same advantages as the common white fiber, but also require no chemical dyeing and bleaching. Therefore, they are environmental friendly as well. Early in 1990s, Bird Brothers successfully bred four colored fibers: green, brown, red (a reddish brown) and mocha (similar to tan). However, so far breeding of naturally colored cotton fiber is limited by the lack of proper germplasm, which is largely limited to brown and green fiber traits. Furthermore, genetic studies on genes controlling pigment production in cotton have been hampered by its complex genome. Therefore, it is necessary to find more cotton germplasm that introduces novel color traits into cotton through transgenic approaches, for example, by modifying anthocyanin biosynthetic pathway.

Biosynthetic pathway leading to anthocyanin pigmentation has been studied extensively in *Zea mays* and genes controlling each step have been characterized (Sharma et al. 2012). Anthocyanin biosynthesis is controlled by two groups of genes. One group is the structural genes which encode the enzymes, catalyzing the biochemical reactions of the formation of anthocyanins. The second group includes genes which encode transcription regulators that control the expression of the structural genes, and determine the spatial and temporal pattern of anthocyanin accumulation. In general, the transcription regulators belong to either the MYB-type C1 family or the basic helix-loop-helix, MYC-type R family (Dooner et al. 1991; Boudet 2007; Kong et al. 2012). Both *C1* and *R* genes have been studied extensively in many plants, such as maize (Kong et al. 2012) and petunia (Quattrocchio et al. 1993; Schwinn et al. 2014). The *C1* gene of maize plays a regulatory role in the production of anthocyanin pigments in the aleurone layer of the endosperm (Cone et al. 1986; Stinard et al. 2009) and the *R* genes control the temporal and spatial distribution of anthocyanin pigments (Ludwig et al. 1989).

The *Lc* (leaf color) gene is a bHLH (basic/helix-loop-helix) anthocyanin regulator of maize, belongs to the *R* gene family, determines the timing

(Ludwig and Wessler 1990), distribution, and amount of anthocyanin pigmentation in maize and is the target for plant pigment engineering (Schwinn et al. 2014). In this family, it also consists of other regulatory genes, such as *Sn* and the B locus genes (Chandler et al. 1989; Dooner et al. 1991; Consonni et al. 1993; Petroni and Tonelli 2011). The *Lc* gene as a transcription factor can regulate the anthocyanin biosynthesis pathway and use anthocyanin-regulating transcription factors to alter anthocyanin production and pigmentation patterns in several plant systems (Boase et al. 1998; Bradley et al. 1998), such as in tomato (Goldsbrough et al. 1996; Bovy et al. 2002), tobacco (Lloyd et al. 1992; Huang et al. 2012), apple (Flachowsky et al. 2010) and petunia (Bradley et al. 1998; Albert et al. 2009). In transgenic petunia plants, the maize *Lc* gene, under the control of the constitutive CaMV 35S promoter, enhanced pigmentation in both vegetative and floral tissues. Additionally, in *Lc*-transgenic alfalfa foliage, red–purple colored anthocyanin accumulated under high light conditions (Ray et al. 2003). These reports illustrated that anthocyanin production in vegetative and reproductive tissues was enhanced and induced by *Lc* transcription factors under high light conditions (Procissi et al. 1997; Albert et al. 2009). However, in several different plant species, anthocyanin biosynthesis was stimulated by *Lc* and C1 together (Han et al. 2009), or in some species *Lc* has no visible effect (Boase et al. 1998; Bradley et al. 1999). In cotton, few studies on *Lc* transformation and expression have been reported so far. Therefore our aims for this study were to determine whether the *Lc* gene alone can active anthocyanin biosynthesis in cotton and characterize the anthocyanin pigmentation response in *Lc*-transformed cotton plants.

For another, anthocyanin in leaves has been correlated with resistance to biotic and abiotic agents, including fungi, herbivores and cold (Chalker-Scott 1999; Gould et al. 2000; Huang et al. 2012). The anthocyanin biosynthetic pathway is interlinked with the biosynthesis of flavonoid compounds, which are important to fertility and defense against insects and pathogens (Dooner et al. 1991; Flachowsky et al. 2010). Red leaves are less frequently attacked than green leaves; in *Lc*-transgenic rice, red leaves exhibited increased resistance to blast infection (Gandikota et al. 2001). Therefore, expression of *Lc* provides us a tool to engineer cotton color and insect resistance. Here, we report our approach to employ *Lc* gene to engineer cotton color and insect resistance.

# 2. MATERIALS AND METHODS

## 2.1 Plant materials

Cotton (*Gossypium hirsutum*) Coker 312 seeds were surface sterilized by soaking for 1 min in 70 % ethanol followed by ~2 h in 10 % $H_2O_2$. They were then washed with sterilized distilled water 3 times. Seeds were allowed to germinate in sterilized water for 16–20 h at 28 °C. The germinated seeds were transferred to half strength MS medium (Murashige and Skoog 1962) containing 15 g/L glucose and 3 g/L Phytagel. The pH was adjusted to 5.8 with 1 M KOH prior to autoclaving. The transformation procedure was carried out as described by Li et al. (2005). Briefly, hypocotyl explants of 5- to 7-day-old seedlings were cut into 0.8–1-cm-long segments and immersed in a bacterial culture suspension with gentle agitation. Then, the co-culturing, selection, tissue culturing, differentiation, germination and rooting processes were performed as described previously (Li et al. 2005). All tissue culture processes before transplanting to pots were performed in a culture room that was maintained at 28 °C with a 16/8 h light/dark photoperiod and light supplied by warm-white fluorescent lamps at an irradiance of 26 $\mu mol/m^2/s$.

## 2.2 Bacterial culture and cotton genetic transformation

*Agrobacteriumtumefaciens* strain LBA4404, which contains a modified plasmid pBI121 harboring a CaMV 35S::Lc::Nos transgene (provided by Dr. Lim Saw Hoon, Malaysia University of Science and Technology, Malaysia), was cultured overnight in LB liquid medium with continuous shaking at 250 rpm, at 28 °C. Bacterial cultures at $OD_{600}$ 0.6–0.8 were collected by centrifugation and resuspended in MS liquid medium to a final $OD_{600}$ of 0.2–0.4. Acetosyringone was added to the culture medium to a final concentration of 100 μM at ~2 h before harvesting. Cotton transformation and tissue culturing were performed as described previously (Li et al. 2005).

## 2.3 Screening of transgenic plants using polymerase chain reaction (PCR)

The analysis of the transgenic lines of the T0 and T1 generations was carried out by amplifying a fragment of the *Lc* gene using PCR. Genomic DNA was extracted from young leaves of greenhouse-grown transgenic and nontransgenic plants using the CTAB method (Li et al. 2002). PCR screening of putative transformants was performed using primers specific to the *Lc* gene. The plasmid was used as a positive amplification control. The sequences of the forward and reverse primers are 5′ATGGCGCTTTCAGCTTCCCGAGT3′ and 5′TCACCGCTTCCCTATAGC TTTG C3′, respectively. The PCR master mix was made according to the PCR kit's protocol. Amplification was accomplished using a Thermal Cycler (Biometro, Germany). The reaction was as follows: incubation at 94 °C for 5 min, and then a step cycle program set to denature at 94 °C for 45 s, to anneal at 58 °C for 45 s, and then extended at 72 °C for 2 min, for a total of 35 cycles, followed by a final extension at 72 °C for 10 min.

## 2.4 Southern blot analysis

For the Southern blot analysis, 50 μg genomic DNA was isolated from PCR-positive plants and digested with 20 units of *BamH*1 at 37 °C overnight in a total volume of 200 μl. Digested DNA was gel fractionated and transferred onto Hybond N+ nylon membrane (Amersham) according to the manufacturer's instructions. The *Lc* probe was labeled with Digoxigenin-11-dUTP using PCR amplification according to the manufacturer's instructions (Roche). Hybridization was performed overnight at 42 °C using DIG-Easy-hybridization solution containing a DIG-labeled *Lc* probe. The hybridized membrane was washed 3 times with a 0.1× SSC plus 0.1 % SDS solution at 68 °C and detection was carried out using an anti-DIG-AP conjugate according to the manufacturer's instructions.

## 2.5 RNA isolation and reverse-transcription (RT)-PCR analysis

The fresh young leaves, petiole and floral tissues, including petal, stigma, anther and sepal, in addition ovule, fiber at the proper development stage, were collected and frozen in liquid nitrogen. Total RNA from these materials was extracted using the hot phenol extraction method (Li et al. 2005). Then, 1 μg of each total RNA sample was reverse transcribed using the Qiagen one-step RT-PCR protocol. Briefly, the RNA template was added to the master mix and reverse transcribed at 50 °C for 30 min. Then, the reverse transcriptase was inactivated, and the cDNA template was denatured at 95 °C for 15 min. PCR amplification was performed using *Lc* gene-specific primers, and a program of 30 cycles of 94 °C for 45 s, 56 °C for 45 s, and 72 °C for 2 min, followed by a final extension at 72 °C for 10 min. The cotton *actin* gene was used as an internal standard to normalize the cDNA concentration. The same amplification was performed using the *actin* primers: Actin-F, 5′TTTGCTGGTGATGATGCTCC3′; and Actin-R, 5′CTC CAATC CAGACACTGTACT3′.

## 2.6 Ovule culture

Cotton bolls from 2-day post-anthesis T1 CaMV 35S::Lc::Nos transgenic plants of two lines, L73 and L143, and wild type (Wt) were surface sterilized in 70 % ethanol for 5–10 min, and then immature ovules were gently transferred onto refined liquid medium as described by Beasley and Ting (1974). Immature ovules from half of one boll were cultured under a 16/8 h light/dark photoperiod, while the other half were cultured in complete darkness in a culture room. Same stage ovules from non-transgenic bolls were cultured under the same conditions. The comparative experiment was repeated six times.

## 2.7 Anthocyanin extraction and measurement

The anthocyanin extraction was performed according to Choung et al. (2001). Air-dried old or fresh leaves were ground to a powder. Then, 5 g of powder was extracted with 100 ml of a 1 % HCl and 40 % methanol mixture in the darkness at 4 °C overnight. Subsequently, the extracts were centrifuged and filtrated through a 0.45 μm filter. The anthocyanin pigment content was determined by a pH differential method commonly used in food technology (Gong et al. 1997). Briefly, 10 ml of extract was adjusted to pH 1.0 and another 10 ml to pH 4.5 with 0.025 M potassium chloride buffer and 0.4 M sodium acetate buffer, respectively. Then, the extracts were analyzed using a UV–VIS spectrophotometer and quantified (Gong et al. 1997). The anthocyanin content, expressed as cyanidine-3-glucoside, was determined according to Lambert–Beer's Law: $A = \varepsilon C\text{Len}$ ($A$, absorbance; $C$ concentration; Len, path length in cm; and $\varepsilon$, molar extinction coefficient) using the molar extinction coefficient of 26,900 Len/mol and molar mass of 449 g/mol.

The color density, polymeric color and the percent of tannin contribution were measured using the bisulfite solution method (Somers and Evans 1974). The metabisulfite is able to bleach the anthocyanin, but not polymeric tannin pigments, and spectrophotometric assay results were calculated according to Cortell et al. (2007). The percent contribution of tannin is the ratio of polymeric color to color density.

## 2.8 Insect bioassay

The 4th leaf from the apical shoot of each line (L73, L74, L143, L144 and Wt) picked for the bioassay was washed thoroughly with sterile distilled water prior to feeding. Both transgenic red leaves and non-transgenic leaves at the same distance from shoot apex were placed in one petri dish and infested with three-third-instar cotton bollworm larvae (*Helicoverpa armigera*) that had been starved for 2 h. After 3 days, the feeding preference was recorded, and 13 replicate samples from each line were used.

In addition, the third-instar larvae were placed into one petri dish with either a piece of transgenic or non-transgenic leaf, observed every day and weighed after 5 days of feeding. The insect weights on different plant lines were determined using the between-column variance method. Larval mass (to an accuracy of 0.1 mg) was measured. To compare herbivore resistance-associated parameters between Wt and *Lc*-transgenic lines a one-way analysis of variance (ANOVA) was performed, and the experiment was replicated three times.

# 3. RESULTS

## 3.1 Generation of transgenic cotton plants

After transformation and regeneration from tissue culture, through somatic embryogenesis, six independent transgenic lines were regenerated on *kanamycin* selective medium. To confirm the presence of the *Lc* transgene in these T0 transgenic lines, genomic DNAs were digested by *BamH*I and fractionated by gel electrophoresis. The blot was probed with a DIG-labeled *Lc* probe. Six PCR positive plantlets were confirmed by Southern hybridization. Southern blot hybridization analysis determined transgene copy numbers from these lines. Lines of L82 and L28 had multiple copies insertion and the other four lines (L73, L74, L143 and L144) had single insertion, respectively. The non-transgenic control (NC) lacks the *Lc* hybridization signal (Fig. 1a). Southern blot analysis was performed on the T1 generation to reveal the *Lc* gene information of integration and separation in T1 generation. In line L143 T1 generation, 5 (L143-1, 2, 5, 6 and 7) of 15 were positive (Fig. 1b).

## *3.2 Lc* gene expressed at the molecular level

To investigate whether the *Lc* transgene is expressed in the transgenic plants, RT-PCR was performed from different tissues and organs. Our results showed that *Lc* was expressed in leaves and floral tissues of the transgenic plants, and no *Lc* transcripts were detected in non-transgenic plants (data not shown). The expression levels of the *Lc* gene varied slightly among tissues and organs (Fig. 2a). Leaves and petioles had relatively higher *Lc* expression levels than the floral tissues. Additionally, the *Lc* expression levels were different among flower parts. The highest expression level was in anther, while it was moderate in stigma, but no signal was detected in petals.

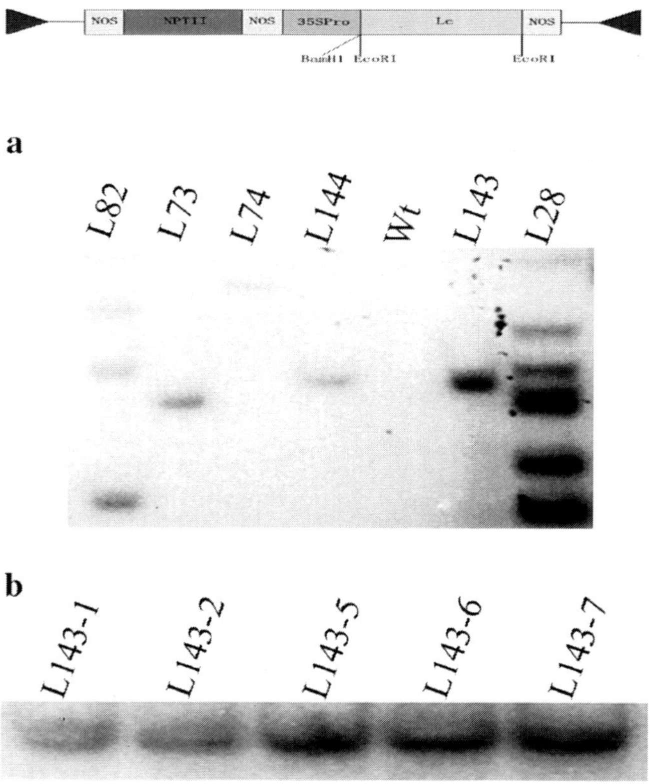


**Figure. 1** Southern blot analysis of the *Lc*-transgenic cotton plants. Genomic DNA from T0 transgenic (**A**) and T1 transgenic plants (**B**) was digested by *BamH*I and fractionated using gel electrophoresis. The blot was probed with a DIG-labeled *Lc* probe. **A** Lanes *1–4* and *6–7* represent six independent transgenic lines as indicated. Lane *5* is a nontransgenic control (Wt). **B** Lanes *1–5* represent five T1 siblings from L143 showing *Lc* hybridization bands

The *Lc* gene was expressed during all the ovule and fiber developmental stages in the transgenic plants. There were high expression levels in the 4–16 DPA (days post-anthesis) ovules, and moderate levels were detected in 1–3 DPA ovules and 18 DPA ovules. To determine whether the *Lc* transgene was expressed only in cotton fiber cells, we isolated RNA from fibers of 7, 8, 13 and 23 DPA ovules and mature fibers, respectively. The test results showed that in fiber, the expression level rose along with fiber development from 7 to 13 days. The expression level decreased to a weak but detectable signal in 23 DPA fibers and was weaker in mature fiber cells (Fig. 2b).

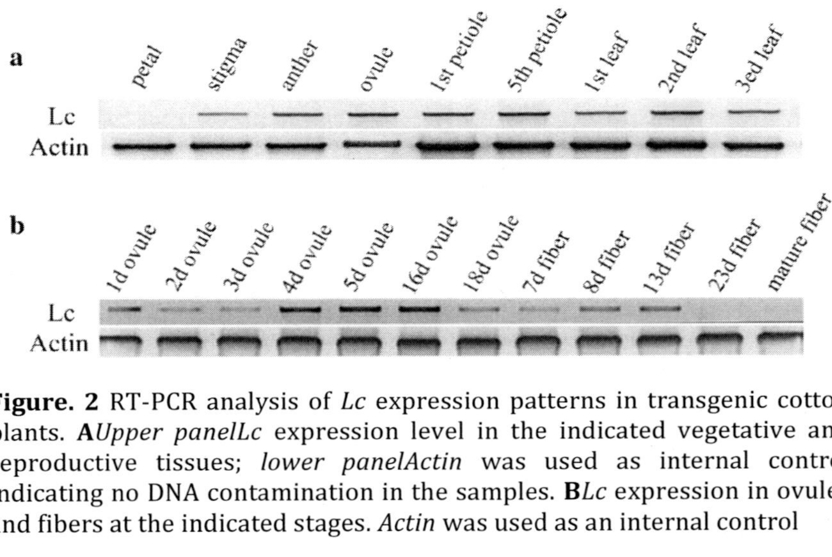


**Figure. 2** RT-PCR analysis of *Lc* expression patterns in transgenic cotton plants. **A***Upper panelLc* expression level in the indicated vegetative and reproductive tissues; *lower panelActin* was used as internal control indicating no DNA contamination in the samples. **B***Lc* expression in ovules and fibers at the indicated stages. *Actin* was used as an internal control

## 3.3 Morphology of *Lc*-transgenic cotton

During the regeneration process, most of the transformants accumulated high levels of anthocyanin pigment in their vegetative and reproductive tissues. In the tissue culture process, we observed that *kanamycin*-resistant calluses and somatic embryos appeared an unusual red color (Fig. 3a) that was never observed in control tissues (Fig. 3b). Generally, anthocyanin pigmentation accumulated in radicles, hypocotyls and cotyledons of transgenic somatic embryos. The purple color present in cotyledons disappeared several days later, and the young root turned red gradually in a culture bottle. The red root phenotype never presented in non-transgenic plants. Juvenile leaves usually displayed red, when the transgenic plantlets were transplanted into soil. Some appeared purple in pigmentation when the leaves unfolded (Fig. 3c), and Wt leaves were green under the same growth conditions (Fig. 3d). With further growth of the regenerated plants, some red patches appeared on the stem's surface (Fig. 3e). The sepal around the flower bud turned red (Fig. 3f) when the flower opened, and the anthers and the stigma also appeared red (Fig. 3g). We observed that the anthers are larger and contain less pollen compared with the Wt. However, no anthocyanin pigmentation was observed in petals at the day of anthesis. We found that the old leaves of transgenic plants turned deep red, and their air-dried leaves were more fragile when squeezed. Non-transgenic senescent leaves remained yellowish and were more flexible after air drying. The L143 and L73 lines were fertile, and their T1 and T2 generations had the same red color characteristics. The T1 generation segregated 3:1 (positive: negative). The positive seedlings had red roots (Fig. 3h-r1, r2) and the negative ones were white when germinated in a bottle under the same culture conditions (Fig. 3h-r3, r4).

**Figure. 3** Phenotypic characterization of *Lc*-transgenic cotton. **A** Red transgenic calluses and somatic embryos (*bar* 2.5 mm). **B** Non-transgenic calluses (*bar* 2.5 mm). **C** Transgenic *purple colored* leaf, indicating anthocyanin accumulation. **D** Non-transgenic leaf. **E** Anthocyanin accumulation on stem (*red patches*). **F** Transgenic flower bud with a red sepal. **G** Transgenic flower with a red anther. **H** T1 siblings showing the

segregation of *Lc*-dependent anthocyanin accumulation in roots. r1 and r2 are positive seedlings with red roots; r3 and r4 are negative seedlings with white roots. **I** Transgenic immature ovules cultured under light for 1 week (*bar* 2 mm). Note red fiber cells. **J** Magnification of J showing a single red fiber cell (*bar* 10 μm). Note red oily liquid gathered in the central big vacuole. **K** Transgenic immature ovules cultured in completely dark for 1 week (*bar* 2 mm). **L** Transgenic leaf abaxial surface toward the sun; its freehand cross section with red palisade mesophyll cells. The epidermal cells are colorless and transparent (*bar* 1 mm). **M** Anthocyanin extract from transgenic dry leaves (e1), transgenic fresh leaves (e2), Wt leaves (e3), and empty control (e4)

## 3.4 Light-induced anthocyanin accumulation in transgenic fibers and leaves

The initiating fibers of transgenic ovules became red (Fig. 3i) after being cultured in vitro under light for 7 days, and a red oil-like liquid accumulated in the large central vacuole of the single fiber cell observed under the microscope (Fig. 3j). Fiber cells of non-transgenic ovules at the same developmental stage showed little color change on the fiber surface under the same light (data not show). When cultured in the dark, the other areas, the fiber no matter from transgenic or non-transgenic bolls, remained completely white (Fig. 3k). Light is also an important factor that affects anthocyanin production in leaves. Freehand cross section of leaves from transgenic plants grown with exposure to normal sunlight for 3 days exhibited anthocyanin accumulation in their palisade mesophyll cells (Fig. 3l). When the upper surface of the leaf was not exposed to sunlight, for example with positioning facing down (due to abnormal growth), the spongy mesophyll cells turned red but the palisade mesophyll cells remained green (data not shown). The leaf epidermis was always maintained colorless and transparent whichever side faced toward the sun (Fig. 3l).

## 3.5 Transgenic plants produce more anthocyanin than non-transgenic plants

We further quantified the amount of anthocyanin in transgenic plants. The leaf extract of transgenic dry or fresh leaves showed more intense red (Fig. 3m-e1, e2) compared with the Wt leaves (Fig. 3m-e3). Figure 3m-e4 shows the empty control.

To quantify the amount of anthocyanin, we employed the pH differential method commonly used in food technology. Measurement results showed that the anthocyanin, exemplified by cyanidin-3-glucoside, increased from 15.41 mg/100 g (Wt) to 53.60 mg/100 g (*Lc*) in the dry leaf extract and from 9.74 mg/100 g (Wt) to 12.80 mg/100 g (*Lc*) in the fresh leaf extract.

Using the bisulfite solution method, anthocyanin degradation indices data showed that the three indices, color density (total anthocyanin), and monomeric and polymeric anthocyanin (condensed tannin) (Cortell et al. 2007) increased when compared with Wt extract. Color density in *Lc*-transgenic leaf rose from 1.62 to 4.03 in the dry leaf extracts; in fresh leaf extract, it increased from 0.71 to 2.76. However, the ratio varied between monomeric anthocyanin and polymeric anthocyanin. In *Lc*-transgenic dry leaf, monomeric anthocyanin accounted for 27.2 % of the total anthocyanin relative to 17.3 % in non-transgenic dry leaf. Polymeric anthocyanin ratio from 67.2 % in Wt varied to 62.4 % of the total anthocyanin. In fresh materials, monomeric anthocyanin percentage decreased from 22.5 to 20.7 %, while the polymeric anthocyanin percentage increased from 50.7 to 51.4 % in transgenic leaf extract (Fig. 4). This might be due to the formation of polymers and anthocyanin degradation that could be occurring in *Lc* leaves during different stages of growth, leading to changes in the anthocyanin ratio.

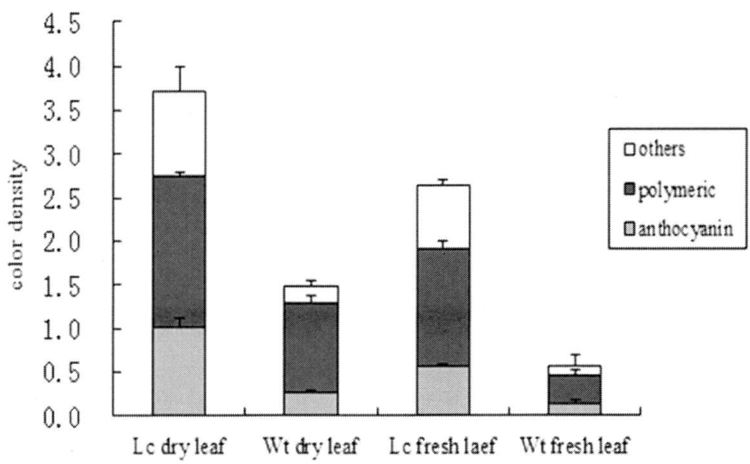


**Figure. 4** Anthocyanin content revealed by UV spectrometric analysis. Relative fractions of monomeric anthocyanin, polymeric anthocyanin (condensed tannins) and other compounds of the total anthocyanin calculated using the bisulfite leaching method. *Values* represent averages ± STEV, and every *bar* represents an average value from three repeated measurements

## 3.6 Transgenic plants are resistant to cotton bollworm

To test whether the CaMV 35S::Lc-transgenic plants were also resistant to cotton bollworm, an insect feeding experiment was performed. Our data showed that the insects fed preferably on non-transgenic leaves when both transgenic and non-transgenic leaves were present in one dish. The non-transgenic leaf was consumed more after 2 days of feeding (Fig. 5a-a1), while the transgenic leaves remained more (Fig. 5a-b1, c1). This trend became more obvious after 4 days of feeding (Fig. 5b-a2, b2, c2). Interestingly, the green part of the leaf was preferentially eaten by larvae compared with red patches on the same leaf.

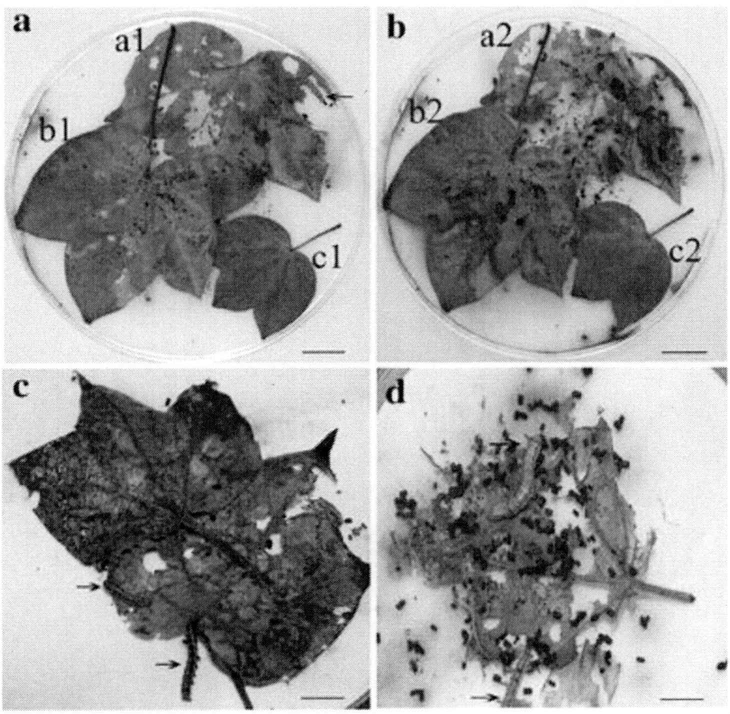


**Figure. 5** Bioassay for cotton bollworm feeding on *Lc* and Wt leaves. **A** Leaf images after 2 days of feeding. *b1* and *c1* are *Lc*-transgenic leaves, *a1* is a Wt leaf. The *arrow points* to worm (*bar* 2.2 cm). **B** Leaf images after 4 days of feeding. *b2* and *c2* are *Lc*-transgenic leaves, *a2* is a Wt leaf (*bar* 2.2 cm). **C** Image showing larvae after feeding 5 days on *Lc*-transgenic red leaves. The *arrows point* to worms (*bar* 2 cm). **D** Image showing larvae after feeding 5 days on Wt leaves. The *arrows point* to worms (*bar* 1.8 cm)

A force-feeding assay is shown in Fig. 5; insect larvae that fed on transgenic leaves after 5 days (Fig. 5c) hardly ate and were immobile, and they were visibly smaller than those fed on non-transgenic leaves those were more active(Fig. 5d). To compare herbivore resistance-associated parameters between Wt and *Lc*-transgenic lines, the differences in larvae weight after feeding on Wt or *Lc*-transgenic leaves for 5 days, 13 groups data produced automatically a bar chart as Fig. 6 using Excel software, and was analyzed using a one-way ANOVA with SPSS statistical software. The analysis results are presented in Table 1; the $F$ test value is 15.792, and $P < 0.01$. This illustrated that there was a significant difference between the larvae weights which fed on transgenic *Lc* leaves or non-transgenic leaves (Table 1). The larvae weight of those fed on non-transgenic leaves was significantly heavier than those fed on *Lc*-transgenic leaves. These data showed that expression of *Lc* confirms transgenic cotton resistance to cotton bollworms.

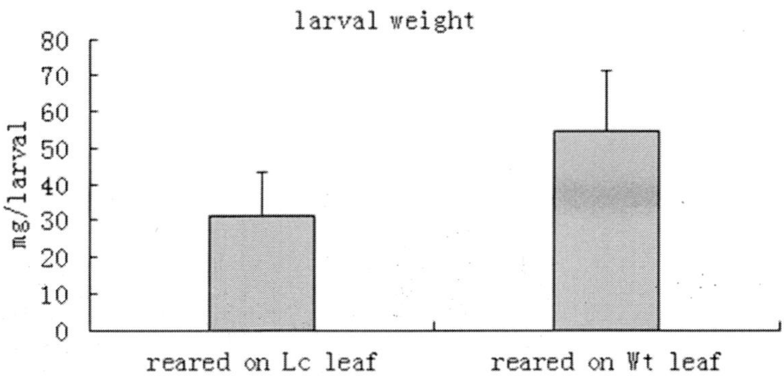


**Figure. 6** Statistical analysis of the larval weights after 3 days of rearing on *Lc*-transgenic and Wt leaves from Fig. 5b values are from 13 samples and represent averages ± STEV. Every *bar* represents the average weight repeated three times ($n = 13$)

**Table 1** Larval weights of cotton bollworms fed on *Lc*-transgenic leaves or Wt leaves were analyzed using a one-way ANOVA test by SPSS

| | Sum of squares | df | Mean square | $F$ | Sig |
|---|---|---|---|---|---|
| Between groups | 3301.885 | 1 | 3301.885 | 15.792 | 0.001* |
| Within groups | 5018.000 | 24 | 209.083 | | |
| Total | 8319.885 | 25 | | | |

Statistical analysis of the larval weights ($n = 13$) after 3 days of rearing on *Lc* and Wt leaves. Average weight measurements were repeated three times ($n = 13$). $P < 0.01$* indicates that there was a significant difference between the two groups

# 4. DISCUSSION

## 4.1 Expression of maize *Lc* promotes anthocyanin production in cotton under light

Our data showed that there were 6 lines of regenerated plants obtained from 16 lines of transgenic calluses. These plants were positive and confirmed by Southern blotting using *Lc* probe. Since there is only one *BamH*1 site in the T-DNA, the number of bands of hybridization is the T-DNA copy number of insertion into plant genome (Fig. 1a). From Fig. 1b, we find that the T-DNA was stably transmitted to the T1 generation and segregated in a 3:1 ratio as a single gene insertion in line of L143. In the positive transgenic plants, the maize *Lc* gene under the CaMV35S promoter enhanced pigmentation in both vegetative and most floral tissues except petals. During tissue culture process, it is evident that the color of the transgenic calluses varies from pink, red, dark red to dark purple due to the amount and the localized production of anthocyanin. Similar phenomenon has been observed in rice tissues (Gandikota et al. 2001). The purple colored calluses became more intense to dark purple during culture leading to no somatic embryos formed from them. Because too much anthocyanin or toxic flavonoids in this kind of calluses was detrimental to cells (Burbulis and Winkel-Shirley 1999; Gandikota et al. 2001). And those with slight pigment, red or pink colored calluses can continuously propagate to form somatic embryos. Purple or red colored calluses never have been seen in control untransformed cultures. Therefore, *Lc* gene could be used as a color-based selection marker for genetic transformation in cotton. In transgenic embryos, color intensity varies among embryos and no consistent pattern was observed. One type of embryos with slight pink color on hypocotyl, cotyledon and radicle can develop to maturity and germinate. Another type of embryos with dark purple around the whole body or most of the embryo were often arrested before the torpedo stage. Might anthocyanin accumulation have adverse effects on hypocotyl formation? Consistent observation has reported in rice that the maize *Lc* gene up-regulated the expression levels of various anthocyanin regulatory genes leading to sterility or abnormal seed development (Li et al. 2013). Anthocyanin production and pigmentation involve multiple steps and often display cell and tissue-specific patterns. In *Lc*-transgenic leaves, anthocyanin accumulates mainly in veins and palisade mesophyll cells rather than epidermis. There was stronger expression of signal in leaves and petioles. And the expression level increased slightly from the 1st leaf to the 3rd leaf. In reproductive tissues, although pigment and signal appeared in anther, sepal and stigma at the same stage in the same cotton flower, neither red color nor transcription signal was observed in the petals. This expression pattern was mainly determined by *Lc* gene in cotton rather than the CaMV 35S promoter which stimulated higher expression levels in all floral parts, including petals in cotton (Sunilkumar et al. 2002). Therefore, it is inferred that the *Lc*-activated pathway is under developmental control, and the transgene expression is regulated both spatially and temporally (Martin and Gerats 1993); like in *Lc*-transgenic tomato fruits, the reddish-purple pigmentation disappears as fruit ripening progresses (Bovy et al. 2002). Another reasoning based on studies in petunia may be that in the petals and anthers anthocyanin synthesis is controlled by two different ternary complexes (Petroni and Tonelli 2011). In the ovule or fiber cells, no regular pattern of transgene expression was found. Although stronger expression signal presented throughout the fiber developmental phase, the fibers still appeared white when the bolls opened. Is the color due to anthocyanin in cotton fibers going to last processing or will it fade when dried? Or anthocyanins are soluble pigment molecules and generally may not be stable in tissues after drying? To resolve this problem, we open the cotton bolls in every development stage, and found that the fiber was white all the way. Therefore, the test of ovule cultured in vitro under light was done. The light induction experiment showed that the white fiber changed to red color, and leaf turned redder, suggesting that light signaling is essential for the *Lc*-activated anthocyanin biosynthesis or the conversion of colorless intermediates into colored pigments in fibers and leaves, because light can influence the accumulation of anthocyanin primarily through the activation of the transcription factors that regulate the flavonoid biosynthetic pathway (Irani and Grotewold 2005). But it is not clear whether the light signaling activates additional genes essential for the conversion of colorless anthocyanin to colored anthocyanin, or enhanced *Lc* expression in fiber cells. Therefore breeding a red fiber germplasm needs to solve this problem of having to increase anthocyanin accumulation in the fiber without having to culture them in vitro or to find another method of induction of the transgene expression.

## 4.2 Enhanced resistance against herbivores with anthocyanin content increased

Anthocyanins is one class of flavonoid compounds, and flavonoids are polyphenolic phytochemicals that constitute a large group of secondary metabolites in plants. Anthocyanin biosynthesis related to flavonoid biosynthesis produces important compounds for the defense against insects and pathogens (Dooner et al. 1991; Moctezuma et al. 2014). Although there have been relatively few reports concerning resistance against herbivores in *Lc*-transgenic plants, red leaves have been suggested by Hamilton and Brown (2001), to function as a visual warning to herbivores. Owing to increased amounts of defensive compounds in red leaves, the leaves containing higher concentrations of anthocyanins incur less natural herbivory. Additionally, the red leaves provide an effective chemical defense against larvae (Cooney et al. 2012) and anthocyanins in red leaf, as secondary metabolites, can function in plant defense against herbivores (Onkokesung et al. 2014). Analogous report showed that red kidney bean *Sayyad* was the most unsuitable host for feeding cotton bollworm (*H. armigera.common*) among common bean *Talash* and white kidney bean *Pak* (Namin et al. 2014). Similarly, transgenic plants that produced more anthocyanin exhibited a higher level of resistance against environmental stresses owing to their altered physiology (Flachowsky et al. 2010). Our study showed that the foreign *Lc* gene was not only expressed in cotton but also increased the anthocyanin content in transgenic leaves, and led to the green leaf changing to red or purple. In feeding experiment, cotton bollworms disfavor consuming red transgenic leaves, but ate green leaves. This is likely due to the higher anthocyanin in fresh leaves, which has been reported in the past to minimize herbivory (Cooney et al. 2012). The statistical analysis of insect weights showed that there was a significant difference between insects who fed on transgenic leaves and those fed on non-transgenic leaves ($P < 0.01$). This is most attributable to the accumulation of anthocyanin in the *Lc*-transgenic leaves. One aspect, cyanidin-3-β-glucoside is an important resistance factor in leaves that acts against the budworm and provides a potential basis for achieving insect resistance (Hedin et al. 1983). Cyanidin chloride and cyanidin-3-glucoside chloride slowed the growth of larvae when added to insect diet (Johnson et al. 2010). In another aspect, condensed tannin (polymeric anthocyanin) is also an important factor in plant–herbivore interactions (Furstenburg and van Hoven 1994; Boeckler et al. 2011). In plants, tannins serve to increase insect resistance by inhibiting several digestive enzymes, including proteases, pectinases, amylases, cellulases and lipases, as well as reducing protein digestibility. And these enzyme-inhibiting components decrease the conversion rate of food in larvae when reared on red kidney bean (Namin et al. 2014). Therefore, the resistance ability against herbivores, which was enhanced in *Lc*-transgenic leaves, was mainly attributable to the increased total anthocyanins.

In conclusion, based on our results, *Lc* alone is sufficient to trigger anthocyanin synthesis in different cell types in cotton under light conditions, and also increased resistance against bollworm. This may provide an alternative control method for cotton bollworms. Additionally, breeding a red fiber cotton variety through the regulation of anthocyanin biosynthesis in cotton fiber may be plausible in the near future.

# ACKNOWLEDGMENTS

The authors thank Dr. Lim Saw Hoon at Malaysia University of Science and Technology, Malaysia, for providing the *Lc* construct. Special thanks to Dr. Yan Hong at Temasek Life Sciences Laboratory, Singapore, and Professor Nam-Hai Chua at Rockefeller University, USA, for their input. This work was supported by a grant from the Shanxi Abroad Returnee Office, Science and Research Project (2010) to Xiaoping Fan, and a grant (2006AA10A108) from the Ministry of Science and Technology of China to Weicai Yang.

# REFERENCES

1. Albert NW, Lewis DH, Zhang H, Irving LJ, Jameson PE, Davies KM (2009) Light induced vegetative anthocyanin pigmentation in Petunia. J Exp Bot 60:2191–2202
2. Beasley CA, Ting IP (1974) Effects of plant growth substances on in vitro fiber development from unfertilized cotton ovules. Am J Bot 61:188–194
3. Boase MR, Bradley JM, Borst NK (1998) Genetic transformation mediated by Agrobacterium tumefaciens of florists' chrysanthemum (Dendranthema 9 Grandiflorum) cultivar "Peach Margaret". In Vitro Cell Dev Biol 34:46–51
4. Boeckler GA, Gershenzon J, Unsicker SB (2011) Phenolic glycosides of the salicaceae and their role as anti-herbivore defenses. Phytochemistry 72:1497–1509
5. Boudet AM (2007) Evolution and current status of research in phenolic compounds. Phytochemistry 68:2722–2735
6. Bovy A, de Vos R, Kemper M, Shijlen E, Almenar PM, Muir S, Collins G, Robinson S, Verhoeyen M, Hughes S, Santos BC, van Tunen A (2002) High-flavonol tomatoes resulting from the heterologous expression of the maize transcription factor genes Lc and C1. Plant Cell 14:2509–2526
7. Bradley JM, Davies KM, Deroles SC, Bloor SJ, Lewis DH (1998) The maize Lc regulatory gene up-regulates the flavonoid biosynthetic pathway of Petunia. Plant J 13:381–392
8. Bradley JM, Deroles SC, Boase MR, Bloor SJ, Swinny E, Davies KM (1999) Variation in the ability of the maize Lc regulatory gene to upregulate flavonoid biosynthesis in heterologous systems. Plant Sci 140:31–39
9. Burbulis IE, Winkel-Shirley B (1999) Interactions among enzymes of the Arabidopsis flavonoid biosynthetic pathway. Proc Natl Acad Sci USA 96:12929–12934
10. Chalker-Scott L (1999) Environmental significance of anthocyanin in plant stress responses. Photochem Photobiol 70:1–9
11. Chandler VL, Radicella JP, Robbins TP, Chen J, Turks D (1989) Two regulatory genes of the maize anthocyanin pathway are homologous: isolation of B utilizing R genomic sequences. Plant Cell 1:1175–1183
12. Choung MG, Baek IY, Kang ST, Han WY, Shin DC, Moon HP, Kang KH (2001) Isolation and determination of anthocyanins in seed coats of black soybean. J Agric Food Chem 49:5848–5851
13. Cone KC, Burr FR, Burr B (1986) Molecular analysis of the maize anthocyanin regulatory locus Cl. Proc Natl Acad Sci USA 83:9631–9635
14. Consonni G, Geuna F, Gavazzi G, Tonelli C (1993) Molecular homology among members of the R gene family in maize. Plant J 3:335–346

15. Cooney LJ, van Klink JW, Hughes NM, Perry NB, Schaefer HM, Menzies IJ, Gould KS (2012) Red leaf margins indicate increased polygodial content and function as visual signals to reduce herbivory in Pseudowintera colorata. New Phytol 194:488–497
16. Cortell JM, Halbleib M, Gallagher AV, Righetti TL, Kennedy JA (2007) Influence of vine vigor on grape (Vitis vinifera L. Cv. Pinot Noir) anthocyanins. 2. Anthocyanins and pigmented polymers in wine. J Agric Food Chem 55(16):6585–6595
17. Dooner HK, Robbins TP, Jorqensen RA (1991) Genetic and developmental control of anthocyanin biosynthesis. Ann Rev Genet 25:173–199
18. Flachowsky H, Szankowski Fischer TC, Richter K, Peil A, Ho¨fer M, Do¨rschel C, Schmoock S, Gau AE, Halbwirth H, Hanke MV (2010) Transgenic apple plants overexpressing the Lc gene of maize show an altered growth habit and increased resistance to apple scab and fire blight. Planta 231:623–635
19. Furstenburg D, van Hoven W (1994) Condensed tannin as antidefoliate agent against browsing by giraffe (Giraffa camelopardalis) in the Kruger National Park. Comp Biochem Physiol 107:425–431
20. Gandikota M, de Kochko A, Chen L, Ithal N, Fauquet C, Reddy AR (2001) Development of transgenic rice plants expressing maize anthocyanin genes and increased blast resistance. Mol Breeding 7:73–83
21. Goldsbrough AP, Tong Y, Yode JI (1996) Lc as a non-destructive visual reporter and transposition excision maker gene for tomato. Plant J 9:927–933
22. Gong ZZ, Yamazaki M, Sugiyama M, Tanaka Y, Saito K (1997) Cloning and molecular analysis of structural genes involved in anthocyanin biosynthesis and expressed in a forma-specific manner in Perilla frutescens. Plant Mol Biol 35:915–927
23. Gould KS, Markham KR, Smith RH, Goris JJ (2000) Functional role of anthocyanin in the leaves of Quintina serrata A Cunn. J Exp Bot 51:1107–1115
24. Hamilton WD, Brown SP (2001) Autumn tree colours as a handicap signal. Proc R Soc Lond B 268:1489–1493
25. Han YJ, Kim YM, Lee JY, Kim SJ, Cho KCh, Chandrasekhar T, Song PS, Woo YM, Kim J (2009) Production of purple-colored creeping bentgrass using maize transcription factor genes Pl and Lc through Agrobacterium-mediated transformation. Plant Cell Rep 28:397–406
26. Hedin PA, Jenkins JN, Collum DH, White WH, Parrott WL, Macgown MW (1983) Cyanidin-3-ß-glucoside, a newly recognized basis for resistance in cotton to the tobacco budworm. Experientia 39:799–801
27. Huang ZA, Zhao T, Fan HJ, Wang N, Zheng SS, Ling HQ (2012) The upregulation of NtAN2 expression at low temperaturre is required for anthocyanin accumulation in juvenile leaves of Lc-transgenic tobacco (Nicotiana tabacum L.). J Genet Genomics 39:149–156
28. Irani NG, Grotewold E (2005) Light-induced morphological alteration in anthocyanin accumulating vacuoles of maize cells. BMC Plant Biol 5:1471–2229
29. Johnson ET, Berhow MA, Dowd PF (2010) Constitutive expression of the maize genes B1 and C1 in transgenic Hi II maize results in differential tissue pigmentation and generates resistance to Helicoverpa zea. J Agric Food Chem 58:2403–2409
30. Kong Q, Pattanaik S, Feller A, Werkman JR, Chai C, Wang Y, Grotewold E, Yuan L (2012) Regulatory switch enforced by basic helix-loop-helix and ACT-domain mediated dimerizations of the maize transcription factor R. Proc Natl Acad Sci USA 109:E2091–E2097
31. Li XB, Cai L, Cheng NH, Liu JW (2002) Molecular characterization of the cotton GhTUB1 gene that is preferentially expressed in fiber. Plant Physiol 130:666–674
32. Li XB, Fan XP, Wang XL, Cai L, Yang WC (2005) The cotton ACTIN1 gene is functionally expressed in fibers and participates in fiber elongation. Plant Cell 17:859–875
33. Li Y, Zhang T, Shen Z, Xu Y, Li JY (2013) Overexpression of maize anthocyanin regulatory gene Lc affects rice fertility. Biotechnol Lett 35:115–119
34. Lloyd AM, Walbot V, Davis RW (1992) Arabidopsis and Nicotiana anthocyanin production activated by maize regulators R and Cl. Science 258:1773–1775
35. Ludwig SR, Wessler SR (1990) Maize R gene family: tissue-specific helix-loop-helix proteins. Cell 62:849–851

36. Ludwig SR, Habera LF, Dellaporta SL, Wessler SR (1989) Lc, a member of the maize R gene family responsible for tissuespecific anthocyanin production, encodes a protein similar to transcriptional activators and contains the myc-homology region. Proc Natl Acad Sci USA 86:7092–7096
37. Martin C, Gerats T (1993) Control of pigment biosynthesis genes during petal development. Plant Cell 5:1253–1264
38. Moctezuma C, Hammerbacher A, Heil M, Gershenzon J, Me´ndez- Alonzo R, Oyama K (2014) Specific polyphenols and tannins are associated with defense against insect herbivores in the tropical oak Quercus oleoides. J Chem Ecol 40:458–467
39. Murashige T, Skoog F (1962) A revised medium for rapid growth and bioassays with tobacco tissue cultures. Physiol Plant 15:473–497
40. Namin FR, Naseri B, Razmjou J (2014) Nutritional performance and activity of some digestive enzymes of the cotton bollworm, Helicoverpa armigera, in response to seven tested bean cultivars. J Insect Sci. doi:10.1093/jis14-1-93
41. Onkokesung N, Reichelt M, van Doorn A, Schuurink RC, van Loon JJ, Dicke M (2014) Modulation of flavonoid metabolites in Arabidopsis thaliana through overexpression of the MYB75 transcription factor: role of kaempferol-3,7-dirhamnoside in resistance to the specialist insect herbivore Pieris brassicae. J Exp Bot 65:2203–2217
42. Petroni K, Tonelli C (2011) Recent advances on the regulation of anthocyanin synthesis in reproductive organs. Plant Sci 181:219–229
43. Procissi A, Dolfini S, Ronchi A, Tonelli C (1997) Light-dependent spatial and temporal expression of pigment regulatory genes in developing maize seeds. Plant Cell 9:1547–1557
44. Quattrocchio F, Wing JF, Leppen HT, Mol J, Koes RE (1993) Regulatory genes controlling anthocyanin pigmentation are functionally conserved among plant species and have distinct sets of target genes. Plant Cell 5:1497–1512
45. Ray H, Yu M, Auser P, Blahut-Beatty L, McKersie B, Bowley S, Westcott N, Coulman B, Lloyd A, Gruber MY (2003) Expression of anthocyanins and proanthocyanidins after transformation of alfalfa with maize Lc. Plant Physiol 132:1448–1463
46. Schwinn KE, Boase MR, Bradley JM, Lewis DH, Deroles SC, Martin CR, Davies KM (2014) MYB and bHLH transcription factor transgenes increase anthocyanin pigmentation in petunia and lisianthus plants, and the petunia phenotypes are strongly enhanced under field conditions. Front Plant Sci 5:603. doi:10. 3389/fpls-2014-00603
47. Sharma M, Chai C, Morohashi K, Grotewold E, Snook ME, Chopra S (2012) Expression of flavonoid 3'-hydroxylase is controlled by P1, the regulator of 3-deoxyflavonoid biosynthesis in maize. BMC Plant Biol 12:196. doi:10.1186/1471-2229-12-196
48. Somers TC, Evans ME (1974) Wine quality: correlations with colour density and anthocyanin equilibria in a group of young red wines. J Sci Food Agric 25:1369–1379
49. Stinard PS, Kermicle JL, Sachs MM (2009) The maize enr system of r1 haplotype-specific aleurone color enhancement factors. J Hered 100:217–228
50. Sunilkumar G, Mohr L, Lopata-Finch E, Emani C, Rathore KS (2002) Developmental and tissue-specific expression of CaMV 35S promoter in cotton as revealed by GFP. Plant Mol Biol 50:463–474

# CHAPTER 13

# Environmental Biotechnology for Bioconversion of Agricultural and Forestry Wastes into Nutritive Biomass

**Marian Petre and Violeta Petre**

*1 Department of Natural Sciences, Faculty of Sciences, University of Pitesti, Romania*

*2 Department of Fruit Growing, Faculty of Horticulture, University of Agronomic Sciences and Veterinary Medicine-Bucharest, Romania*

## 1. INTRODUCTION

The cellulose is the most widely distributed skeletal polysaccharide and represents about 50% of the cell wall material of plants. Beside hemicellulose and lignin, cellulose is a major component of agricultural wastes and municipal residues. The cellulose and hemicellulose comprise the major part of all green plants and this is the main reason of using such terms as "cellulosic wastes" or simply "cellulosics" for those materials which are produced especially as agricultural crop residues, fruit and vegetable wastes from industrial processing, and other solid wastes from canned food and drinks industries.

The cellulose biodegradation using fungal cells is essentially based on the complex interaction between biotic factors, such as the morphogenesis and physiology of fungi, as the cellulose composition and its complexness with hemicellulose and lignin (Andrews & Fonta, 1988; Carlile & Watkinson, 1996).

An efficient method to convert cellulose materials, in order to produce unconventional high-calorie foods or feeds, is the direct conversion by cellulolytic microorganisms. Theoretically, any microorganism that can grow as pure culture on cellulose substrata, used as carbon and energy sources, should be considered a potential organism for "single-cell protein" (SCP) or "protein rich feed" (PRF) producing.

# 2. BIOTECHNOLOGY OF MYCELIA BIOMASS PRODUCING THROUGH SUBMERGED BIOCONVERSION OF AGRICULTURAL CROP WASTES

The submerged cultivation of mushroom mycelia is a promising method which can be used in novel biotechnological processes for obtaining pharmaceutical substances of anticancer, antiviral, immuno-modulating, and anti-sclerotic action from fungal biomass and cultural liquids and also for the production of liquid spawn (Breene, 1990).

The researches that were carried out to get nutritive supplements from the biomass of Ganoderma lucidum species (Reishi) have shown that the nutritive value of its mycelia is owned to the huge protein content, carbohydrates and mineral salts. Lentinula edodes species (Shiitake) is a good source of proteins, carbohydrates (especially polysaccharides) and mineral elements with beneficial effects on human nutrition (Wasser & Weis, 1994; Mizuno et al., 1995).

It is well known the anti-tumor activity of polysaccharide fractions extracted from mycelia of Pleurotus ostreatus, known on its popular name as Oyster Mushroom (Mizuno et al., 1995; Hobbs, 1996).

The main purpose of this research work consists in the application of biotechnology for continuous cultivation of edible and medicinal mushrooms by submerged fermentation in agro-food industry which has a couple of effects by solving the ecological problems generated by the accumulation of plant wastes in agro-food industry through biological means to valorise them without pollutant effects as well as getting fungal biomass with high nutritive value which can be used to prepare functional food (Carlile & Watkinson, 1996; Moser, 1994).

The continuous cultivation of medicinal mushrooms was applied using the submerged fermentation of natural wastes of agro-food industry, such as different sorts of grain by-products as well as winery wastes that provided a fast growth as well as high biomass productivity of the investigated strains (Petre & Teodorescu, 2012; Petre & Teodorescu, 2011).

## 2.1. Materials and methods

Ganoderma lucidum (Curt. Fr.) P. Karst, Lentinula edodes (Berkeley) Pegler and Pleurotus ostreatus (Jacquin ex Fries) Kummer were used as pure strains. The stock cultures were maintained on malt-extract agar (MEA) slants, incubated at 25°C for 5-7 d and then stored at 4°C. The seed cultures were grown in 250-ml flasks containing 100 ml of MEA medium (20% malt extract, 2% yeast extract, 20% agar-agar) at 23°C on rotary shaker incubator at 100 rev.min$^{-1}$ for 7 d (Petre & Petre, 2008; Petre et al., 2007).

The fungal cultures were grown by inoculating 100 ml of culture medium using 3-5% (v/v) of the seed culture and then cultivated at 23-25°C in rotary shake flasks of 250 ml. The experiments were conducted under the following conditions: temperature, 25°C; agitation speed, 120 rev. min$^{-1}$; initial pH, 4.5–5.5.

After 10–12 d of incubation the fungal cultures were ready to be inoculated aseptically into the glass vessel of a laboratory-scale bioreactor (Fig. 1).

For fungal growing inside the culture vessel of this bioreactor, certain special culture media were prepared by using liquid nutritive broth, having the following composition: 15% cellulose powder, 5% wheat bran, 3% malt extract, 0.5% yeast extract, 0.5% peptone, 0.3% powder of natural argillaceous materials. After the steam sterilization at 121oC, 1.1 atm., for 15 min. this nutritive broth was transferred aseptically inside the culture vessel of the laboratory scale bioreactor shown in figure 1.

The culture medium was aseptically inoculated with activated spores belonging to *G. lucidum*, *L. edodes* and *P. ostreatus* species. After inoculation into the bioreactor vessel, a slow constant flow of nutritive liquid broth was maintained inside the nutritive culture medium by recycling it and adding from time to time a fresh new one.

The submerged fermentation was set up at the following parameters: constant temperature, 23°C; agitation speed, 80-100 rev. min$^{-1}$; pH level, 5.7–6.0 units; dissolved oxygen tension within the range of 30-70%. After a period of submerged fermentation lasting up to 120 h, small fungal pellets were developed inside the broth (Petre & Teodorescu, 2010; Petre & Teodorescu, 2009).

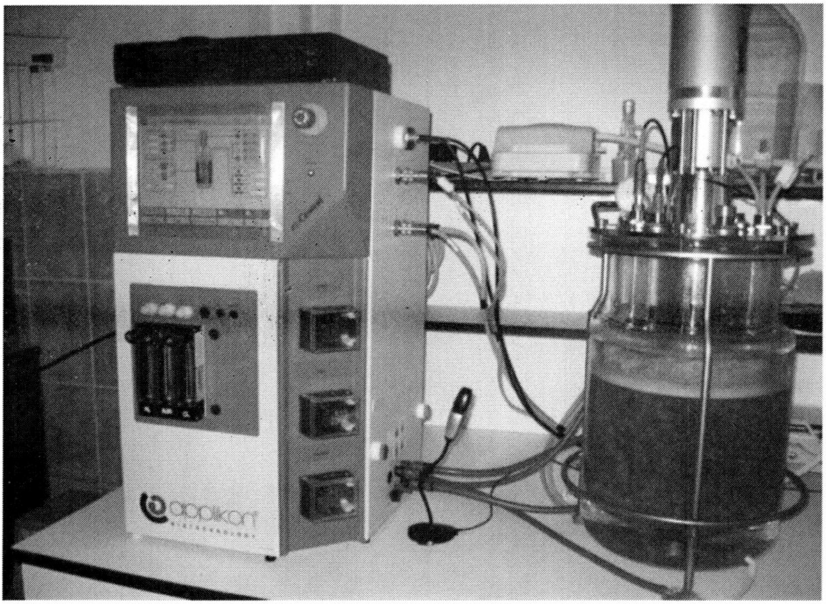


**Figure 1**. Laboratoy-scale bioreactor for submerged cultivation of edible and medicinal mushrooms

The experimental model of biotechnological installation, represented by the laboratory scale bioreactor shown in figure 1, was designed to be used in submerged cultivation of the mentioned mushroom species that were grown on substrata made of wastes resulted from the industrial processing of cereals and grapes (Table 1).

**Table 1**. The composition of compost variants used in mushroom cultures

| Variants of culture substrata | Composition |
|---|---|
| S1 | Mixture of winery wastes and wheat bran 2.5% |
| S2 | Mixture of winery wastes and barley bran 2.5% |
| S3 | Mixture of winery wastes and rye bran 2.5% |
| Control | Pure cellulose |

## 2.2. Results and discussion

The whole process of mushroom mycelia growing lasts for a single cycle between 5-7 days in case of *L. edodes* and between 3 to 5 days for *G. lucidum* and *P. ostreatus*. All experiments regarding the fermentation process were carried out by inoculating the growing medium volume (15 L) with secondary mycelium inside the culture vessel of the laboratory-scale bioreactor (see Fig. 1).

The strains of these fungal species were characterized by morphological stability, manifested by its ability to maintain the phenotypic and taxonomic identity. Observations on morphological and physiological characters of these two tested species of fungi were made after each culture cycle, highlighting the following aspects:

- sphere-shaped structure of fungal pellets, sometimes elongated, irregular, with various sizes (from 7 to 12 mm in diameter), reddish-brown colour of *G. lucidum* specific culture (Fig. 2a);
- globular structures of fungal pellets, irregular with diameters of 5 up to 10 mm or mycelia congestion, which have developed specific hyphae of *L. edodes* (Fig. 2b);
- round-shaped pellets with diameter measuring between 5 and 15 mm, having a white-cream colour and showing compact structures of *P. ostreatus* mycelia (Fig. 2c).

The experiments were carried out in three repetitions. Samples for analysis were collected at the end of the fermentation process, when pellets formed specific shapes and characteristic sizes. For this purpose, fungal biomass was washed repeatedly with double distilled water in a sieve with 2 mm diameter eye, to remove the remained bran in each culture medium (Petre at al., 2005a).

Biochemical analyses of fungal biomass samples obtained by submerged cultivation of edible and medicinal mushrooms were carried out separately for the solid fraction and extract fluid remaining after the separation of fungal biomass by pressing and filtering. Also, the most obvious sensory characteristics (color, odor, consistency) were evaluated and presented at this stage of biosynthesis taking into consideration that they are very important in the prospective view of fungal biomass using as raw matarials for nutraceuticals producing. In each experimental variant the amount of fresh biomass mycelia was analyzed.

Percentage amount of dry biomass was determined by dehydration at 70° C, until constant weight. The total protein content was investigated by using the biuret method, whose principle is similar to the Lowry method, being recommended for the protein content ranging from 0.5 to 20 mg/100 mg sample (Bae et al., 2000; Lamar et al., 1992).

The principle method is based on the reaction that takes place between copper salts and compounds with two or more peptides in the composition in alkali, which results in a red-purple complex, whose absorbance is read in a spectrophotometer in the visible domain (λ 550 nm). In addition, this method requires only one sample incubation period (20 min) eliminating the interference with various chemical agents (ammonium salts, for example).

In table 2 are presented the amounts of fresh and dry biomass as well as the protein contents for each fungal species and variants of culture media.

According to registered data, using a mixture of wheat bran 2.5% and winery wastes the growth of *G. lucidum* biomass was stimulated, while the barley bran led to increased growth of *L. edodes* mycelium and *G. lucidum* as well.

In contrast, the dry matter content was significantly higher when using barley bran 2.5% mixed with winery wastes for both species used. Protein accumulation was more intense when using barley bran compared with those of wheat bran and rye bran, at both mushroom species.

The sugar content of dried mushroom pellets collected after the biotechnological experiments was determined by using Dubois method. The mushroom extracts were prepared by immersion of dried pellets inside a solution of NaOH pH 9, in the ratio 1:5. All dispersed solutions containing the dried pellets were maintained 24 h at the precise temperature of 25 oC, in full darkness, with continuous homogenization to avoid the oxidation reactions.

After the removal of solid residues by filtration the samples were analyzed by the previous mention method (Wasser & Weis, 1994).

The nitrogen content of mushroom pellets was analyzed by Kjeldahl method. All the registered results are related to the dry weight of mushroom pellets that were collected at the end of each biotechnological culture cycle (Table 3).

Comparing all the registered data, it could be noticed that the correlation between the dry weight of mushroom pellets and their sugar and nitrogen contents is kept at a balanced ratio for each tested mushroom species.

From these mushroom species that were tested in biotechnological experiments *G. lucidum* (variant III) showed the best values concerning the sugar and total nitrogen content. On the very next places, *L. edodes* (variant I) and *G. lucidum* (variant II) could be mentioned from these points of view.

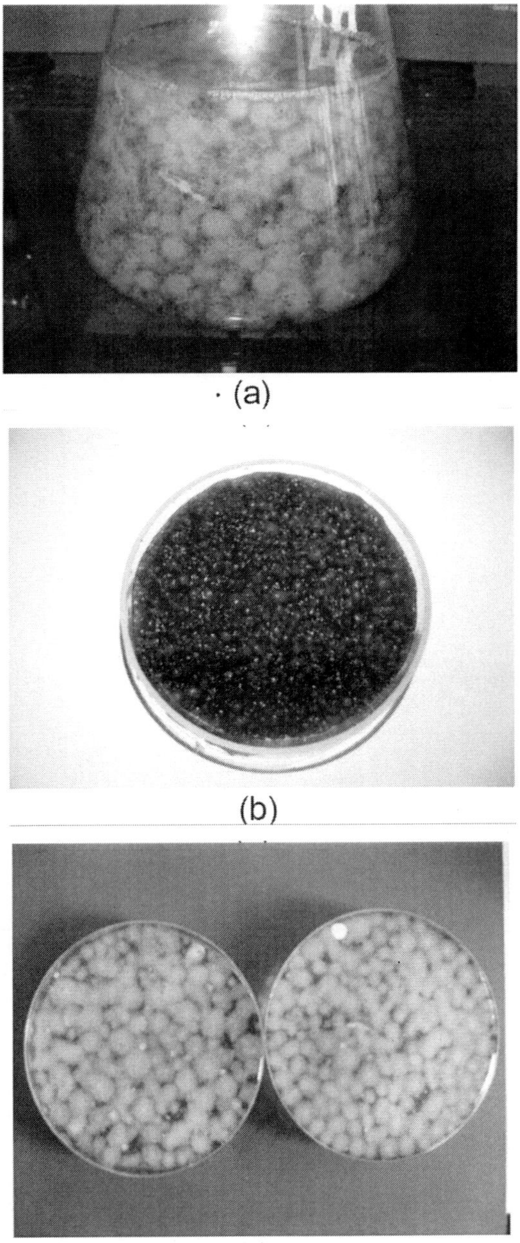


**Figure 2**. Fungal pellets of G. lucidum, b. Fungal pellets of L. Edodes,. c. Fungal pellets of P. ostreatus

**Table 2**. Fresh and dry biomass and protein content of G. lucidum, L. edodes and P. ostreatus mycelia grown by submerged fermentation

| Mushroom species | Culture variants | Fresh biomass (g) | Dry biomass (%) | Total proteins (g % d.w.) |
|---|---|---|---|---|
| *G. lucidum* | I | 25.94 | 9.03 | 0.67 |
| *G. lucidum* | II | 22.45 | 10.70 | 0.55 |
| *G. lucidum* | III | 23.47 | 9.95 | 0.73 |
| | Control | 5.9 | 0.7 | 0.3 |
| *L. edodes* | I | 20.30 | 5.23 | 0.55 |
| *L. edodes* | II | 23.55 | 6.10 | 0.53 |
| *L. edodes* | III | 22.27 | 4.53 | 0.73 |
| | Control | 4.5 | 0.5 | 0.2 |
| *P. ostreatus* | I | 21.50 | 5.73 | 0.63 |
| *P. ostreatus* | II | 23.95 | 7.45 | 0.55 |
| *P. ostreatus* | III | 23.25 | 4.79 | 0.75 |
| | Control | 4.7 | 0.5 | 0.3 |

**Table 3**. The sugar and total nitrogen contents of dried mushroom pellets

| Mushroom species | Culture variants | Mushroom pellets d. w. (%) | Sugar content of dried pellets (mg/ml) | Kjeldahl nitrogen of dried pellets (%) |
|---|---|---|---|---|
| *G. lucidum* | I | 17.64 | 4.93 | 5.15 |
| *G. lucidum* | II | 14.51 | 3.70 | 5.35 |
| *G. lucidum* | III | 20.16 | 5.23 | 6.28 |
| | Control | 0.7 | 0.45 | 0.30 |
| *L. edodes* | I | 19.67 | 4.35 | 6.34 |
| *L. edodes* | II | 17,43 | 3.40 | 5.03 |
| *L. edodes* | III | 15.55 | 4.75 | 6.05 |
| | Control | 0.5 | 0.45 | 0.35 |
| *P. ostreatus* | I | 19.70 | 5.15 | 6.43 |
| *P. ostreatus* | II | 14.93 | 4.93 | 6.25 |
| *P. ostreatus* | III | 15.63 | 5.10 | 5.83 |
| | Control | 0.55 | 0.50 | 0.35 |

The registered results concerning the sugar and total nitrogen contents have higher values than those obtained by other researchers (Bae et al., 2000; Jones, 1995; Moo-Young, 1993). The nitrogen content in fungal biomass is a key factor for assessing its nutraceutical potential, but the assessing of differential protein nitrogen compounds requires additional investigations.

# 3. LABORATORY-SCALE BIOTECHNOLOGY OF EDIBLE MUSHROOM PRODUCING ON GROWING COMPOSTS OF APPLE AND WINERY WASTES

The agricultural works as well as the industrial activities related to apple and grape processing have generally been matched by a huge formation of wide range of cellulosic wastes that cause environmental pollution effects if they are allowed to accumulate in the environment or much worse they are burned on the soil (Petre, 2009; Verstrate & Top, 1992).

The solid substrate fermentation of plant wastes from agro-food industry is one of the challenging and technically demanding biotechnology that is known so far (Petre & Petre, 2008; Carlile & Watkinson, 1996).

The major group of fungi which are able to degrade lignocellulose is represented by the edible mushrooms of Basidiomycetes Class. Taking into consideration that most of the edible mushrooms species requires a specific micro-environment including complex nutrients, the influence of physical and chemical factors upon fungal biomass production and mushroom fruit bodies formation were studied by testing new biotechnological procedures (Petre & Petre, 2008; Moser, 1994; Beguin & Aubert, 1994; Chahal & Hachey, 1990).

The main aim of research was to find out the best biotechnology of recycling the apple and winery wastes by using them as a growing source for edible mushrooms and, last but not least, to protect the environment (Petre et al., 2008; Smith, 1998; Raaska, 1990).

## 3.1. Materials and methods

Two fungal species of Basidiomycetes group, namely *Lentinula edodes* (Berkeley) Pegler (folk name: Shiitake) as well as *Pleurotus ostreatus* (Jacquin ex Fries) Kummer (folk name: Oyster Mushroom) were used as pure mushroom cultures isolated from the natural environment and now being preserved in the local collection of the University of Pitesti.

The stock cultures were maintained on malt-extract agar (MEA) slants (20% malt extract, 2% yeast extract, 20% agar-agar). Slants were incubated at 25°C for 120-168 h and stored at 4°C. The pure mushroom cultures were expanded by growing in 250-ml flasks containing 100 ml of liquid malt-extract medium at 23°C on rotary shaker incubators at 110 rev. $min^{-1}$ for 72-120 h. To prepare the inoculum for the spawn cultures of *L. edodes* and *P. ostreatus* the pure mushroom cultures were inoculated into 100 ml of liquid malt-yeast extract culture medium with 3-5% (v/v) and then maintained at 23-25°C in 250 ml rotary shake flasks.

After 10–12 d of incubation the fungal cultures were inoculated aseptically into glass vessels containing sterilized liquid culture media in order to produce the spawn necessary for the inoculation of 10 kg plastic bags filled with compost made of winery and apple wastes.

These compost variants were mixed with other needed natural ingredients in order to improve the enzymatic activity of mushroom mycelia and convert the cellulose content of winery and apple wastes into protein biomass. The best compositions of five compost variants are presented in Table 4.

**Table 4**. The composition of five compost variants used in mushroom culture cycles

| Compost variants | Compost composition |
|---|---|
| S1 | Winery and apple wastes (1:1) |
| S2 | Winery wastes + wheat bran (9:1) |
| S3 | Winery wastes and rye bran (9:1) |
| S4 | Apple wastes and wheat bran (9:1) |
| S5 | Apple wastes + rye bran (9:1) |
| Control | Poplar, beech and birch sawdust (1:1:1) |

In this way, the whole bags filled with compost were steam sterilized at 121oC, 1.1 atm., for 30 min. In the next stage, all the sterilized bags were inoculated with liquid mycelia, and then, all inoculated bags were transferred into the growing chambers for incubation. After 10-15 d, on the surface of sterilized plastic bags filled with compost, the first buttons of mushroom fruit bodies emerged. For a period of 20-30 d there were harvested between 1.5–3.5 kg of mushroom fruit bodies per 10 kg compost of one bag (Petre et al., 2012; Oei, 2003; Stamets, 1993; Wainwright, 1992; Ropars et al., 1992).

## 3.2. Results and discussion

To increase the specific processes of winery and apple wastes bioconversion into protein of fungal biomass, there were performed experiments to grow the mushroom species of *P. ostreatus* and *L. edodes* on the previous mentioned variants of culture substrata (see Table 1).

During the mushroom growing cycles the specific rates of cellulose biodegradation were determined using the direct method of biomass weighing the results being expressed as percentage of dry weight (d.w.) before and after their cultivation (Stamets, 1993; Wainwright, 1992).

In order to determine the evolution of the total nitrogen content in the fungal biomass there were collected samples at precise time intervals of 50 h and they were analyzed by using Kjeldahl method. The registered results concerning the evolution of total nitrogen content in *P. ostreatus* biomass are presented in figure 3 and the data regarding *L. edodes* biomass could be seen in figure 4.

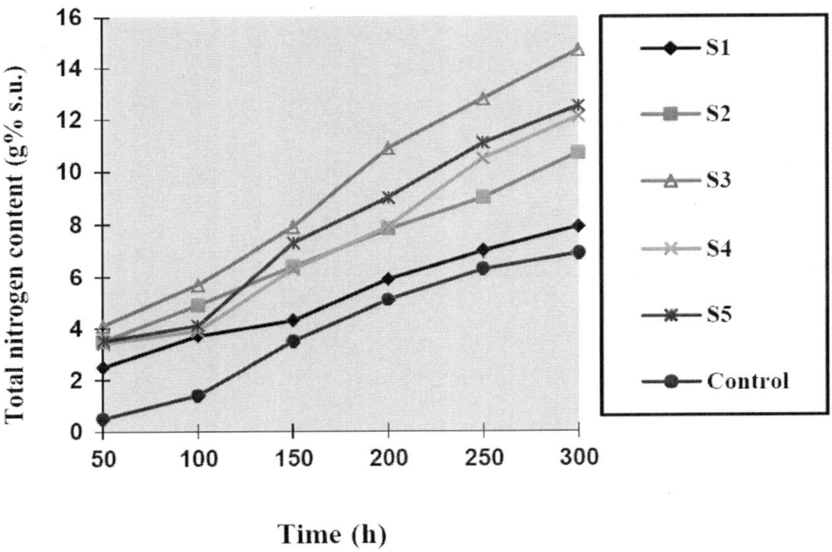


**Figure 3**. The evolution of total nitrogen content in P. ostreatus biomass

During the whole period of fruit body formation, the culture parameters were set up and maintained at the following levels, depending on each mushroom species:

- air temperature, 15–17oC;
- the air flow volume, 5–6m3/h;
- air flow speed, 0.2–0.3 m/s;
- the relative moisture content, 80–85%;
- light intensity, 500–1,000 luces for 8–10 h/d.

According to the registered results of the performed experiments the optimal laboratory-scale biotechnology for edible mushroom cultivation on composts made of marc of grapes and apples was established (Fig. 5).

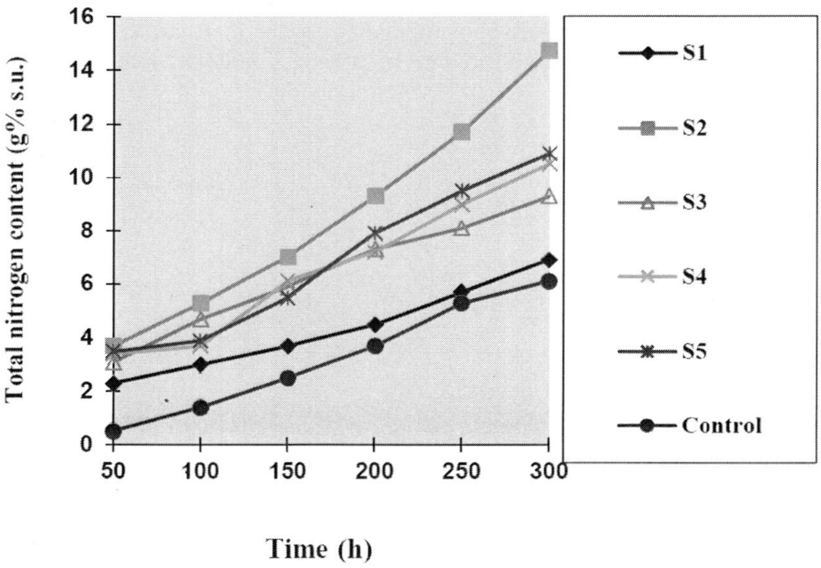


**Figure 4.** The evolution of total nitrogen content in L. edodes biomass

As it is shown in figure 5, two technological flows were carried out simultaneously until the first common stages of the inoculation of composts with liquid mushroom spawn followed by the mushroom fruit body formation.

The whole period of mushroom growing from the inoculation to the fruit body formation lasted between 30–60 d, depending on each fungal species used in experiments.

The registered data revealed that by applying such biotechnology, the winery and apple wastes can be recycled as useful raw materials for mushroom compost preparation in order to get significant mushroom production.

In this respect, the final fruit body production of these two mushroom species was registered as being between 20–28 kg relative to 100 kg of composts made of apple and winery wastes.

# 4. BIOTECHNOLOGY OF FORESTRY WASTES RECYCLING AS GROWING COMPOSTS FOR EDIBLE AND MEDICINAL MUSHROOM CULTURES

The most part of wastes produced all over the world arise from industrial, agricultural and domestic activities. These wastes represent the final stage of the technical and economical life of products (Verstraete & Top 1992).

As a matter of fact, the forestry works as well as the industrial activities related to forest management and wood processing have generally been matched by a huge formation of wide range of waste products (Beguin & Aubert 1994, Wainwright 1992).

Many of these lignocellulosic wastes cause serious environmental pollution effects, if they are allowed to accumulate in the forests or much worse to be burned for uncontrolled domestic purposes. So far, the basis of most studies on lignocellulose-degrading fungi has been economic rather than ecological, with emphasize on the applied aspects of lignin and cellulose decomposition, including biodegradation and bioconversion (Carlile & Watkinson 1996).

In this respect, the main aim of this work was focused on finding out the best way to convert the wood wastes into useful food supplements, such as mushroom fruit bodies, by using them as growing sources for the edible and medicinal mushrooms (Smith, 1998).

## 4.1. Materials and methods

### 4.1.1. Fungal species and culture media

According to the main purpose of this work, three fungal species from Basidiomycetes, namely *Ganoderma lucidum* (Curt.:Fr.) P. Karst, *Lentinus edodes* (Berkeley) Pegler and *Pleurotus ostreatus* (Jacquin ex Fries) Kummer were used as pure mushroom cultures during all experiments. The stock mushroom cultures were maintained by cultivating on malt-extract agar

(MEA) slants. After that, they were incubated at 25° C for 5-7 d and then stored at 4° C. These pure mushroom cultures were grown in 250-ml flasks containing 100 ml of MEA medium (20% malt extract, 2% yeast extract) at 23°C on rotary shaker incubators at 110 rev min $^{-1}$ for 5-7 d.

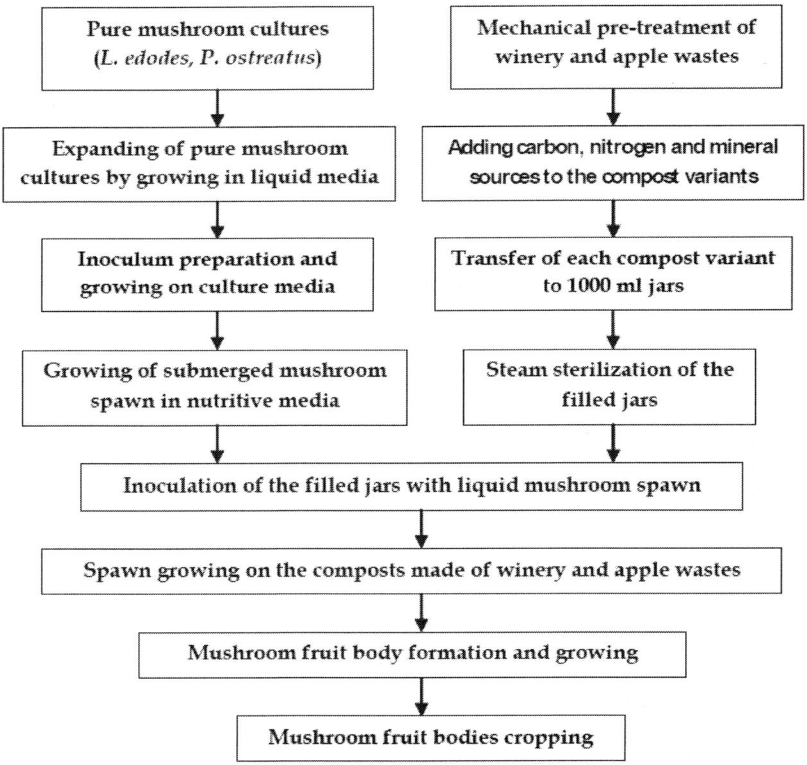


**Figure 5**. Scheme of laboratory-scale biotechnology for edible mushroom producing on winery and apple wastes

### 4.1.2. Methods used in experiments

#### *4.1.2.1. Preparation of submerged mycelia inoculum*

The pure mushroom cultures for experiments were prepared by inoculating 100 ml of culture medium with 3-5% (v/v) of the seed culture and then cultivated at 23-25°C in rotary shake flasks of 250 ml. The experiments were conducted under the following conditions:

- temperature, 25°C;
- agitation speed, 90-120 rev min-1;
- initial pH, 4.5–5.5.

The seed culture was transferred to the fungal culture medium and cultivated for 7–12 d (Petre et al., 2005a; Glazebrook et al., 1992).

#### *4.1.2.2. Incubation of mushroom cultures*

The experiments were performed by growing all the previous mentioned fungal species in special culture rooms, where all the culture parameters were kept at optimal levels in order to get the highest production of fruit bodies. The effects of culture compost composition (carbon, nitrogen and mineral sources) as well as other physical and chemical factors (such as: temperature, inoculum size and volume and incubation time) on mycelial net formation and especially, on fruit body induction were investigated (Petre & Petre, 2008).

All the culture composts for mushroom growing were inoculated using liquid inoculum with the age of 5–7 days and the volume size ranging between 3-7% (v/w). During the period of time of 18–20 d after this inoculation, all the fungal cultures had developed a significant biomass on the culture substrata made of wood wastes, such as: white poplar and beech wood sawdusts. These woody wastes were used as main ingredients to prepare natural composts for mushroom growing. The optimal temperatures for incubation and mycelia growth were maintained between 23–25°C. The whole period of mushroom growing from the inoculation to the fruit body formation lasted between 30–60 days, depending on each fungal species used in experiments (Petre & Teodorescu, 2010).

#### *4.1.2.3. Preparation of mushroom culture composts*

The lignocellulosic materials were mechanical pre-treated to breakdown the lignin and cellulose structures in order to induce their susceptibility to the enzyme actions during the mushroom growing. All these pre-treated lignocellulosic wastes were disinfected by steam sterilization at 120o C for 60 min (Petre et al., 2005b; Leahy & Colwell 1990).

The final composition of culture composts was improved by adding the following ingredients: 15-20% grain seeds (wheat, rye, rice) in the ratio 2:1:1, 0.7–0.9% CaCO3, 0.3–0.5% $NH_4H_2PO_4$, each kind of culture medium composition depending on the fungal species used to be grown. As control samples for each variant of culture composts used for the experimental growing of all these fungal species were used wood logs of white poplar and beech that were kept in water three days before the experiments and after that they were steam sterilized to be disinfected.

### *4.1.2.4. Preparation of mushroom spawn*

3000 g of white poplar sawdust and 1500 g of beech sawdust were mixed with cleaned and ground rye grain, 640 g of $CaCO_3$, 50 g of $NH_4H_2PO_4$ and 3550 ml of water, in order to obtain the growth substratum for mushroom spawn. The ingredients of such smal compost were mixed and then they were sterilized at 121° C, for 20 min. and allowed to cool until the mixture temperature decreased below 35° C. The spawn mixture was inoculated with 100-200 ml of liquid fungal inoculums and mixed for 10 min. to ensure complete homogeneity. Sterile polyethylene bags, containing microporus filtration strips, were filled with the smal composts and incubated at 25° C, until the spawn fully colonized the whole composts. At this point the spawn may be used to inoculate the mushroom growing substrate or alternatively it may be stored for up to 6 months at 4° C before use (Chahal & Hachey, 1990).

All the culture composts were inoculated using inoculum with the age of 5–7 d and the volume size ranging between 3-7% (v/w). The optimal temperatures for incubation and mycelia growth were maintained between 23–25°C. The whole period of mushroom growing from the inoculation to the fruit body formation lasted between 30–50 days.

### *4.1.2.5. Mushroom cultivation*

The experiments were carried out inside such in vitro growing rooms, where the main culture parameters (temperature, humidity, aeration) were kept at optimal levels to get the highest production of mushroom fruit bodies (Moser, 1994).

In order to find a suitable carbon source for the mycelia growth and consequently for fungal biomass synthesis, the pure cultures of *P. ostreatus* (Oyster Mushroom), as well as *L. edodes* (Shiitake) and *G. lucidum* (Reishi) were cultivated in different nutritive culture media containing various carbon sources, and each carbon source was added to the basal medium at a concentration level of 1.5% (w/v) for 7-12 d (Raaska, 1990).

To investigate the effect of nitrogen sources on mycelia growth and fungal biomass production, the pure cultures of these two fungal species were cultivated in media containing various nitrogen sources, where each nitrogen source was added to the basal medium at a concentration level of 10 g/l. At the same time, malt extract was one of the better nitrogen sources for a high mycelia growth. Peptone, tryptone and yeast extract are also known as efficient nitrogen sources for fungal biomass production by using the pure cultures of such fungal species (Chang & Hayes, 1978). In comparison with organic nitrogen sources, inorganic nitrogen sources gave rise to relatively lower mycelia growth and fungal biomass production (Bae et al., 2000).

The influence of mineral sources on fungal biomass production was examined at a standard concentration level of 5 mg. In order to study the effects of initial pH correlated with the incubation temperature upon fruit body formation, *G. lucidum, P. ostreatus and L. edodes* were cultivated on substrates made of wood wastes of white poplar and beech at different initial pH values (4.5–6.0). The experiments were carried out for 6 days at 25°C with the initial pH 5.5. Similar observations were made by Stamets (1993), during the experiments. $K_2HPO_4$ could improve the productivity through its buffering action, being favourable for mycelia growth. The experiments were carried out between 30-60 days at 25°C.

## 4.2. Results and discussion

The effects of carbon, nitrogen and mineral sources as well as other physical and chemical factors on mycelial net formation and especially, on fruit body induction were investigated by adding them to the main composts made of white poplar and beech sawdusts in the ratio 2:1. For the experimental growing of all these fungal species white poplar and beech logs were used as control samples.

### 4.2.1. The effect of carbon sources upon mushroom mycelia growth

When the cells were grown in the maltose medium, the fungal biomass production was the highest among the tested variants. Data presented in the following table are the means ± S.D. of triple determinations (Table 5).

**Table 5**. The effect of carbon sources upon the mycelia growth of pure mushroom cultures on white poplar and beech composts

| Carbon source (g/l) | Fresh Fungal Biomass Weight (g/l) | | | Final pH | | |
|---|---|---|---|---|---|---|
| | *G. lucidum* | *L. edodes* | *P. ostreatus* | *G. l* | *L. e* | *P. o* |
| Glucose | 27±0.10 | 41±0.05 | 43±0.03 | 5.5 | 5.3 | 5.1 |
| Maltose | 27±0.14 | 45±0.12 | 49±0.05 | 5.8 | 5.4 | 5.3 |
| Sucrose | 25±0.23 | 35±0.03 | 37±0.09 | 5.1 | 5.1 | 5.7 |
| Xylose | 26±0.07 | 38±0.07 | 35±0.07 | 5.3 | 5.5 | 5.9 |

What is very important to be noticed is that the maltose has a significant effect upon the increasing of mycelia growth and fungal biomass synthesis. The experiments were carried out for 12 days at 25 °C with the initial pH 5.5 (Petre, 2002).

### 4.2.2. The effect of nitrogen sources upon mushroom mycelia growth

Among five nitrogen sources examined, rice bran was the most efficient for mycelia growth and fungal biomass production. The experiments were carried out for 12 days at 25 °C with the initial pH 5.5 (Table 6).

**Table 6**. The effect of nitrogen sources upon the mycelia growth of pure mushroom cultures on white poplar and beech composts

| Nitrogen sources (1%, w/v) | Fresh Fungal Biomass Weight (g/l) | | | Final pH | | |
|---|---|---|---|---|---|---|
| | *G. lucidum* | *L. edodes* | *P. ostreatus* | *G. l* | *L. e* | *P. o* |
| Rice bran | 37±0.21 | 57±0.05 | 73±0.23 | 5.5 | 5.5 | 5.1 |
| Malt extract | 36±0.12 | 55±0.03 | 69±0.20 | 5.3 | 5.2 | 5.7 |
| Peptone | 35±0.03 | 41±0.12 | 57±0.15 | 4.6 | 4.9 | 5.3 |
| Tryptone | 36±0.15 | 38±0.07 | 55±0.17 | 5.1 | 5.3 | 5.9 |
| Yeast extract | 37±0.20 | 30±0.01 | 61±0.14 | 4.3. | 5.1 | 5.1 |

Data presented in table 6 are the means ± S.D. of triple determinations.

### 4.2.3. The effect of mineral sources upon mushroom mycelia growth

Among the various mineral sources examined, $K_2HPO_4$ yielded good mycelia growth as well as fungal biomass production and for this reason it was recognized as a favourable mineral source (Table 7). Data presented in table 7 are the means ± S.D. of triple determinations

**Table 7**. The effect of mineral source upon mycelia growth of pure mushroom cultures on white poplar and beech composts

| Mineral Sources (5 mg) | Fresh Fungal Biomass Weight (g/l) | | | Final pH | | |
|---|---|---|---|---|---|---|
| | *G. lucidum* | *L. edodes* | *P. ostreatus* | *G. l* | *L. e* | *P. o* |
| $KH_2PO_4$ | 37±0.15 | 45±0.07 | 53±0.12 | 5.5 | 5.3 | 5.9 |
| $K_2HPO_4$ | 45±0.07 | 57±0.05 | 59±0.07 | 5.1 | 5.1 | 5.7 |
| $MgSO_4 \cdot 5H_2O$ | 35±0.25 | 55±0.09 | 63±0.28 | 5.6 | 5.4 | 6.1 |

### 4.2.4 the influence of initial ph and temperature upon mushroom fruit body formation

The optimal pH and temperature levels for fungal fruit body production were 5.0–5.5 and 21–23°C (Table 8).

To find the optimal incubation temperature for mycelia growth, these fungal species were cultivated at different temperatures ranging from 20-25°C, and, finally, the optimum level of temperature was found at 23°C, being correlated with the appropriate pH level 5.5, at it is shown in Table 8. All data presented in the previous table are the means ± S.D. of triple determinations

**Table 8**. The effects of initial pH and temperature upon mushroom fruit body formation on white poplar and beech composts

| Initial pH (pH units) | Initial temperature (t°) | Final Weight of Fresh Mushroom Fruit Bodies (g / kg substratum) | | |
|---|---|---|---|---|
| | | *G. lucidum* | *L. edode* | *P. ostreatus* |
| 4.5 | 18 | 175±0.23 | 191±0.10 | 180±0.02 |
| 5.0 | 21 | 193±0.15 | 203±0.05 | 297±0.14 |
| 5.5 | 23 | 198±0.10 | 195±0.15 | 351±0.23 |
| 6.0 | 26 | 181±0.12 | 179±0.12 | 280±0.03 |
| 6.5 | 29 | 173±0.09 | 105±0.23 | 257±0.15 |

### 4.2.5. The influence of inoculum age and inoculum volume upon mushroom fruit body formation

Amongst several fungal physiological properties, the age and volume of mycelia inoculum may play an important role in fungal hyphae development as well as in fruit body formation (Petre & Teodorescu, 2012).

To examine the effect of inoculum age and inoculum volume, mushroom species *G. lucidum, P. ostreatus* and *L. edodes* were grown on substrates made of vineyard wastes during different time periods between 30 and 60 days, varying the inoculum volume (5 - 7 v/w).

All the experiments were carried out at 25°C and initial pH 5.5. As it is shown in Tables 9 and 10, the inoculum age of 120 h as well as an inoculum volume of 6.0 (v/w) have beneficial effects on the fungal biomass production.

**Table 9**. The effect of inoculum age upon mushroom fruit body formation on white poplar and beech composts

| Inoculum age (h) | Final Weight of Fresh Mushroom Fruit Bodies (g /kg substratum) | | |
|---|---|---|---|
| | *G. lucidum* | *L. edodes* | *P. ostreatus* |
| 264 | 123±0.14 | 128±0.05 | 135±0.23 |
| 240 | 141±0.10 | 150±0.28 | 157±0.17 |
| 216 | 154±0.12 | 195±0.90 | 193±0.15 |
| 192 | 155±0.23 | 221±0.25 | 215±0.05 |
| 168 | 169±0.37 | 235±0.78 | 241±0.07 |
| 144 | 210±0.20 | 248±0.03 | 259±0.12 |
| 120 | 230±0.15 | 253±0.05 | 264±0.21 |
| 96 | 215±0.09 | 230±0.15 | 253±0.10 |
| 72 | 183±0.05 | 205±0.23 | 210±0.05 |

**Table 10**. The effect of inoculum volume upon mushroom fruit body formation on white poplar and beech composts

| Inoculum Volume (v/w) | Final Weight of Fresh Mushroom Fruit Bodies (g /kg substratum) | | |
|---|---|---|---|
| | *G. lucidum* | *L. edodes* | *P. ostreatus* |
| 7.0 | 234±0.12 | 215±0.20 | 220±0.05 |
| 6.5 | 245±0.15 | 248±0.23 | 251±0.20 |
| 6.0 | 253±0.1 | 257±0.07 | 280±0.15 |
| 5.5 | 243±0.12 | 235±0.03 | 247±0.07 |
| 5.0 | 255±0.23 | 215±0.15 | 235±0.03 |

From all these fungal species tested, *P. ostreatus* was registered as the fastest mushroom (25–30 days), then *L. edodes* (35–45 days) and eventually, *G. lucidum* as the longest mushroom culture (40–50 days).

The registered data revealed that the white poplar and beech wood wastes have to be used as substrates for mushroom growing only after some mechanical pre-treatments (such as grinding) that could breakdown the whole lignocellulose structure in order to be more susceptible to the fungal enzyme action (Chahal, 1994).

Due to their high content of carbohydrates and nitrogen, the variants of culture composts supplemented with wheat grains at the ratio 1:10 and rice grains at the ratio 1:5 as well as a water content of 60% were optimal for the fruit body production of *P. ostreatus* and, respectively, L. edodes. The mushroom culture of G. lucidum does not need such supplements (Ropars et al., 1992; Lamar et al., 1992).

So far, lignocellulose biodegradation made by mushroom species of Ganoderma genus had been little studied, mostly because of their slow growth, difficulty in culturing as well as little apparent biotechnological potential. Only, Stamets (1993) reported a few experimental data concerning the cultivation of such fungal species in natural sites and he noticed its slowly growing.

In spite of these facts, some strains of G. lucidum were grown in our experiments on culture substrates made of wood wastes of white poplar and beech mixed with rye grains at the ratio 1:7 and a water content of 50%.

Higher ratio of rye grains might lead to an increase of total dry weight of fruit body, but also could induce the formation of antler branches and smaller fruit bodies than those of the control samples.

The final fruit body mushroom production ranged between 15 and 20 kg relative to 100 kg of compost made of wood, depending on the specific strains of those tested mushroom species.

# 5. CONCLUSIONS

1. The cereal by-products and winery wastes used as substrata for growing the fungal species G. lucidum, L. edodes and P. ostreatus by controlled submerged fermentation showed optimal effects on the mycelia development in order to get high nutritive biomass.
2. The dry matter content of fungal biomass produced by submerged fermentation of barley bran was higher for both tested species.
3. The protein accumulation is more intense when using barley bran compared with those of wheat and rye, at both fungal species.
4. G. lucidum (variant III) registered the best values of sugar and total nitrogen contents, being followed by L. edodes (variant I)
5. The winery and apple wastes can be recycled as useful raw materials for mushroom compost preparation in order to get significant mushroom fruit body production and protect the natural environment surrounding apple juice factories as well as wine making industrial plants.
6. By applying the biotechnology of recycling the grape and apple wastes can be produced between 20–28 kg of mushroom fruit bodies relative to 100 kg of composts made of winery and apple wastes.
7. From all these fungal species tested in experiments, P. ostreatus was registered as the fastest mushroom culture (25–30 days), then L. edodes (35–45 days) and finally, G. lucidum as the longest mushroom culture (40–50 days).
8. The registered data revealed that when the cells were grown in the maltose medium, the fungal biomass production was the highest among the tested variants.

9. From five nitrogen sources examined, rice bran was the most efficient for mycelia growth and fungal biomass production
10. Among the various mineral sources examined, K2HPO4 yielded good mycelia growth as well as fungal biomass production and for this reason it was as a favourable mineral source.
11. The inoculum age of 120 h as well as an inoculum volume of 6.0 (v/w) have beneficial effects on the fungal biomass production and the optimal pH and temperature levels for fungal fruit body production were 5.0–5.5 and 21–23° C.
12. The final fruit body mushroom production ranged between 15 and 20 kg relative to 100 kg compost made of wood, depending on the specific strains of those tested mushroom species.

## ACKNOWLEDGEMENTS

The authors express their highest respect and deepest gratefulness for the professional competence and outstanding scientific contribution which were proven by Dr. Paul Adrian during so many research works.

## REFERENCES

1. Bae, J. T, Sinha, J, Park, J. P, Song, C. H, & Yun, J. W. (2000). Optimization of submerged culture conditions for exo-biopolymer production by *Paecilomyces japonica*. *Journal of Microbiology and Biotechnology*, 1017-7825, 10, 482-487.
2. Beguin, P, & Aubert, J. P. (1994). The biological degradation of cellulose. *FEMS Microbiol. Rev.*, , 13, 25-58.
3. Breene, W. M. (1990). Nutritional and medicinal values of mushrooms. *J. Food Prot.*, , 53, 833-894.
4. Carlile, M. J, & Watkinson, S. C. (1996). Fungi and biotechnology. In: *The Fungi*, M.J. Carlile, S.C. Watkinson (Eds.), Academic Press, 0-12159-960-4England, 253-264.
5. Chahal, D. S. (1994). Biological disposal of lignocellulosic wastes and alleviation of their toxic effluents. In: *Biological Degradation and Bioremediation of Toxic Chemicals*, G.R. Chaudry (Ed.), Chapman & Hall, 978-0-41262-290-8London, England, 347-356.
6. Chahal, D. S, & Hachey, J. M. (1990). Use of hemicellulose and cellulose system and degradation of lignin by *Pleurotus sajor-caju* grown on corn stalks. *Am. Chem. Soc. Symp.*, , 433, 304-310.
7. Chang, S. T, & Hayes, W. A. Eds.), ((1978). The Biology and Cultivation of Edible Mushrooms. Academic Press, New York, , 147-156.
8. Jones, K. (1995). *Shiitake- The Healing Mushroom*. Healing Arts Press, Rochester, 0-89281-499-3USA
9. Glazebrook, M. A, Vining, L. C, & White, R. L. (1992). Growth morphology of *Streptomyces akiyoshiensis* in submerged culture: influence of pH, inoculum, and nutrients. *Canadian Journal of Microbiology*, , 38, 98-103.
10. Hobbs, C. (1996). Medicinal mushrooms. Santa Cruz, Botanika Press, , 251-270.
11. Lamar, R. T, Glaser, J. A, & Kirk, T. K. (1992). White rot fungi in the treatment of hazardous chemicals and wastes. In: Leatham, G.F. (ed.), *Frontiers in industrial mycology*, Chapman & Hall, New York, , 223-245.
12. Leahy, J. G, & Colwell, R. R. (1990). Microbial Degradation of Hydrocarbons in the Environment, *Microbial Rev.*, , 54, 305-315.

13. Moo-young, M. (1993). Fermentation of cellulose materials to mycoprotein foods, *Biotechnology Advances*, 0734-9750, 11(3), 469-482.
14. Moser, A. (1994). Sustainable biotechnology development: from high-tech to ecotech. *Acta Biotechnologica*, 0138-4988, 12(2), 10-15.
15. Oei, P. (2003). Mushroom Cultivation. 3rd Edition, Backhuys Publishers, 9-05782-137-0The Netherlands
16. Petre, M, & Teodorescu, A. (2012). Biotechnology of Agricultural Wastes Recycling Through Controlled Cultivation of Mushrooms. In: Advances in Applied Biotechnology (M. Petre Editor), InTech Open Access Publisher, 978-9-53307-820-5, 3-23.
17. Petre, M, Teodorescu, A, & Giosanu, D. (2012). Advanced Biotechnological Procedures of Mushroom Cultivation. In: Mushroom Biotechnology and Bioengineering, M. Petre and M. Berovic Editors, CD Press, Bucharest, 978-6-06528-146-2, 1-21.
18. Petre, M, & Teodorescu, A. (2011). Recycling of Vineyard and Winery Wastes as Nutritive Composts for Edible Mushroom Cultivation. Proc. of the International Conference on Advances in Materials and Processing Technologies AMPT, American Institute of Physics (978-0-73540-871-5
19. Petre, M, & Teodorescu, A. (2010). Handbook of submerged cultivation of eatable and medicinal mushrooms. CD Press, 978-6-06528-087-8Bucharest, Romania
20. Petre, M. & Petre, V. (2008). Environmental Biotechnology to Produce Edible Mushrooms by Recycling the Winery and Vineyard Wastes. *Journal of Environmental Protection and Ecology*, Vol. 9, No.1, pp. 88-95, ISSN: 1311-5065
21. Petre, M, & Petre, V. (2008). Environmental Biotechnology to Produce Edible Mushrooms by Recycling the Winery and Vineyard Wastes. *Journal of Environmental Protection and Ecology*, 1311-5065, 9(1), 88-95.
22. Petre, M, Peng, M-X, & Mao, L-X. (2005a). The influence of culture conditions on fungal pellets formation by submerged fermentation of *Cordyceps sinensis* (*Paecilomyces hepiali*)- *Cs 4. Acta Edulis Fungi*, Suppl.), , 12, 345-353.
23. Petre, M, Teodorescu, A, & Dicu, G. (2005b). The Growing Effect of Vineyard and Winery Wastes on the Production of Mycelia and Fruit Bodies of Edible and Medicinal Fungi. *International Journal of Medicinal Mushrooms*, 1521-9437, 7(3), 444-445.
24. Petre, M, Zarnea, G, Adrian, P, & Gheorghiu, E. (2001). Biocontrol of cellulose waste pollution by using immobilized filamentous fungi. In: Environmental Monitoring and Biodiagnostics of Hazardous Contaminants (Healy, M., Wise, D.L. Moo-Young, M, eds), Kluwer Academic Publishers, The Netherlands, 079236869, 227-241.
25. Raaska, L. (1990). Production of *Lentinus edodes* mycelia in liquid media: Improvement of mycelial growth by medium modification. *Mushroom Journal of the Tropics*, , 8, 93-98.
26. Ropars, M, Marchal, R, Pourquie, J, & Vandercasteele, J. P. (1992). Large scale enzymatic hydrolysis of agricultural lignocellulosic biomass. *Biores. Technol.*, , 42, 197-203.
27. Smith, J. (1998). Biotechnology. 3rd Edition. Cambridge University Press, 0-52144-911-1England
28. Stamets, P. (2000). Growing Gourmet and Medicinal Mushrooms. Ten Speed Press, 1-58008-175-4Toronto, Canada
29. Verstraete, W, & Top, E. (1992). Holistic Environmental Biotechnology. Cambridge University Press, 0-52142-078-4England
30. Wainwright, M. (1992). An Introduction to Fungal Biotechnology. Wiley, Chichester, , 103-115.
31. Wasser, S. P, & Weis, A. L. (1994). Therapeutic effects of substances occurring in higher Basidiomycetes mushrooms: a modern perspective. *Critical Reviews in Immunology*, 1040-8401, 19, 65-96.

# CHAPTER 14

# Enhanced resistance to bacterial and fungal pathogens by overexpression of a human cathelicidin antimicrobial peptide (hCAP18/LL-37) in Chinese cabbage

**Yu-Jin Jung1, Soon-Youl Lee2, Yong-Sun Moon3, Kwon-Kyoo Kang1**

*1 Department of Horticulture, Hankyong National University, 67 Sukjong-dong, Ansung, Kyonggi-do 456-749, South Korea*

*2 Korea School of Applied Life Sciences Genetic Informatics Center, GRRC Hankyong National University, 67 Sukjong-dong, Ansung, Kyonggi-do 456-749, South Korea*

*3 Department of Horticulture, Yeungnam University, 214-1 Daedong, Gyeongsan, Gyeongbuk 712-749, South Korea*

## ABSTRACT

The human cathelicidin antimicrobial protein hCAP18, which includes the C-terminal peptide LL-37, is a multifunctional protein. As a possible approach to enhancing the resistance to plant disease, a DNA fragment coding for hCAP18/LL-37 was fused at the C-terminal end of the leader sequence of endopolygalacturonase-inhibiting protein under the control of the cauliflower mosaic virus 35S promoter region. The construct was then introduced into *Brassica rapa*. LL-37 expression was confirmed in transgenic plants by reverse transcription-polymerase chain reaction and western blot analysis. Transgenic plants exhibited varying levels of resistance to bacterial and fungal pathogens. The average size of disease lesions in the transgenic plants was reduced to less than half of that in wild-type plants. Our results suggest that the antimicrobial LL-37 peptide is involved in wide-spectrum resistance to bacterial and fungal pathogen infection.

**Keywords:** Antimicrobial peptide, Cathelicidin, Disease resistance, Transgenic Chinese cabbage

# 1. INTRODUCTION

Bacterial and fungal plant pathogens severely affect crop productivity. For example, *Xanthomonas campestris* pv. *campestris* and *Pectobacterium carotovorum* subsp. *carotovorum*, which cause black rot and soft rot, respectively, are present worldwide and severely damage plants and reduce their yields, especially in cruciferous plants (Boman 2003). Therefore, the development of cruciferous plants that are resistant to black and soft rot diseases has been a major goal of researchers for several decades. Strategies based on transgenic approaches to enhance plant disease resistance involve the use of genes associated with plant defense pathways (Makandar et al. 2006; Zhang et al. 2007) and genes encoding plant or fungal hydrolytic enzymes (Bieri et al. 2003), defense-related transcription factors (Chen and Chen 2002; Sohn et al. 2006) and antimicrobial peptides (Alan et al. 2004).

A large number of antimicrobial peptides from different organisms have been characterized (Simmaco et al. 1998). The human cathelicidin antimicrobial protein hCAP18 is the only member of the mammalian cathelicidin family of proteins that is present in humans (Gudmundsson et al. 1996). The holoprotein consists of a conserved prodomain, a cathelin domain, and the non-conserved C-terminal peptide LL-37, which is enzymatically cleaved after secretion (Sorensen et al. 2001; Yamasaki et al. 2006). Its precursor molecule, an 18 kDa human cationic antimicrobial protein (hCAP-18), is secreted by activated neutrophil granulocytes. After release, the helical C-terminal end of this precursor comprising 37 amino acids is cleaved off, thereby forming the functional antimicrobial peptide LL-37 (Sorensen et al. 2001). Since LL-37 is the only human antimicrobial peptide that is active at physiological or elevated salt concentration conditions, there is a significant interest in using this peptide for pharmaceutical applications (De Smet and Contreras 2005; Reddy et al. 2004; Travis et al. 2000).

In the present study, we report the transgenic expression of human cathelicidin antimicrobial peptide carrying the substitution Met37Leu in Chinese cabbage. The expression of this peptide in cabbage plants significantly inhibited the growth of *Pectobacterium carotovorum* subsp. *carotovorum* on the plant leaves, and it conferred resistance to several fungal pathogens. These results further support the assignment of a defense role to LL-37 and highlight its plant biotechnological potential.

# 2. MATERIALS AND METHODS

## 2.1 Expression vector construction

The leader sequence of the gene encoding *Phaseolus vulgaris* endopolygalacturonase-inhibiting protein (PGIP) (GenBank Accession No. X64769) was fused upstream of an LL-37-coding DNA fragment to cause the extracellular localization of the mature protein. The PGIP signal peptide (87 bp) was amplified by PCR with primers A linked with *Bam*HI site (5′-CCGGATCCATGACTCAATTCAATATCCCA-3′) and B (5′-AGAGAGTGC AGTT CTCAA-3′). The coding region of *LL-37* (111 bp) was substituted with Met-LL37-Leu and amplified by PCR from *pFALL37* DNA using primers C (5′-ATGCTGC TGGGTGATTTCTTC-3′) and D with *Sac*I site (5′-CGAGAGCT CCTAGGACTCTGTCC TGGG-3′). The two products were ligated into pBlueScript-SK (Stratagene, La Jolla, CA, USA) at the *Bam*HI and *Sac*I restriction sites. The generated *LL-37* fragment was further amplified using primers A and D. For *Agrobacterium* transformation, the PCR product was subcloned into pBI121 binary vector driven by cauliflower mosaic virus 35S (CaMV35S) promoter (Gelvin 1998). The Ti plasmid vector construct pBI–LL37 was confirmed by DNA sequencing (ABI 377 DNA sequencer; Perkin-Elmer, Cypress, CA, USA).

## 2.2 Plant transformation and regeneration

The prepared construct was transformed into Chinese cabbage using the protocol described in Min et al. (2007). A total of 168 hypocotyls from in vitro grown seedlings of Chinese cabbage (*Brassica rapa* cv. Osome) were inoculated with *Agrobacterium tumefaciens* strain LBA4404 carrying either pBI-LL37 or pBI121. Green shoots that developed in the selective medium were transferred to a rooting medium containing 100 mg $L^{-1}$ kanamycin and 500 mg $L^{-1}$carbenicillin. Rooted shoots were screened by PCR for the presence of the transgene before transfer to plastic pots.

## 2.3 Estimation of transformants and generation of homozygous lines

Self-pollinated seeds obtained from $T_0$ plants were sown into plastic pots in the greenhouse. Two weeks after germination, seedlings were sprayed with 400 mg $L^{-1}$ kanamycin in water; they were sprayed again 2 days later. Three days after the second spray, the ratios of green seedlings to bleached seedlings were determined, and the results were analyzed by a Chi-square test for goodness of fit to the ratios 3:1, 15:1, or 63:1. In order to obtain transformants homozygous for the *LL-37* gene, kanamycin-resistant $T_1$ progenies were grown to produce selfed $T_2$ seeds. $T_2$ lines that had no bleached segregants after kanamycin sprays were assumed to be homozygous for the *LL-37* and *nptII* genes.

## 2.4 Molecular analysis of transformants

PCR analyses were conducted to detect the presence of *LL-37*- or *nptII*-specific fragments. Primers 35SF (5′-TCCACTGACGTAAGGGATGA-3′) and LL-37R (5′-CGAGAGCTCCTAGGACTCTGTCCTGGG-3′), which amplified a fragment of size approximately 750 bp, including sequences from the 3′ end of the 35S promoter, signal peptide (SP), and part of the *LL-37* gene, were used to screen for putative *LL-37* transformants. Putative transformants were screened using primers *nptII*F (5′-TCGGCTATGACTGGGCACAACAGC-3′) and *nptII*R (5′-AAGA AGGCGATAGAAGGCGATGCG-3′), which amplified a 722-bp *nptII*-specific fragment. Genomic DNA was isolated from young leaves of Chinese cabbage plants using a DNeasy Plant kit (Qiagen, Germantown, MD, USA.) as per the manufacturer's instructions. PCR was performed and the reaction conditions were followed (1 cycle of 94°C for 1 min; 30 cycles of 94°C for 30 s, 55°C for 30 s, and 72°C for 1 min; and 1 cycle of 72°C for 10 min). The reaction products were electrophoresed on a 2% (w/v) agarose gel with 1× TAE buffer and visualized by staining with ethidium bromide.

In order to analyze gene expression in transgenic plants by RT-PCR, total RNA from wild-type and transgenic plants was reverse transcribed using AMV reverse transcriptase (Roche, USA) with oligo (dT) primers for 1 h at 42°C. The expression level of actin mRNA was used as a quantitative control.

Western blot analysis was performed by following standard molecular techniques (Sambrook et al. 1998). Briefly, for western blotting, 200 ng of purified protein or synthetic peptide was electroblotted onto a polyvinylidene difluoride (PVDF) membrane. Blots were probed with 1:1,000 dilution of polyclonal Hbt Ab antiserum raised in rabbit and then with 1:10,000 dilution of goat anti-rabbit immunoglobulin-peroxidase conjugate (Vector, Burlingame, CA, USA). The blots were then developed using the enhanced chemiluminescence (ECL) developing system (GE Health Care, USA). The Low Range BioRad 161-0304 markers (BioRad, Hercules, CA, USA) were used as the molecular size markers.

## 2.5 Pathogen inoculation

*P. carotovorum* subsp. *carotovorum* KACC 10057 obtained from the Korean Agricultural Culture Collection (http://kacc.rda.go.kr) was grown in Luria–Bertani (LB) medium until the absorbance at 600 nm ($A_{600}$) was 0.2, which corresponds to a concentration of approximately 2 × $10^8$ CFU/mL. Three different concentrations of the culture ($10^4$, $10^6$, and $10^8$ CFU/mL) with 10 mM $MgCl_2$ were inoculated by syringe infiltration. The inoculated plants were transferred to a growth chamber and incubated at 28°C under continuous light. They were examined for 12–96 h after inoculation. Lesion length (cm) and disease index (DI) were recorded for each individual plant; the disease index ranged from 0 to 6 on the basis of the development of the disease lesions: 0, no lesion; 1, lesion size 0.1–0.5 cm; 2, 0.5–1.5 cm; 3, 1.5–3.5 cm; 4, 3.5–5.5 cm; 5, 5.5–8.5 cm; and 6, over 8.5 cm or plant dead.

The fungi *Fusarium oxysporum* f. sp. *Lycopersici* (KACC 40032), *Colletotrichum higginsianum* (KACC 40807), and *Rhizoctonia solani* (KACC 40107) were inoculated on plant leaves by placing 10 μl of an aqueous suspension containing $10^6$ spores/mL on the leaves. Plants were maintained in highly humidified conditions (100% RH) at 25°C with 16 h of light in a growth chamber.

## 2.6 Evaluation of in vitro inhibition assays

Total and extracellular fluids were extracted by the methods described in Alan et al. (2004). Protein concentrations in the leaflet fluids were determined by the Bradford assay (Bradford 1976). The in vitro inhibition assays were evaluated by the protocol from Alan et al. (2004). Briefly, the experiments tested whether total fluid (TF) and extracellular fluid (EF) from three homozygous lines possessed antimicrobial activity. Assays were also performed using TF and EF from wild-type plant and LB medium as controls. A volume of 248 μL of TF, EF, and LB was mixed with 2.5 μL of $10^8$ CFU/mL *P.carotovorum* subsp. *carotovorum* in Eppendorf tubes and incubated on a shaker for 4 h. The samples remaining in the tubes were mixed in the ratio 1:9 with LB, returned to the shaker, and incubated overnight at 37°C. The bacterial growth in these tubes was determined by a spectrophotometer at 600 nm.

# 3. RESULTS

The sequence of mature peptide of Met-LL37-Leu (GenBank accession No. NM-004345) is shown in Fig. 1a and the construct of binary pBI121 in Fig. 1b. Morphological characteristics of non-transformed Chinese cabbage (wt) and transgenic plants are shown in Fig. 1c. Transformation experiments with Agrobacterium carrying pBI-LL37 yielded 17 independent kanamycin-resistant putative transformants. Three out of 17 transformants showed abnormal phenotypes and eliminated for further experiment. PCR analysis with 35SF and LL-37R primers revealed that 11 of 14 transformants contained the expected amplified product (Fig. 2a). The PCR-positive plants (T0) were transferred to the greenhouse where they were further observed for their phenotypic characters and were grown for recovery of self-pollinated seeds. Most transformants were phenotypically similar to non-transformed plants. Segregation analysis was indicated that T1 transformants with a single copy showed 3:1 kanamycin resistance. Four of them were randomly selected for selfing, and named lines B11, B12, B13, and B14 (T2). Ten plants per line after selfing were screened on kanamycin to get homozygotics (T3). Additionally, we confirmed the integration of the genes in four homozygous lines (T3; B21, 22, 23, and 24 from T2; B11, 12, 13, and 14) by PCR using the 35S-specific, nptII-specific, and LL-37-specific primer sets (Fig. 2b). All of these contained the expected amplified product.

## 3.1 Expression of *LL-37* in homozygous lines

In order to confirm *LL-37* expression in the transformants, we performed quantitative RT-PCR using total RNA from 4 homozygous lines. The results of RT-PCR indicated that LL-37 RNA was expressed in the homozygous lines, and the expression levels were similar across all the transformants (Fig. 3). Western blot analysis also confirmed that the four homozygous lines expressed the 4-kDa peptide at varying levels, whereas the control plants did not show the LL-37-specific band (Fig. 4). Therefore, we concluded that the antimicrobial *LL-37* gene was stably integrated into the genome of the transformants and was transcribed into mRNA that yielded the LL-37 peptide.

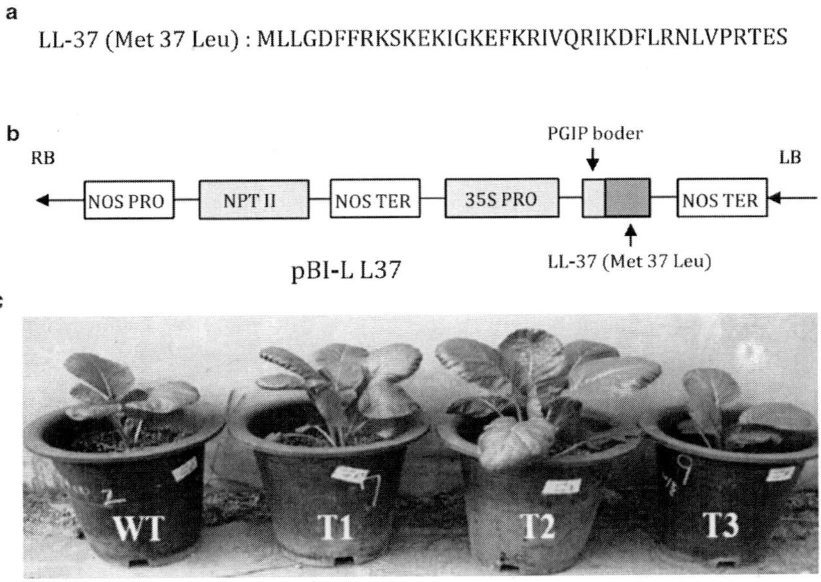


**Figure 1** Construction of a binary plant expression vector pBI121 for the transformation of Chinese cabbage plants. **a** Sequence of the peptide (GenBank Accession No. NM-004345). **b** Schematic diagram of the expression construct pBI-LL37. **c** Morphological characteristics of non-transformed Chinese cabbage (wt) and transgenic plants.

## 3.2 Increased resistance of the transgenic plants to soft rot

The resistance to several rot pathogens was evaluated for the four homozygous lines harboring the human *LL-37* gene by comparing the severity of the disease to that seen with the nontransgenic plants. Transgenic and control plants were inoculated with conidia, and the sizes of the disease lesions were determined. In order to test the resistance of the transgenic plants, we first cultivated the cabbage soft rot pathogen *P.carotovorum* subsp. *carotovorum* in LB medium for 1–2 days. We tested the resistance of the plants to three different concentrations of the bacterial pathogen: $10^4$, $10^6$, and $10^8$ CFU/mL. We challenged the leaf body and leaf vein with $10^4$ and $10^6$ CFU/mL of the bacterial culture, respectively; subsequently, the pathogenesis at the leaf was examined after 24, 48, and 72 h of inoculation. The control cabbage plants were susceptible to the bacterial pathogen and their leaves became softer with visible lesions 12 h after inoculation. After 48 h, the symptom spread all over the leaves, and the plants died owing to softening 72 h after inoculation of the pathogen. On the other hand, the softening symptom was not observed in three of the four transgenic lines, B21, B22, and B24, by 72 h after bacterial inoculation (Fig. 5). When $10^8$ CFU/mL of the bacterial culture was inoculated, the leaves of the control plants showed the softening symptom 12 h after inoculation; the condition worsened and the leaves became amorphous after 24 h of inoculation. The transgenic lines only showed mild softening of the leaves around the area of inoculation (Fig. 6). Therefore, it could be concluded that transgenic plants showed significantly increased resistance to the bacterial pathogen *P. carotovorum subsp. carotovorum*, which causes soft rot.

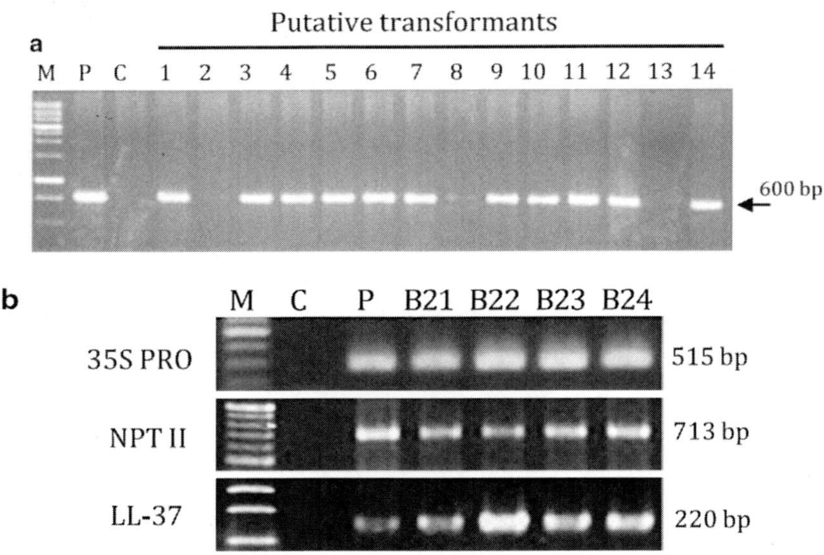


**Figure 2** PCR amplification of the foreign genes in the transgenic Chinese cabbage lines. **a** PCR analysis of 14 putative LL-37 Chinese cabbage lines (T0) using the 35SF (5′-ATGGAGTCAAAGATTCAAATAGAG-3′) and LL-37R (5′-CGAGAGCTCCTAGGACTCTGTCCTGGG-3′) primers. *Lanes*: *M* 1 kb ladder, *P* positive control (pBI LL-37), *C* non-transformed plant, *1–14* putative transformants. **b** Four T3 (B21, 22, 23, 24) of single transgenic homozygous lines from T2 (B11, 12, 13, 14) were reconfirmed by PCR after genetic segregation events by kanamycin resistance. Amplification products of the 35S promoter, *NptII*, and *LL-37* were separated using a 1.5% agarose gel. *Lanes*: *M* DNA ladder, *C* non-transformed plant, *P* pBI*LL-37*, *B21–B24* independent T3 transgenic homozygous lines

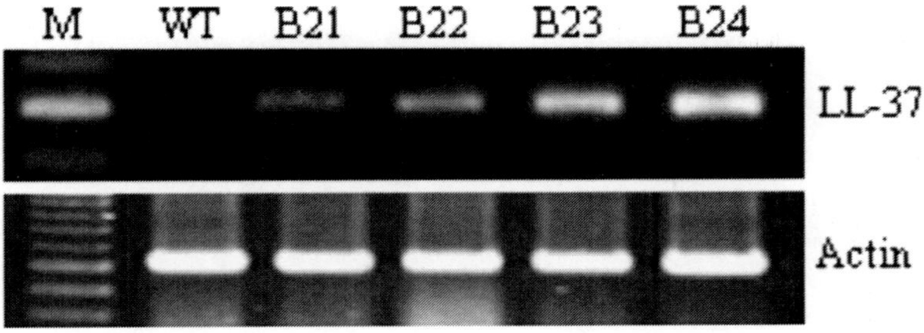


**Figure 3** Analysis of the *LL-37* gene expression in the transgenic homozygous lines by using reverse transcription polymerase chain reaction. Total RNA was isolated from each plant, and 0.5 μg of this RNA was amplified with LL-37-specific primers. The Chinese cabbage actin gene was amplified as a loading control. The amplification products of the *LL-37* gene were separated using a 1.5% agarose gel. *Lanes*: *M* DNA ladder, *WT* wild-type plant, *B21–B24* independent transgenic homozygous lines

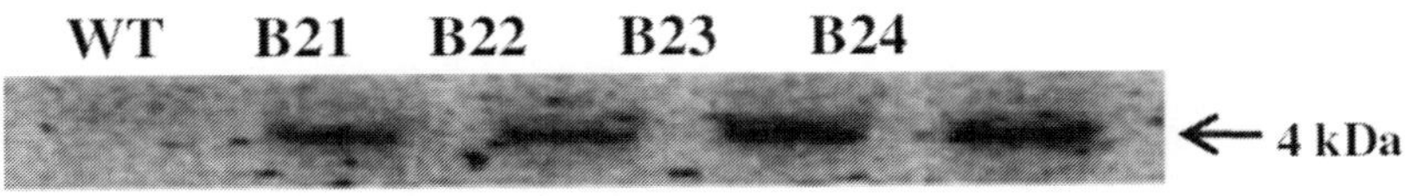


**Figure 4** Western blot analysis of transgenic homozygous lines. *Lanes*: *WT* wild-type plant, *B21–B24*, transgenic homozygous lines containing the *LL-37* gene, *arrow* indicates 4 kDa protein

## 3.3 Increased resistance of the transgenic plants to fungal pathogens

The leaves of the transgenic plants were challenged with several fungal pathogens such as *F. oxysporum, C. higginsianum,* and *R. solani* in order to test the resistance of the transgenic plants to fungal pathogens. Three of the four transgenic homozygous lines, B21, B22, and B23, showed higher resistance to *F. oxysporum* than the control plants (Fig. 7). In particular, B21 showed remarkably high resistance to *F. oxysporum.* After being challenged with *C. higginsianum* and *R. solani*, the control plants showed typical symptoms and died after 6 days of inoculation, while the transgenic plants were resistant, as shown in Fig. 7. Therefore, we can conclude that the LL-37 peptide expressed in the transgenic plants acts as an effective antimicrobial peptide.

## 3.4 Bacterial inhibition assays in transgenic leaflets

We collected TF and EF from the leaflets of three transgenic homozygous lines and a non-transformed line in order to study their effects on the activity of LL-37. The inhibition of the bacterial (*P.carotovorum* subsp. *carotovorum* subsp. *carotovorum*. Disease manifestations were scored 3 days after inoculation. *RG* water, *WT* wild-type plant, *B21–B24* LL-37 transgenic homozygous lines. Disease lesions: *0*, no lesion; *1*, lesion size 0.1–0.5 cm; *2*, 0.5–1.5 cm; *3*, 1.5–3.5 cm; *4*, 3.5–5.5 cm; *5*, 5.5–8.5 cm; and *6*, over 8.5 cm or plant dead *carotovorum*) growth by TF and EF was evaluated using the spectrophotometric method. As expected, bacteria grew normally when they were incubated in LB medium only and in TF or EF obtained from the non-transformed line, while the growth was completely inhibited when EF obtained from the three transgenic lines was added to the LB medium (Fig. 8). Bacteria also grew well in the TF obtained from the three transgenic lines, although their growth was lower than that in LB medium alone.

**Figure. 5** Temporal development of soft rot in Chinese cabbage ("Osome") plants when the leaves were inoculated with *Pectobacteriumcarotovorum* subsp. *carotovorum*. *B21–B24* Transgenic homozygous lines

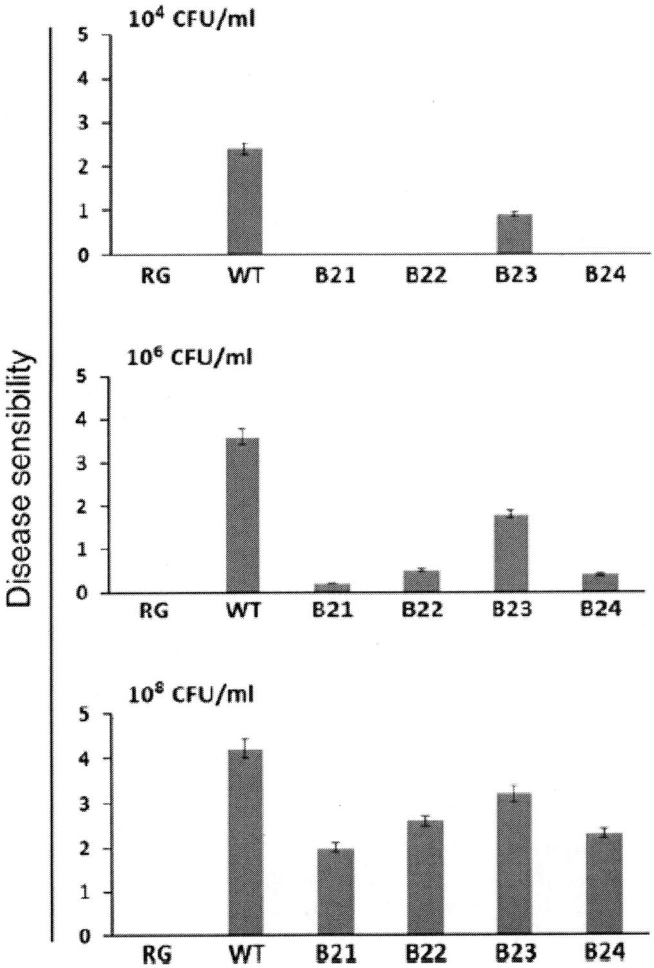


**Figure 6** Disease development in transgenic homozygous lines containing the *LL-37* gene, which were inoculated with $10^4$, $10^6$, or $10^8$ CFU/mL of *P.*

## 4. DISCUSSION

In response to a microbial attack, plants activate a complex series of responses that lead to the local and systemic induction of a broad spectrum of antimicrobial defenses (Kim and Martin 2004; Kunkel 2002). When induced defense responses are rapidly and coordinately triggered during a given plant–pathogen interaction, plants become broadly resistant to diseases. These defense responses include the strengthening of mechanical barriers, oxidative burst, and production of antimicrobial compounds (Hammond-Kosack and Parker 2003; Park 2005). Some research has been performed to bolster plant defenses against bacteria and fungi by genetically engineering plants to express antimicrobial peptides (Lee et al. 2008; Prasad et al. 2008).

In this study, we have chosen a variant of the antimicrobial peptide LL-37 as an interesting candidate for transgenic plant expression; this variant is designed to target the peptide into extracellular spaces. Extracellular targeting was originally intended to prevent possible deleterious effects of the peptide in plant cells. Moreover, secretion into extracellular spaces allows the plant-produced peptide to come into direct contact with pathogens growing and multiplying extracellularly before attacking the cells (Ponti et al. 2003). We developed homozygous Chinese cabbage lines stably expressing LL-37, which did not cause adverse effects on the plant phenotypes. These transgenic homozygous lines were tested with four important pathogens of this crop and found to inhibit the growth of the bacterial pathogen causing Chinese cabbage rot and also that of three fungal pathogens.

The results of our pathogenicity assays suggest that the expression of LL-37 provides a moderate level of resistance against a bacterial pathogen (*P. carotovorum* subsp. *carotovorum*) at the inoculum concentration of $10^4$ CFU/mL. However, the extent of disease suppression provided by LL-37 expression was reduced as the inoculum concentrations were increased to $10^6$ and $10^8$ CFU/mL. Moreover, we observed reduction in the survival of *P. carotovorum* subsp. *carotovorum* cells incubated with EF from LL-37-expressing Chinese cabbage cell lines. However, bacteria grew normally when they were incubated in LB medium only and in TF or EF obtained from a non-transformed cell line. Therefore, leaves appeared to express LL-37 at sufficient levels in the extracellular spaces to retard bacterial multiplication and hence decrease disease severity. The functionality of this transgene and the presence of antimicrobial activity in the EF indicates that the LL-37 peptide was properly targeted to the extracellular space even with a foreign plant signal peptide. Furthermore, the mammalian peptide was not subjected to a processing step in the foreign plant cell environment that rendered it inactive.

Until now, studies involving the enhancement of resistance to various bacterial, fungal, and oomycete pathogens by the expression of antimicrobial peptides have been reported for rice, tobacco, poinsettia, banana, and more host species (Chakrabarti et al. 2003; Liang et al. 2002; Smith et al. 1998). However, progress on identifying the defense mechanisms in Chinese cabbage (*B. rapa*), an important vegetable crop in Asia, has been very slow. Previously, we reported an enhancement in the resistance to bacterial soft rot by the expression of the bromelain gene in Chinese cabbage (Jung et al. 2008). Here, we demonstrate that expression of the human LL-37 peptide has antimicrobial activity toward both bacterial and fungal pathogens of Chinese cabbage. Since pathogens have the ability to overcome gene-for-gene host defense mechanisms in the field by undergoing mutations in the cognate avirulence genes, any transgenic model based on such resistance-conferring genes may easily be evaded by pathogens. On the other hand, this problem is less likely to occur in transgenic plants overexpressing genes with more general antimicrobial activity. Therefore, the expression of the human LL-37 peptide is expected to confer durable resistance (i.e., field resistance) to a wide variety of pathogens infecting Chinese cabbage plants.

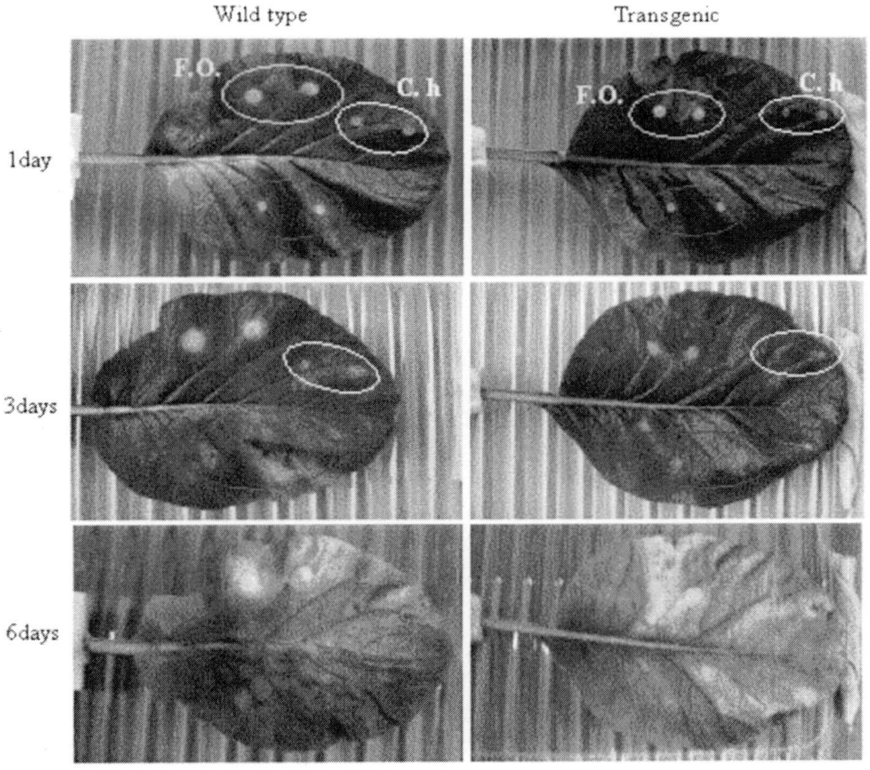


**Figure 7** Enhanced disease resistance of transgenic homozygous lines to several fungal pathogens. F.O., *Fusarium oxysporum*; C.h., *Colletotrichum higginsianum*; R.S., *Rhizoctonia solani*. Disease manifestations assessed until 6 days after inoculation

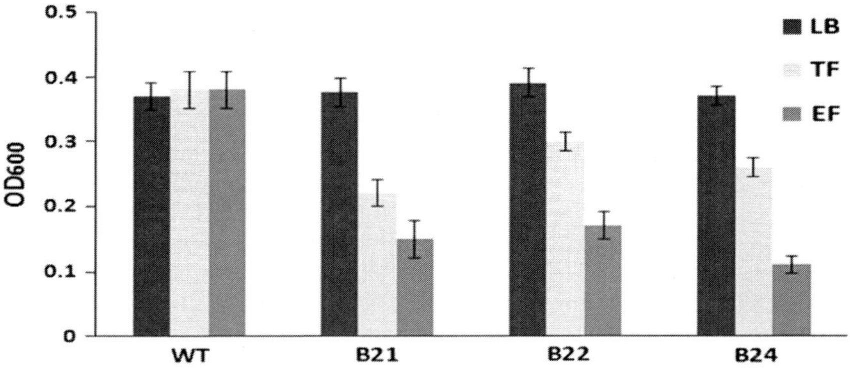


**Figure 8** Bacterial inhibition assays of total fluid (*TF*) and extracellular fluid (*EF*) obtained from the leaflets of transgenic homozygous lines; *LB* Luria–Bertani medium

In order to obtain a high level of resistance as observed in the case of R gene-mediated resistance, the overexpression of multiple antifungal proteins with different functions may be necessary. In a *R. solani* infection assay, tobacco plants coexpressing the barley transgenes (a class II chitinase, a class II β-1,3-glucanase, and a type I ribosome-inactivating protein) were reported to elicit significantly enhanced protection against fungal attack as compared to that of the corresponding isogenic lines expressing a single barley transgene at a similar level (Brogue et al. 1991). Thus, other antimicrobial genes such as the bromelain gene (Jung et al. 2008) could be stacked with the human *LL-37* gene by crossing different transgenic lines and this may be able to strongly and durably inhibit the growth of pathogens.

## ACKNOWLEDGMENTS

We thank to Dr. Philip Becraft for critical reviews and comments. This work was supported by a grant from the Next-Generation BioGreen 21 Program (No. PJ008085), Rural Development Administration, Republic of Korea.

# REFERENCES

1. Alan AR, Blowers A, Earle ED (2004) Expression of a magainin-type antimicrobial peptide gene (MSI-99) in tomato enhances resistance to bacterial speck disease. Plant Cell Rep 22:388–396
2. Bieri S, Potrykus I, Fütterer J (2003) Effects of combined expression of antifungal barley seed proteins in transgenic wheat on powdery mildew infection. Mol Breed 11:37–48
3. Boman HG (2003) Antibacterial peptides: basic facts and emerging concepts. J Intern Med 254:197–215
4. Bradford MM (1976) A rapid and sensitive method for the quantification of microgram quantities of protein utilizing the principle of protein-dye binding. Anal Biochem 72:248–254
5. Brogue K, Chet I, Holliday M, Cressman R, Biddle P, Knowlton S, Mauvais CJ, Broglie R (1991) Transgenic plants with enhanced resistance to the fungal pathogen Rhizoctonia solani. Science 254:1194–1197
6. Chakrabarti A, Ganapathi TR, Mukherjee PK, Bapat VA (2003) MSI-99, a magainin analogue, imparts enhanced disease resistance in transgenic tobacco and banana. Planta 216:587–596
7. Chen C, Chen Z (2002) Potentiation of developmentally regulated plant defense response by AtWRKY18, a pathogen-induced *Arabidopsis* transcription factor. Plant Physiol 129:706–716
8. De Smet K, Contreras R (2005) Human antimicrobial peptides: defensins, cathelicidins and histatins. Biotechnol Lett 27:1337–1347
9. Gelvin SB (1998) The introduction and expression of transgenes in plants. Curr Opin Biotech 9:227–232
10. Gudmundsson GH, Agerberth B, Odeberg J, Bergman T, Olsson B, Salcedo R (1996) The human gene FALL39 and processing of the cathelin precursor to the antibacterial peptide LL-37 in granulocytes. Eur J Biochem 238:325–332
11. Hammond-Kosack KE, Parker JE (2003) Deciphering plant-pathogen communication: fresh perspectives for molecular resistance breeding. Curr Opin Biotech 14:177–193
12. Jung YJ, Choi CS, Park JH, Kang HW, Choi JE, Nou IS, Lee SY, Kang KK (2008) Overexpression of the pineapple fruit bromelain gene (BAA) in transgenic Chinese cabbage (*Brassica rapa*) results in enhanced resistance to bacterial soft rot. Electron J Biotechnol 11:1–8
13. Kim YJ, Martin GB (2004) Molecular mechanisms involved in bacterial speck disease resistance of tomato. Plant Pathol J 20:7–12
14. Kunkel BN (2002) Brooks cross talk between signaling pathways in pathogen defense. Curr Opin Plant Biol 5:325–331
15. Lee SC, Hwang IS, Choi HW, Hwang BK (2008) Involvement of the pepper antimicrobial protein *CaAMP1* gene in broad spectrum disease resistance. Plant Physiol 143:1004–1020
16. Liang H, Catranis CM, Maynard CE, Powell WA (2002) Enhanced resistance to the poplar fungal pathogen, *Septoria musiva*, in hybrid poplar clones transformed with genes encoding antimicrobial peptides. Biotechnol Lett 24:383–389
17. Makandar R, Essig JS, Schapaugh MA, Trick HN, Shah J (2006) Genetically engineered resistance to *Fusarium* head blight in wheat by expression of *Arabidopsis NPR1*. Mol Plant Microbe Interact 19:123–129
18. Min BW, Cho YN, Song MJ, Noh TK, Kim BK, Chae WK, Park YS, Choi YD, Harn CH (2007) Successful genetic transformation of Chinese cabbage using phosphomannose isomerase as a selection marker. Plant Cell Rep 26:337–344
19. Park JM (2005) The hypersensitive response: a cell death during disease resistance. Plant Pathol J 21:99–101

20. Ponti D, Mangoni ML, Mignogna G, Simmaco M, Barra D (2003) An amphibian antimicrobial peptide variant expressed in *Nicotiana tabacum* confers resistance to phytopathogens. Biochem J 370:121–127
21. Prasad BD, Jha S, Chattoo BB (2008) Transgenic *indica* rice expressing *Mirabilis jalapa* antimicrobial protein (Mj-AMP2) shows enhanced resistance to the rice blast fungus *Magnaporthe oryzae*. Plant Sci 175:364–371
22. Reddy KV, Yedery RD, Aranha C (2004) Antimicrobial peptides: premises and promises. Int J Antimicrob Ag 24:536–547
23. Sambrook J, Fritsch EF, Maniaatis T (1998) Molecular cloning: a laboratory manual, 2nd edn. Cold Spring Harbor Laboratory Press, Cold Spring Harbor, New York
24. Simmaco M, Mignogna G, Barra D (1998) Antimicrobial peptides from amphibian skin: what do they tell us? Biopolymers 47:435–450
25. Smith FD, Gadoury DM, Vaneck JM, Blowers A, Sanford JC, Van der Meij J, Eisenreich R (1998) Enhanced resistance to powdery mildew in transgenic poinsettia conferred by antimicrobial peptides. Phytopathology 88:S83
26. Sohn KH, Lee SC, Jung HW, Hong JK, Hwang BK (2006) Expression and functional roles of the pepper pathogen-induced transcription factor RAV1 in bacterial disease resistance, and drought and salt stress tolerance. Plant Mol Biol 61:897–915
27. Sorensen OE, Follin P, Johnsen AH, Calafat J, Tjabringa GS, Hiemstra PS, Borregaard N (2001) Human cathelicidin, hCAP-18, is processed to the antimicrobial peptide LL-37 by extracellular cleavage with proteinase 3. Blood 97:3951–3959
28. Travis SM, Anderson NN, Forsyth WR, Espiritu C, Conway BD, Greenberg EP, McCray PB, Lehrer RI, Welsh MJ, Tack BF (2000) Bactericidal activity of mammalian cathelicidin-derived peptides. Infect Immun 68:2748–2755
29. Yamasaki K, Schauber J, Coda A, Lin H, Dorschner RA, Schechter NM, Bonnart C, Descargues P, Hovnanian A, Gallo RL (2006) Kallikrein-mediated proteolysis regulates the antimicrobial effects of cathelicidins in skin. Fed Am Soc Exp Biol J 20:2068–2080
30. Zhang X, Dai Y, Xiong Y, DeFraia C, Li J, Dong X, Mou Z (2007) Overexpression of *Arabidopsis MAP kinase kinase 7* leads to activation of plant basal and systemic acquired resistance. Plant J 52:1066–1079

# CHAPTER 15

# Biotechnological Approaches for the Control of Insect Pests in Crop Plants

**Jackie Stevens, Kerry Dunse, Jennifer Fox, Shelley Evans and Marilyn Anderson**

*La Trobe Institute for Molecular Sciences, La Trobe University, Melbourne, , Australia*

## 1. INTRODUCTION

Each year billions of dollars are spent worldwide on insect control in agriculture [1]. Despite this expenditure, up to 40% of a crop can be lost to insect damage, particularly in developing countries [2]. Some of the most damaging insect species belong to the Lepidoptera, the second largest insect order comprised of moths and butterflies. The larval stage of moths cause major damage to an array of economically valuable crops including cotton, tobacco, tomato, corn, sorghum, lucerne, sunflower, pulses, and wheat [3]. Until recently, broad spectrum chemical insecticides have been the primary control agent for agricultural pests, with about 40% targeted to the control of lepidopteran insects [4]. Over the years the widespread use of pesticides has led to pesticide resistant insects, a reduction in beneficial insect populations and harmful effects to humans and the environment [5-8]. These problems have led researchers to develop different insect control strategies using both synthetic and natural molecules that are more environmentally friendly.

One such approach has been the use of transgenic plants expressing plant defence molecules. Genetic modification can potentially provide a much larger array of novel insecticidal genes that are otherwise beyond the scope of conventional breeding. The first transgenic plant that expressed an insecticidal gene was produced in 1987. The transgenic tobacco plant produced cowpea trypsin inhibitor at levels of up to 1% of the soluble protein and had enhanced protection against the lepidopteran pest *Heliothis virescens* [9,10]. The gene encoding the cowpea trypsin inhibitor was subsequently transferred into rice [11] and potato [12,13], but did not provide sustainable insect protection and was thus not commercially viable. Commercial development of insecticidal genes has focused on the *Bacillus thuringiensis* (Bt) toxins [14,15]. In 1987, genes encoding the Bt endotoxins were also transformed into tobacco and tomato plants [16-18]. Since the commercialisation of biotech crops in 1996, farmers have adopted the technology at such a dramatic rate, that in 2011, 16.7 million farmers in 29 counties planted 160 million hectares of the biotech crops. This has led to a reduction in chemical pesticide use of 443 million Kg and an additional financial gain for farmers of US $78 billion in the last 15 years [19]. In India alone, Bt-cotton has increased cotton yields by up to 60%, and has reduced insecticide sprays by around half. This in turn has lead to an income increase of up to US $11.9 billion per annum [19]. The reliance of a worldwide industry on one insect resistance trait has led to real concerns about the development of Bt-resistant insects [20], especially since at least four cases of field based resistance have already been documented [21-23]. This in turn has led to a search for new insecticidal proteins and their encoding genes that have commercial potential for plant protection [8,24]. They include -amylase inhibitors [25,26], vegetative insecticidal protein [27,28], chitinases [29] and protease inhibitors [30,31], as well as several other proteins directed to targets in the insect gut (Table 1).

**Table 1**. Biotechnological approaches for the control of lepidopteran insects with transgenes

| Transgene | Source and Mode of Action | Example of use |
|---|---|---|
| *Bacillus thuringiensis* (Bt) endotoxin | See section "The *Bacillus thuringiensis* endotoxin" | See section "The *Bacillus thuringiensis* endotoxin" |
| Vegetative insecticidal protein (VIP) | VIPs are produced by *Bacillus cereus* and *Bacillus thuringiensis*. They have similar activity to endotoxins from Bt. Vip1/Vip2 are toxic to coleopteran insects and Vip3 is toxic to lepidopteran insects [32]. | VIP was highly toxic to *Agrotis* and *Spodoptera* species. VIP induced gut paralysis, complete lysis of the gut epithelial cells and resulted in larval mortality [33]. |
| | | *Agrotis ipsilon* and *Spodoptera frugiperda* larvae suffered gut |

| | | |
|---|---|---|
| | | paralysis, disruption of midgut epithelial cells and mortality on Vip3A [34]. |
| | | Vip3A was toxic to *A. ipsilon* and *S. frugiperda*. Larvae of *Ostrinia nubilalis* and *Danaus plexippus* were insensitive [35]. |
| | | Vip3Aa14 was toxic to *Spodoptera litura* and *Plutella xylostella*. Larvae of *Helicoverpa armigera* and *Pieris brassicae* were insensitive [27]. |
| | | VIP3Ac1 had insecticidal activity against larvae of *S. frugiperda, Helicoverpa zea* and *Trichoplusia ni,* but low activity against *Bombyx mori* and *O. nubilalis.* The chimeric protein Vip3AcAa was insecticidal to *O. nubilalis* [28]. |
| | | Vip3LB resulted in growth inhibition of *Spodoptera littoralis* when incorporated into a semi solid artificial diet [36]. |
| Biotin binding proteins (avidin and streptavidin) | Biotin is an essential vitamin for insects. It functions as a covalently-bound cofactor in various carboxylases, which have major roles in gluconeogenesis, lipogenesis, amino acid and fatty acid catabolism, and the citric acid cycle. | Avidin and streptavidin increased mortality in four Lepidoptera; *Epiphyas postvittana, Planotortrix octo, Ctenopseustis obliquana* and *Phthorimaea operculella* when incorporated into artificial diets [37]. |
| | Avidin is a water-soluble tetrameric glycoprotein from chicken egg, which binds strongly to biotin. Streptavidin is a homologous protein found in the culture supernatant of *Streptomyces avidinii.* | Transgenic plants with leaves expressing avidin in the vacuole halted growth and caused mortality in *H. armigera* and *S. litura* larvae [38]. |
| | | Transgenic tobacco plants expressing either avidin or streptavidin increased |

| | | |
|---|---|---|
| | | mortality of the potato tuber moth (*P. operculella*). Similarly, transgenic apple expressing either avidin or streptavidin increased mortality and decreased growth of the lightbrown apple moth (*E. postvittana*) [39]. |
| | | Transgenic tobacco expressing avidin reduced *S. litura* larval mass [40]. |
| | | Transgenic tobacco expressing three variants of biotin binding proteins in the vacuole increased mortality of *P. operculella* larvae [41]. |
| Chitinase (enzyme) | Chitinase catalyses the hydrolysis of chitin, which is one of the vital components of the lining of the digestive tract in insects and is not present in plant and higher animals. | Transgenic tobacco plants expressing *M. sexta* chitinase caused a reduction in survival and growth of *H. virescens,* but not *M. sexta* larvae [42]. |
| | | *Lacanobia oleracea* larvae exposed to diet containing recombinant *L. oleracea* chitinase had a reduction in weight gain and consumption compared to control-fed larvae [43]. |
| | | Transgenic rapeseed(*Brassica napus*) expressing *M. sexta* chitinase and scorpion insect toxin increased mortality and reduced growth of *Plutella maculipenis* [44]. |
| | | Oral injection of *B. mori* chitinase (*Bm*-CHI) caused high mortality in Japanese pine beetle, *Monochamus alternates* (Coleoptera). The peritrophic membrane chitin was degraded by *Bm*-CHI, but the midgut epithelium was not affected [29]. |
| Cholesterol | Cholesterol oxidase is a | Cholesterol oxidase from |

| | | |
|---|---|---|
| oxidase (enzyme) | bacterial enzyme that catalyzes the oxidation of cholesterol and other 3-hydroxysterols, resulting in production of the corresponding 3-hydroxysterols and hydrogen peroxide. Functions by damaging midgut membranes. | *Streptomyces* caused stunting of *H. virescens*, *H. zea* and *Pectinophora gossypiella* when incorporated into an artificial diet [45]. |
| | | Cholesterol oxidase expressing tobacco leaves that were incorporated in artificial diets caused mortality and severe stunting of neonate *Anthonomus grandis* larvae [46]. |
| Lipoxygenases (enzyme) | Dioxygenase enzymes are widely distributed in plants and catalyse the hydroperoxidation of cis-cis-pentadiene moieties in unsaturated fatty acids. Functions by damaging midgut membranes. | Lipoxygenase from soybean retards the growth of *Manduca sexta* when incorporated into artificial diet [47]. |
| Alpha-amylase inhibitors | Alpha-amylase inhibitors block starch digestion. Widespread in microorganisms, plants and animals, [25,26]. | Development of pea weevil larvae (*Bruchus pisorum*; Coleoptera) was blocked at an early stage after ingestion of transgenic peas expressing an alpha-amylase inhibitor from the common bean (*Phaseolus vulgaris*) [48]. |
| Alpha-amylase inhibitors | Alpha-amylase inhibitors block starch digestion. Widespread in microorganisms, plants and animals, [25,26]. | Alpha-amylase inhibitor protects against predation by certain species of bruchids (Coleoptera: Bruchidae) and the tomato moth, *L. oleracea* (Lepidoptera) [49]. |
| | | Alpha-amylase inhibitor 1, from the common bean (*P. vulgaris*), provided complete protection against pea weevil (*B. pisorum*; Coleoptera) in transgenic peas. Whereas alpha-amylase inhibitor 2 delayed maturation of larvae [50]. |
| | | The alpha-amylase activity in *Tecia solanivora* larvae was inhibited by alpha-amylase |

| | | |
|---|---|---|
| | | inhibitor from amaranth seeds [51] |
| Protease inhibitors | See section Protease inhibitors for the control of insect pests | See Table 2 |
| Lectins | Multivalent carbohydrate-binding proteins. Some bind to midgut epithelial cells, disrupting their function, causing breakdown of nutrient transport, and absorption of potentially harmful substances [25,52]. | Lectin from soybean seed inhibited larval growth of *M. sexta* [47]. |
| | | Wheatgerm agglutinin was toxic when fed to *O. nubilalis*. Formation of the peritrophic membrane was disrupted in the anterior midgut microvilli [53]. |
| | | *O. nubilalis* growth was strongly inhibited by wheat germ agglutinin (WGA), whereas *M. sexta* was not affected. In *O. nubilalis* larvae, WGA caused hypersecretion of unorganized peritrophic membrane in the anterior midgut lumen, disintegration of microvilli and cessation of feeding [54] |
| | | The snowdrop lectin (*Galanthus nivalis*, agglutinin, GNA) reduced *L. oleracea* larval biomass and slowed larval development when in an artificial diet or expressed in potato plants [55]. |
| | | Transgenic potato expressing snowdrop lectin (*G. nivalis* agglutinin; GNA) reduced development of *L. oleracea* larvae. Transgenic plants were significantly less damaged [56]. |
| | | Transgenic tobacco plants expressing leaf (ASAL) and bulb (ASAII) agglutinins from *Allium sativum* retarded *S. littoralis* larval development and growth [57]. |

| | | |
|---|---|---|
| | | The *Moringa oleifera* lectin (cMoL) reduced *Anagasta kuehniella* larval growth and increased development time and pupal mortality when incorporated into an artificial diet [58] |
| Trypsin-modulating ostatic factor (TMOF) | A peptide that blocks trypsin biosynthesis in mosquitoes (*Aedes aegypti;* Diptera [Aea-TMOF]) and fleshflies (Sarcophaga; Diptera) [59]. | Injection or oral ingestion of Aea-TMOF caused inhibition of trypsin biosynthesis and larval growth in *H. virescens.* Mortality of *H. virescens* increased when fed transgenic tobacco plants expressing Aea-TMOF [60]. |
| Isopentenyl-transferase gene (ipt) | Microorganism-derived gene from *Agrobacterium tumefaciens*. Codes for a key enzyme in the cytokinin-biosynthetic pathway. | Ipt expressed in tobacco and tomato decreased leaf consumption by *M. sexta* and reduced survival of the peach potato aphid, *Myzus persicae* (Hemiptera) [61]. |
| RNAi constructs: 1) Vacuolar ATPase | Nutrient uptake by midgut cells is energized by the electrical difference created by the K+ pump. The K+ pump also regulates midgut lumen pH and determines the potassium concentration in blood, epithelial cells and midgut lumen [62]. The primary motor for transport is a vacuolar-type proton ATPase. | Transgenic corn plants expressing dsRNA of a V-ATPase from *Diabrotica virgifera* (western corn rootworm [WCR], Coleoptera) showed significant reduction in WCR feeding and plant damage [63]. |
| 2) Cytochrome P450 monooxygenase | Cytochrome P450 monooxygenase permits insects to tolerate otherwise inhibitory concentrations of the cotton metabolite, gossypol. | *H. armigera* fed on plants expressing cytochrome P450 dsRNA had retarded growth. Growth inhibition was more dramatic in the presence of gossypol [64]. |
| 3) Hemolin | Recognition of microbial infection is an essential first step in immunity in insects. Induction of this protective effect is associated with up-regulation of microbial pattern recognition protein | Pupae of the giant silkmoth (*Hyalophora cecropia*) were injected with hemolin dsRNA and developed normally into moths. After mating, no larvae emerged from the eggs which had malformed embryos [65]. |

| | | |
|---|---|---|
| | genes such as hemolin. | Prior infection of *M. sexta* larvae with non-pathogenic *E. coli*, elicited effective immunity against subsequent infection by the lethal pathogen *Photorhabdus luminescens*. Injection of hemolin dsRNA left the insect more susceptible to *P. luminescens* infection than insects that had not experienced prior infection with *E. coli* [66]. |

## 1.1. *Helicoverpa* species

*Helicoverpa* species (Figure 1) are polyphagous pests of at least 181 plant species from 49 families including cotton, corn, soybeans, tobacco and chick-pea [67-69]. They are one of the most serious pests in cotton-producing countries like Australia, India and China, causing enormous economic problems [70,71].

The lepidopteran species, *H. armigera*, progresses though four stages of development; egg, six larval instars, pupal and adult. The time frame of each of these stages varies with environmental conditions. Over the warmer months, the life cycle can be completed in 30-40 days and each female moth can lay from 500-3000 eggs.

One of the reasons these pests are so damaging is the larva's feeding preference for plant structures that are high in nitrogen, principally reproductive structures and growing points such as cotton buds and bolls, corn ears, tobacco buds, and sorghum heads. Damage to these structures has a direct influence on yield [67]. *H. armigera* larvae are foliar feeders at the early instar stage and shift to developing seeds or bolls at later stages [72]. *H. armigera* is a major problem in Australia because it has developed resistance to many of the chemical insecticides that have been used for its control [68,73]. Unlike other lepidopteran species, *H. armigera* larvae don't migrate far from their original host plant, consequently their populations in agricultural areas are exposed to consistent selection pressure, leading to greater resistance to insecticides [5].

In the 1995/96 growing season, transgenic cotton known as Ingard that expressed the *Cry1Ac* gene became commercially available in Australia [71]. To preserve the susceptibility of lepidopterans to Bt toxins, a conservative resistance management plan was imposed, where planting of Ingard cotton was restricted to 30% of the cotton production area per farm [71]. The average amount of insecticide used per hectare was 44% lower on Ingard cotton compared to conventional cotton [71]. In the 2004/05 growing season, Ingard cotton was replaced by Bollgard II, which expressed both the *Cry1Ac* and *Cry2Ab* genes [71]. Restrictions were not placed on this new variety and Bollgard II cotton comprised around 80% of the total cotton area planted in Australia during the 2004/05 and 2005/06 seasons [71] and 95% of the total cotton area in the 2010/2011 season [19]. This reduced the average amount of insecticide used per hectare by 85% compared to conventional cotton [71]. So far, there have been no reported field failures of Bollgard II due to resistance. However, while alleles that confer resistance to Cry1Ac in *H. armigera* are rare in the field, alleles that confer resistance to Cry2Ab are more common.

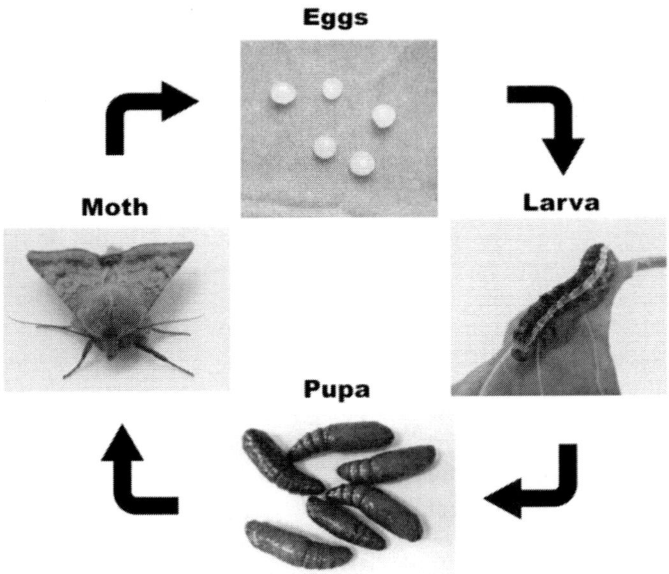


**Figure 1.** *Helicoverpa armigera* life cycle

## 2. THE USE OF GENETICALLY MODIFIED PLANTS FOR CONTROL OF LEPIDOPTERAN INSECTS

As mentioned previously, insects are responsible for major crop losses worldwide. In addition to direct impacts on yield, insects also reduce yields by making crops more susceptible to disease causing pathogens [8]. Last decade, most control measures focused on the use of chemical pesticides, a curative pest control strategy that was useful for rapid control of certain pest outbreaks. However, excessive and indiscriminate large-scale use of pesticides has led to development of pesticide-resistant insects [74]. Additionally, the long-term and extensive use of synthetic chemicals has led to concerns regarding their impact on food safety, associated human health and the environment [8]. As the use of pesticides for prevention of insect-associated losses cannot be overlooked in agriculture, there is a greater need to develop alternative or additional technologies which would allow a more selective use of pesticides and provide sustainable crop protection [52]. To achieve this objective, it is necessary to enhance the resistance of plants to pests and pathogens through integrated pest management (IPM) programs. They will need to consist of a combination of control strategies including (A) the use of natural biocontrol factors such as pathogens, predators or parasites [75]; (B) various preventive pest control strategies including crop rotation, intercropping, and cultivation of pest-resistant varieties of plants [8] and (C) genetic control via the release of sterile insects and also the use of natural insecticides. The latter includes secondary metabolites [52,76], viruses [77,78] and transgenes.

As the products of most transgenes are ingested by the insect pest and therefore act through the gut, most of the focus has been on transgene encoded proteins that target the insect midgut and/or the peritrophic membrane to disrupt digestion or nutrition [53,54,79-81]. Generally, the detrimental effects on larval and insect growth result from limited assimilation of nutrients [82-85]. Furthermore, any severe delay in growth and development, in a natural setting, lengthens the period in which the larvae are vulnerable to natural predators such as mice, spiders and predaceous insects [30,86,87]. The use of transgenic plants that express insecticidal agents thus reduces the population of insect pests and reduces the usage of chemical insecticides. This extends the useful life of the insecticides and also reduces the ecological damage they may cause [61]. As with any new method of insect control, the impact of transgenic plants on non-target and beneficial insects, particularly pollinators such as honey bees, needs to be assessed [88-90]. Table 1 lists a number of biotechnology approaches tested on lepidopteran insects. Since the discovery that dsRNA can silence genes, RNA interference (RNAi) has been developed as an effective tool for regulating gene expression in plants and animals. RNA interference or gene silencing has been used to inhibit virus replication and spread in transgenic plants and has potential to be developed commercially for disease control [91]. The use of RNAi for insect control is less well developed. Insect genes can be down-regulated by injection of dsRNA or by oral administration of high concentrations of exogenously supplied dsRNA as part of an artificial diet, but a much more efficient method of delivering dsRNA is needed before RNAi technology can be used to control pests in the field [64,65]. To date, the most successful transgenes for insect control have been the genes encoding insecticidal toxins from the soil bacterium *Bacillus thuringiensis.*

## 2.1. The bacillus thuringiensis endotoxins

The use of genes encoding endotoxins from *Bacillus thuringiensis* is now a well-established technology for producing transgenic plants with enhanced resistance to the larvae of lepidopteran insect pests [92]. Bt cotton was first released for commercial production in the USA in 1996 and subsequently grown in several countries including Argentina, Australia, China, Colombia, Indonesia, Mexico, South Africa, and India [93]. Since then other transgenic crop species producing Bt toxins have been commercialized including maize, tomato and potato (http://cera-gmc.org). The adoption of Bt crop varieties by farmers has been rapid reflecting the benefits of these crops such as reduced insecticide use, lower production costs and higher yields [94]. Only two Bt crops are grown in Australia (Table 2). In the most recent season (2011/2012) approximately 80% of the cotton grown in Australia was Bollgard II ® [95].

*B. thuringiensis*, a Gram-positive soil bacterium, produces a proteinaceous parasporal crystalline inclusion during sporulation [96]. There are two main categories of Bt toxins: Cry and Cyt. These two groups are classified further by a detailed nomenclature system that describes groups Cry1 to Cry55 and Cyt1 to Cyt2 [97-99]. The Cry toxins are divided into three larger families that are not related phylogenetically. The largest Cry family is the three domain family, and genes from this family are present in the majority of commercialised Bt crops [100].

The larvae of insect orders primarily affected by Bt toxins are Lepidoptera (butterflies and moths), Diptera (mosquitoes) and Coleoptera (larval and adult beetles) [101]. However, Bt toxins are not toxic to people, wildlife, or most beneficial insects [102,103] and therefore the opportunities for biological control are great. The effect of Bt toxins on a range of lepidopteran insects has been studied including: *Bombyx mori* [104], *Helicoverpa armigera* [105], *Heliothis virescens* [106,107], *Manduca sexta* [108,109], *Ostrinia nubilalis* [110-113], *Plutella xylostella* [114,115], *Sesamia nonagrioides* [115], *Spodoptera exigua* [116], *Spodoptera frugiperda* [117] and *Spodoptera littoralis* [118]. The Cry toxins produced in Bt crops generally target lepidopteran pests, although some also target coleopteran pests [100]. The first commercialised Bt crops contained only one Cry toxin, but second generation Bt crops have between two to six different toxins [100].

**Table 2.** Bt crops grown in Australia

| Trade name | Crop | Bt protein | Company | Year released |
|---|---|---|---|---|
| Ingard ® | cotton | Cry1Ac | Monsanto | 1996 |
| Bollgard II ® | cotton | Cry1Ac, Cry2Ab | Monsanto | 2003 |

This table lists the transgenic crops in Australia producing Bt proteins.

## 2.2. Mechanism of action

The Bt toxin mechanism of action is described by two models: The pore formation model and the signal transduction model. The initial steps of both models are the same. Upon ingestion by insects the crystalline inclusion is solubilised in the midgut [119]. Most target insects have a high gut pH [120] that is crucial for the efficacy of Bt toxins since most Bt-protoxins are only soluble above pH 9.5 [121]. The 130 kDa protoxins are activated by insect gut proteases, which typically cleave from both the C- and N-termini resulting in a 43-65 kDa protease-resistant active core [122-125].

The pore formation model has been the accepted mode of action for 20 years and is supported by numerous publications [96,126-128]. In this model the activated toxins bind to the primary receptors in the brush border membrane of the midgut epithelium columnar cells [14]. The major receptors for Cry toxins in lepidopterans are cadherin-like proteins [129-133]. The binding site of Cry toxins varies depending on the structure of the Cry toxin [105,110]. Binding to cadherin facilitates further proteolytic cleavage of the toxin and promotes the formation of oligomers [128,134]. The toxins then interact with secondary receptors in the midgut larval membrane. These secondary receptors are GPI-anchored proteins; either aminopeptidases or alkaline phosphatases [119,128,131,135]. Following secondary receptor binding, the toxin inserts into the membrane and creates pores [128]. These pores lead to the disruption of membrane integrity and cause an electrolyte imbalance that ultimately leads to death by starvation or septicaemia [136,137]. It is likely that there are more receptors involved in Bt toxicity since insects lacking the cadherin receptor are still killed by modified Bt toxins [138,139].

An alternative model for the Bt toxin mechanism of action proposes that Cry toxins trigger a signalling cascade pathway [140,141]. This model differs from the pore formation model in that it does not involve toxin oligomerisation, secondary receptors or the formation of pores in the membrane. Instead, in this model, binding to the cadherin receptor initiates a Mg2+ dependent signal cascade pathway that includes a guanine nucleotide-binding protein, adenylyl cyclase, and protein kinase A which ultimately results in cell death.

## 2.3. Resistance of lepidopteran insects to bt toxins

More recently there have been reports of field resistance to Bt crops in pink bollworm (*Pectinophore gosspiella* [142,143]), cotton bollworm (*Helicoverpa* spp [144-147]), armyworm (*Spodoptera frugiperda*[22]) and western corn rootworm (*Diabrotica virgifera virgifera* [148]. Some insects collected from the field have Bt resistance that has been characterized in the laboratory. However, there is debate about the relevance of this laboratory resistance in the field [149]. A decrease in field performance of Bt corn against *S. frugiperda* was observed in Puerto Rico [150] and against *Busseola fusca* in South Africa [23,151]. In southeastern US problems with control of *H. zea* on Bt cotton have also been reported [144-146].

The most common mechanism of resistance is the disruption of binding of Bt toxin to receptors in the midgut membrane. This disruption may be caused either by mutations in the receptor that blocks binding (reviewed in [20]) or changes in expression of the receptors [152,153]. Mutations in cadherin genes are responsible for Bt resistance in *Heliothis virescens* [154], *Helicoverpa armigera* [155] and *Pectinophora gossypiella* [156]. Another resistance mechanism associated with an ABC transporter locus has been reported in three lepidopteran spp (*H. virescens, P. xylostella and T. ni* [157]). Resistance to Bt in *Ostrinia nubialis* is due to reduced midgut protease activity resulting in less activation of the protoxins [111,158,159].

## 2.4. Management of resistance to bt crops

There are two main strategies for management of insect resistance to Bt crops: Refuge and pyramiding. The main approach for delaying evolution of resistance to Bt crops is the refuge strategy [21]. Farmers are mandated to maintain an abundance of host non-Bt crops as a refuge surrounding their Bt crops. The theory behind this strategy is that any Bt resistant larvae that arise on the Bt crops will mate with susceptible individuals from neighbouring non-Bt crops. As long as inheritance of resistance remains recessive the offspring will be susceptible to Bt crops [160-162]. This strategy is then combined with several other mandatory farming practices that include control of volunteer and ratoon plants that arise post-harvest, planting within a defined period of time to restrict the exposure of the Bt crop to the insect pests, restricted use of foliar Bt and the cultivation of crop residues [95]. The other major strategy to combat the evolution of Bt resistance is gene pyramiding. For example, the development of second generation Bt cotton that has at least two Bt toxins such as the Monsanto Bollgard II cotton variety, but up to six Bt toxins [100]. Another resistance management strategy which is still in the research phase of development is the use of insecticidal genes with completely different modes of action such as proteinase inhibitors. The success of combining multiple Bt genes for resistance management is contingent on the individual toxins having different targets to prevent cross resistance developing [163-165]. Binding studies with various Cry toxins have been used to identify toxins with different binding sites in the lepidopteran midguts [105,166,167]. This information can be used to design combinations of Cry toxins that complement each other to delay the development of resistance to Bt crops.

In addition to the resistance management plan for Bollgard cotton outlined above, farmers also use integrated pest management (IPM) systems as a sustainable approach to control all pests. IPM systems deploy a tactical combination of biotechnological, chemical, biological and cultural control methods to avoid pest problems [168]. Some of the major IPM strategies and tools include maintenance of beneficial insect populations, ensuring healthy plant growth, managing weed hosts and monitoring pest populations and plant damage regularly. All these additional practices lead to better control of insect populations in general and therefore helps prevent the development of resistance in insect populations to Bt.

# 3. PROTEASE INHIBITORS FOR THE CONTROL OF INSECT PESTS

Protease inhibitors are one component of a plant's natural defence mechanism against herbivores and pathogens [169]. Plants protect themselves directly by constitutively expressing protease inhibitors [170] and by inducing protease inhibitors in response to mechanical wounding or insect attack [169,171]. They may also release volatile compounds after insect damage that function as potent attractants for predators of insect herbivores [172]. The release of volatile compounds after wounding, such as methyl jasmonate also triggers the production of proteinase inhibitors in neighbouring unwounded plants essentially prearming the local population against insect attack [173].

## 3.1. Mechanism of action of protease inhibitors on lepidopteran insects

Protease inhibitors when incorporated into artificial diets or expressed in transgenic plants increase mortality [174] and reduce the growth and development of larvae from many insect pest species including Coleoptera [175,176], Orthoptera [177] and Lepidoptera [178,179](Table 2). The mechanisms by which ingested PIs mediate their effects on insect physiology differs between insect species [180]. Proteinase inhibitors bind to insect digestive proteases, preventing proteolysis which blocks digestion of protein [181]. This effectively starves the larvae of protein and essential amino acids required for insect growth, development and reproduction [182-185]. To compensate for this inhibition, several insect species increase production of proteases to swamp the ingested PIs [186,187]. This in turn can lead to a limitation in bioavailability of essential amino acids for protein synthesis, impairment of growth and development, and potentially death [182,186]. The loss of the sulphur-containing amino acids (cysteine and methionine) is critical because the sulfydryl content in trypsin and chymotrypsin is high and reprocurement of the sulphur-containing amino acids is difficult since cysteine and methionine are in relatively low concentrations in the diet, especially if the food source is plant material [186]. Broadway and colleagues confirmed this hypothesis in bioassays with *Spodoptera exiqua* where the weight-reducing effects obtained with soybean trypsin inhibitor were eliminated when the diets were supplemented with methionine [186].

## 3.2. Pis in transgenic plants for plant protection: success and failure

Several groups have reported enhanced protection of plants against lepidopteran pests after transformation with genes encoding PIs (Table 3). Despite this substantial body of work, defense strategies based on PI expression in plants have not resulted in any commercial application so far [61,214,215]. This relates to two distinct problems: (1) the levels of PI-expression in transgenic plants and (2) the pest's capacity to react to PI consumption. Most problems arise from the use of a single transgene producing a PI that targets only one protease or one class of protease in the insect midgut.

**Table 3.** Serine protease inhibitors that have been tested for their effect on growth and development of lepidopteran larvae

| Protease inhibitor | Protease family | Proteases inhibited | Transformed plant | Insect species used in bioassay | Effect of PI on larval growth |
|---|---|---|---|---|---|
| *Arabidopsis thaliana* serpin 1 [AtSerpin 1] | alpha-1-peptidase inhibitor | Chymotrypsin | *Arabidopsis* | *Spodoptera littoralis* | 38% biomass reduction after feeding for 4 days [188] |
| Barley trypsin inhibitor [BTI] | Cereal trypsin inhibitor | Trypsin | Tobacco | *Spodoptera exigua* | 29% reduction in survival [189] |
| | | | Wheat | *Sitotroga cerealella* | No effect on growth or mortality [190] |
| Bovine pancreatic trypsin inhibitor [BPTI] | Kunitz (animal) | Trypsin, chymotrypsin, plasmin, kallikreins | Tobacco | *Spodoptera exigua* | Reduced trypsin activity; induced leucine aminopeptidase and carboxypeptidase A activities; chymotrypsin, elastase, and carboxypeptidase B proteases not affected [190] |
| | | | Sugarcane | *Scirpophaga excerptalis* | Significant reduction in weight [191] |
| Bovine spleen trypsin inhibitor [SI] | Kunitz (animal) | Trypsin, chymotrypsin | Tobacco | *Helicoverpa armigera* | Reduced survival and growth [192] |
| Cowpea trypsin inhibitor [CpTI] | Bowman-Birk | Trypsin | Tobacco | *Heliothis virescens* | Increased mortality [9] |
| | | | Tobacco | *Helicoverpa zea* | Increased mortality [193] |
| | | | Rice | *Chilo suppressalis-Sesamia inferens* | Growth not monitored [11] |

| | | | | | |
|---|---|---|---|---|---|
| | | | Potato | *Lacanobia oleracea* | 45% biomass reduction [13] |
| | | | Tobacco | *Spodoptera litura* | 50% biomass reduction [194] |
| | | | Potato | *Lacanobia oleracea* | Decreased weight and delayed development [12] |
| Giant taro proteinase inhibitor [GTPI] | Kunitz (plant) | Trypsin, chymotrypsin | Tobacco | *Helicoverpa armigera* | Decreased growth, no increase in mortality [195] |
| Mustard trypsin inhibitor 2 [MTI-2] | Brassicaceae proteinase inhibitor | Trypsin, chymotrypsin | Tobacco, *Arabidopsis* and oilseed rape | *Spodoptera littoralis* | Increased mortality; surviving larvae up to 39% smaller after 10 days [187] |
| | | | | *Mamestra brassicae, Plutella xylostella, Spodoptera littoralis* | *P. xylostella:* 100% mortality on *Arabidopsis*; high mortality & delayed development on oilseed rape. *M. brassicae:* increased mortality & weight of survivors on *Arabidopsis* and tobacco, no effect on oilseed rape. *S. littoralis*: delay in development on oilseed rape [178]. |
| | | | Tobacco | *Spodoptera littoralis* | No effect on growth; reduction in fertility [196] |
| | | | Oilseed rape | *Plutella xylostella* | Reduction in survival and weight [30] |

| | | | | | |
|---|---|---|---|---|---|
| *Nicotiana alata* protease inhibitor [NaPI] | Proteinase inhibitor II | Trypsin, chymotrypsin | Tobacco | *Helicoverpa punctigera* | Decreased weight; increased mortality [197] |
| | | | Tobacco and peas | *Helicoverpa armigera* | Increased mortality; delayed growth [198] |
| | | | 'Royal Gala' apple | *Epiphyas postvittana* | Larval and pupal weights reduced; developmental abnormalities [31] |
| | | | Cotton | *Helicoverpa armigera* | A higher number of cotton bolls were recorded in plants expressing NaPI and a PotI inhibitor from potato, StPin1A [199]. |
| Potato inhibitor II [Pin II, PPI- II, Pot II, PI-II] | Proteinase inhibitor II | Trypsin, chymotrypsin, oryzin, subtilisin, elastase | Tobacco | *Manduca sexta* | Growth retarded [200] |
| | | | Tobacco | *Chrysodeixis eriosoma, Spodoptera litura, Thysanoplusia orichalcea* | *C. eriosoma* larvae grew slower; *S. litura and T. orichalcea* growth either unaffected or enhanced [201] |
| | | | Tobacco | *Spodoptera exigua* | Growth not affected [202] |
| | | | Rice | *Sesamia inferens* | Decreased weight [74] |
| | | | *Brassica napus* | *Plutella xylostella* | Lowered growth rates however more plant tissue consumed [203] |
| | | | Tomato | *Heliothis obsoleta* | Increased mortality and decreased weight on homozygous |

| | | | | | |
|---|---|---|---|---|---|
| | | | | | plants expressing PI-II and potato carboxypeptidase inhibitor (PCI), opposite effect on hemizygous plants [204] |
| *Solanum americanum* proteinase inhibitor [SaPIN2a] | Proteinase inhibitor II | Trypsin, chymotrypsin | Tobacco | *Helicoverpa armigera, Spodoptera litura* | Reduction in larval weight and pupation rate [205] |
| Soybean Kunitz trypsin inhibitor [SBTI, SKTI] | Kunitz (plant) | Trypsin, chymotrypsin, kallikrein, plasmin | Poplar | *Clostera anastomosis, Lymantria dispar* | Mortality and growth not significantly affected [206] |
| | | | Potato | *Lacanobia oleracea* | Survival and growth decreased by 33% and 40% respectively after 21 days [174] |
| | | | Tobacco | *Spodoptera litura* | Increased mortality and delayed development [207] |
| | | | Tobacco | *Helicoverpa armigera* | Development unaffected [208] |
| | | | Tobacco and potato | *Spodoptera littoralis* | High mortality on tobacco and up to 50% weight reduction on potato [209] |
| | | | Sugarcane | *Diatraea saccharalis* | Increased mortality; retarded growth [210] |
| Soybean Bowman-Birk | Bowman-Birk | Trypsin, chymotrypsin | Sugarcane | *Diatraea saccharalis* | Growth severely retarded [210] |

| | | | | | |
|---|---|---|---|---|---|
| trypsin inhibitor [SBBI] | | | | | |
| Sweet potato trypsin inhibitor [SWTI, Sporamin ] | Kunitz (plant) | Trypsin | Cauliflower | *Plutella xylostella, Spodoptera litura* | Increased mortality [42] |
| | | | Tobacco | *Spodoptera litura* | Growth and survival severely retarded [211] |
| | | | Tobacco | *Helicoverpa armigera* | Increased mortality and delayed growth and development in larvae on plants expressing sporamin and a phytocystatin from taro, CeCPI [212] |
| | | | Brassica | *Plutella xylostella* | Survival rate and body mass was significantly lower in larvae fed plants expressing sporamin and chitinase [213] |
| Tomato inhibitor I [Tom1] | Proteinase inhibitor I | Chymotrypsin subtilisin, trypsin | Tobacco | *Manduca sexta* | Little effect on growth [200] |
| Tomato inhibitor II [TPI-II] | Proteinase inhibitor II | Chymotrypsin trypsin, subtilisin | Tobacco | *Manduca sexta* | Growth retarded [200] |

This table lists plant and non-plant serine protease inhibitors expressed in transgenic plants that have been tested inbioassays with lepidopteran larvae. The major enzymes targeted by each PI are given, however other enzymes may beweakly inhibited or have not been tested.The first problem of inadequate levels of PI expression is best exemplified by studies with *P. xylostella*, the diamondback moth. When larvae of the diamondback moth consumed transgenic plants expressing the chymotrypsin and trypsin specific potato type II proteinase inhibitor, Pot II, they suffered lower growth rates. However, this did not confer an advantage to the plants because the larvae consumed more tissue to compensate for their decrease in metabolism [13,203]. As a result, the insects maintained population growth rates similar to those of larvae on non-transgenic plants. Growth enhancement has been reported after PI ingestion in insects from a number of orders [201,216]. Larvae that consumed tobacco leaves expressing low levels of mustard trypsin inhibitor 2 (MTI-2) developed faster, had an increased mean weight and caused more damage to leaves compared to control larvae on non-transgenic tobacco [187]. The increase in leaf surface consumption observed with plants expressing low levels of MTI-2 may have resulted from a decrease in available protein due to the presence of MTI-2 and/or to an increase in gut proteolytic capacity induced by PI consumption [187].

The second problem, the pest's capacity to react to PI consumption, is exemplified by the observation that several PIs that are potent inhibitors of insect proteases *in vitro* fail to produce any deleterious effect when fed to larvae [187]. Several mechanisms have been reported for this lack of effect (Figure 2). For example, the complement of proteolytic enzymes in the insect midgut can be altered after PI ingestion [183,214,217]. This could involve a switch to enzymes of different substrate specificity, but the same mechanistic class. For example, production of a chymotrypsin-like enzyme rather than a trypsin-like protease [195,218]. Another mechanism used to detoxify the PIs is degradation via endogenous proteases within the insect midgut [214,219]. Insects that feed regularly on a particular host plant are generally not affected by the PIs produced by the host. For example the PIs from chickpea, a host plant for *H. armigera,* are rapidly degraded by the *H. armigera* gut proteases [219,220]. Similarly, single domain cystatins from potato multicystatin are degraded when fed to larvae of *Diabrotica* spp (Coleoptera). Sometimes non-host PIs are also rapidly degraded. Human stefin A, a potent inhibitor of human cysteine proteases, was degraded by cystatin-insensitive proteases in the gut of Colorado potato beetle (*Leptinotarsa decemlineata*) and black vine weevil (*Otiorynchus sulcatus*) [221]. Another anti-PI mechanism is the production of midgut inhibitor-resistant serine proteases [182,222-224]. Some insect larvae adapt to the presence of PIs by replacing the inhibited enzymes with other PI-resistant proteases and can exhibit increased ingestion rates and faster development than larvae fed on control diets lacking PIs [202,204,225,226]. Some classic examples of this phenomenon are as follows. Soybean Kunitz trypsin inhibitor (SKTI) is normally an effective inhibitor of protease activity in gut extracts from *H. armigera* larvae, this insect is not seriously affected by ingestion of this PI because it responds to chronic ingestion of SKTI by increasing activity of an SKTI-resistant trypsin [227]. Similarly, growth and development of *S. exigua* larvae was not impacted when fed leaves from tobacco plants transformed with the chymotrypsin/trypsin specific potato proteinase inhibitor II (Pot II) [202]. Analysis of the trypsin activity in the gut of these insects demonstrated that only 18% of the trypsin activity of insects reared on these transgenic plants was inhibited by Pot II, whereas 78% of the trypsin activity in the gut of insects reared on control plants was Pot II-inhibitable [202]. The larvae had compensated for the loss of the PI-inhibitable trypsin by a 2.5-fold induction of new activity that was resistant to inhibition by Pot II [202]. Another observation of induction of PI-resistant enzymes was made by Markwick and coworkers who reported that the trypsin in three species of leaf rollers (Tortricidae) that had fed on diets containing SKTI was less inhibited by bovine pancreatic trypsin inhibitor (BPTI) compared to the trypsin in control larvae [228]. These responses have been reported for lepidopteran species that have ingested PIs in native plants, transgenic plants, and artificial diets [195,229]. In summary, potent inhibition of an insect digestive enzyme *in vitro* by a particular PI is not a good prediction that the PI will be useful when expressed as a transgene for crop protection. That is, expression and regulation of midgut serine proteases in herbivorous insects is tightly regulated and is heavily influenced by the levels and the nature of ingested PIs [230,231]. The mechanism by which changes in protease levels and protease isoforms is regulated in response to PI ingestion is still unknown for phytophagous insects. An overview of the effects of PIs on insects is presented in Figure 2.

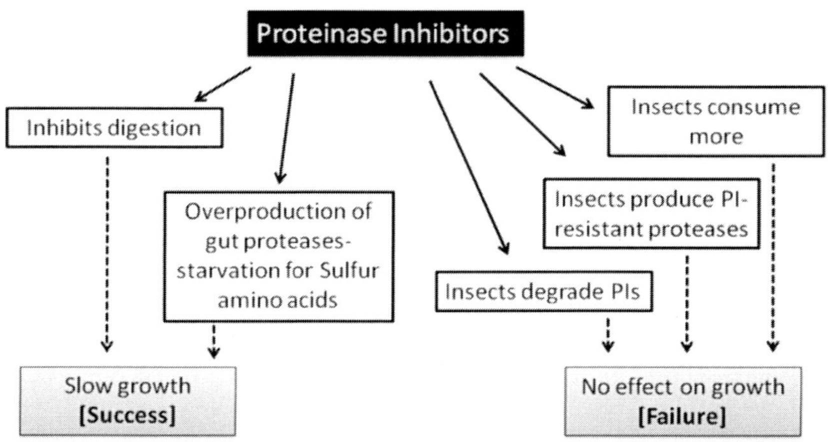


**Figure 2.** Outline of the various effects of ingested PIs on insect pests leading to success or failure in plant protection

## 3.3. Proteinase inhibitors from nicotiana alata as defence molecules against insect pests

Female reproductive tissues and wounded leaves of the ornamental tobacco, *Nicotiana alata* amass high levels of serine proteinase inhibitors for protection against insect pests and pathogens [232]. These serine proteinase inhibitors (NaPI) belong to the Potato type II family (Merops family I20) which have only been described in the Solanaceae. The NaPI precurser protein (NaProPI; 43 kDa), is composed of an ER signal peptide (29 amino acids), six repeated domains each with a potential PI-reactive site, and a 25 residue C-terminal domain that is essential for vacuolar targeting (VTS) [232-234] ( Figure 3).

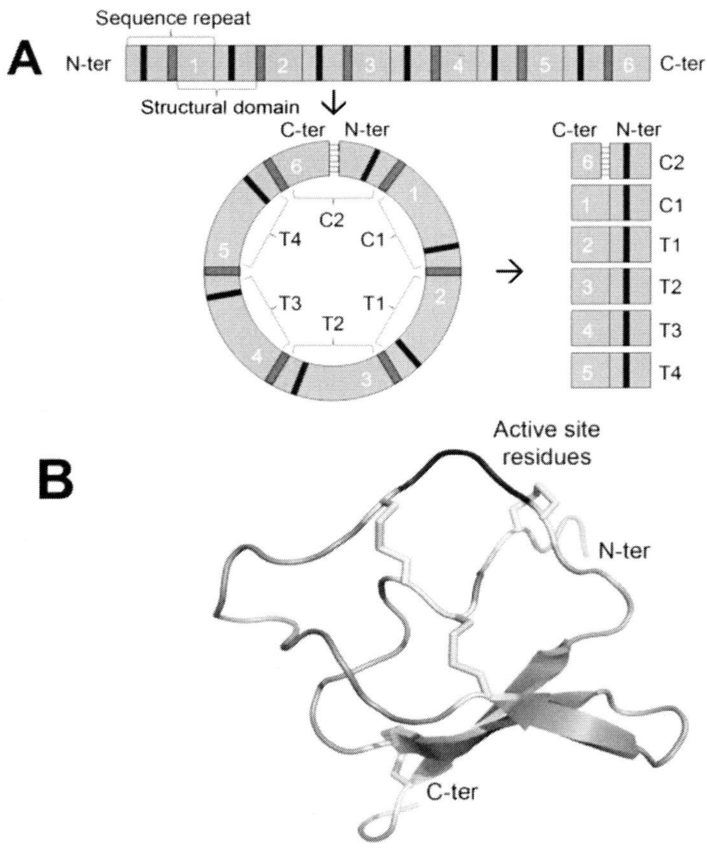


**Figure 3.** Diagrammatic representation of the domain organisation of NaProPI and the structure of the T1 inhibitor domain

(A) The NaPI precursor protein, shown as a linear gene product, forms a circular 'bracelet' structure that is 'clasped' by three disulphide bonds (yellow) between the N- and C-terminal repeats. Each repeat (labeled 1-6) contains a protease-reactive site (black), which is specific for either chymotrypsin (C1 and C2) or trypsin (T1-4). The six linker regions (red), with sequence EEKKN, are cleaved to release the six active inhibitor domains. The N-terminal signal sequence and the C-terminal vacuolar targeting signal have been omitted for clarity. Figure adapted from Scanlon *et al.* [235]. (B) Ribbon view of T1 showing the major secondary structural element, a triple stranded β-sheet (green) and the cysteines involved in disulfide bonds (yellow). The reactive site residues (black) are positioned between two cysteines that anchor the reactive loop to the central coil [236]. The other five inhibitors have the same structure [236-238].

Processing of NaProPI in the secretory pathway removes the ER signal peptide and VTS, and releases six PIs [232,239]. Processing of the six repeat NaProPI occurs at sites located within, rather than between, these repeated regions [232,239]. Complete removal of the linker sequence (Glu-Glu-Lys-Lys-Asn) contained within each repeated region [239], generates five contiguous inhibitors, a chymotrypsin inhibitor (C1) and four trypsin inhibitors (T1-T4), and two flanking peptides from the N- and C-termini. The flanking peptides form a novel two-chain chymotrypsin inhibitor (C2) that can only be formed if NaPI adopts a circular structure (Figure 3; [240]). The peptides have very similar amino acid sequences [239]. The three-dimensional structures of C1, C2, T1, T2, T3 and T4 have been determined by NMR spectroscopy [234,236,240]. A triple stranded -sheet is the dominant secondary structural feature; several -turns and a short region -helix are also present (Figure 3B; [238]). The reactive site is located on an exposed loop which has a higher degree of mobility than other regions of the protein (Figure 3B). This is a common feature of PIs and is thought to allow the inhibitor to adapt to slightly different enzymes [239].

Atkinson and colleagues suggested NaPIs may be involved in deterring insects from feeding on stigmas or in protecting the stigma from pathogen invasion since the related type-II PIs from potato and tomato are effective against proteases of fungal, bacterial and insect origin [232,241]. The PIs from *N. alata* inhibit the digestive gut proteases from five insect orders *in vitro* and display significant inhibitory activity against the midgut proteases of *H. punctigera* and *T. commodus* [197,198]. Significant mortality was recorded when *H. punctigera* larvae were fed transgenic tobacco [197] or transgenic peas [198] expressing the NaPI precursor. More recently, the response of *Helicoverpa* larvae to ingestion of NaPI has been more thoroughly characterized. Following ingestion of NaPI, all surviving *Helicoverpa punctigera* larvae produced high levels of a chymotrypsin that was resistant to inhibition by NaPI [199]. However this NaPI-resistant chymotrypsin was strongly inhibited by a potato type 1 inhibitor which is also produced by solanaceous plants, but belongs to a different class of serine proteinase inhibitors. When presented to *H. armigera* larvae in an artificial diet the combination of NaPI and the potato type I inhibitor had a much more dramatic effect on growth and development of the larvae compared to either of the inhibitors alone (Figure 4).

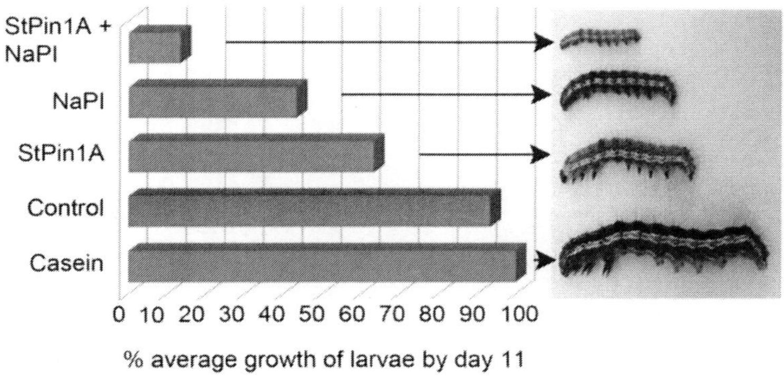


**Figure 4.** Percentage of Helicoverpa larval growth on day 11.

Neonates were transferred to cotton-leaf based artificial diets containing 0.3% of PIs (NaPI, StPin1A) and growth (mg) measured every 2$^{nd}$ day until day 11. Day 11, the % average weights compared to casein control are shown with representative larvae from each treatment (adapted from [199])

This laboratory result was then translated to transgenic plants in the field. Transgenic cotton plants expressing both PI classes, NaPI and StPin1A performed better than transgenic cotton plants expressing either PI alone. The improved performance of the transgenic cotton plants with both PIs was measured by an increase in cotton boll number per plant and increased yield of lint at the end of the cotton growing season (Figure 5)[199].

## 3.4. Commercialisation of pis and strategies to avoid resistance

Since the first transgenic plants appeared almost two decades ago, this technology has contributed to the development of new approaches for crop protection [25]. There are numerous reports showing that expression of PIs in transgenic plants confers resistance to the intended target insects (see Table II; reviewed in [61,215,242,243]). However, many of the candidate genes that have been used in genetic transformation of crops have limited application because they do not have broad spectrum activity against the major insect pests or are only mildly effective against the target pests [52]. To overcome the development of insect resistance to transgenic plants expressing PIs, it is necessary to develop PIs that have broad activity against most or all of the proteases that the insects use for digestion. Several strategies have been proposed.

**Figure 5.** A higher number of cotton bolls were produced on field grown transgenic cotton producing NaPI and StPin1A (A) compared to Coker (B) the control non-transgenic parent

### 3.4.1. Selecting second generation protease inhibitors from novel sources

PI-resistant proteases probably result from the selection pressure imposed on insects when they encounter high endogenous PI levels in certain host plants [170]. Such selection for PI-resistant proteases does not occur for PIs from non-host plants. Therefore, one approach to obtain better inhibitors for a particular insect pest is to search for PIs in plant species that are unrelated to the plant that is the normal host for that pest [10,74,170]. Another approach is to select PIs from synthetic libraries of mutant inhibitors for insect control [170].

### 3.4.2. Use of multiple inhibitors

Another strategy for controlling resistance development is to use at least two inhibitors that have different targets. This can be achieved by producing chimeric proteins, gene stacking (pyramiding) or the use a single inhibitors that have dual targets. Some examples of bifunctional inhibitors are alpha-amylase/trypsin inhibitors [8] and trypsin/ carboxypeptidase A inhibitors [244]. Similarly, expression of a fusion protein composed of a cystatin and a serine PI has been used to control certain nematode pathogens in transgenic plants [245]. Oppert and colleagues [246] demonstrated synergism between soybean Kunitz trypsin inhibitor and the cysteine protease inhibitor L-trans-epoxysuccinyleucylamide [4-guanidino] butane (E64) in artificial diet bioassays with *Tribolium castaneum* (red flour beetle, Coleoptera).

Transgenic tobacco plants expressing both a Bt-toxin and a cowpea trypsin inhibitor (CpTI) were more protected from *H. armigera* damage compared to transgenic tobacco expressing the Bt-toxin alone [247]. The enhanced insecticidal activity was attributed to enhanced stability of the Bt-toxin when the gut protease activity had been lowered [248,249]. In a separate set of experiments, *H. armigera* and *S. litura* larvae that consumed leaves from transgenic tobacco expressing avidin (from chicken egg white) that had been painted with Cry1Ba protein died significantly faster than larvae given either of the two treatments alone [38]. When used together in bioassays with artificial diet, the different and complementary action of Pot I (a chymotrypsin inhibitor) and CPI (a carboxpeptidase inhibitor) also resulted in a synergistic effect at reducing the growth rate of *Cydia pomonella* (codling moth) larvae [250]. However, the protective effects observed with PI gene constructs have not been sufficient to lead to a serious attempt at commercialising these transgenic crops.

# 4. SUMMARY

The usefulness of insect-resistant transgenic plants has been widely demonstrated with the highly successfully implementation of crops that produce the Bt toxin. The current fear is that although Bt toxin has defended crops in the field for nearly 10 years now, the discovery of Bt resistance in *H. zea* populations in crop fields in the USA [251] and Bt resistance in populations of *D. virgifera* found in corn fields [148] might lead to widespread development of resistance to the Bt toxin. We have reported that two structurally different PIs that target different enzymes greatly improved the protection of transgenic cotton plants in the field. This supports the general consensus in the literature that no single insect trait will provide sustainable crop protection and that stacking of multiple insect traits that target different mechanisms should be employed.

# REFERENCES

1. Krattiger AF (1996) Insect resistance in crops: A case study of *Bacillus thuringiensis* (*Bt*)and its transfer to developing countries: The International Agricultural Service for theAcquisition of Agribiotech Applications (ISAAA).
2. Oerke EC (2006) Crop losses to pests. J Agric Sci 144: 31-43.
3. Srinivasan A, Giri A, Gupta V (2006) Structural and functional diversities inlepidopteran serine proteases. Cell Mol Biol Lett 11: 132-154.
4. Brooke E, Hines E (1999) Viral biopesticides for *Heliothine* control-fact of fiction?Today's Life Science 11: 38-45.
5. Fitt GP (1994) Cotton Pest Management: Part 3. An Australian Perspective. Annu RevEntomol 39: 532-562.
6. Gatehouse AM, Hilder VA, Powell KS, Wang M, Davison GM, Gatehouse LN, DownRE, Edmonds HS, Boulter D, Newell CA, et al. (1994) Insect-resistant transgenic plants:choosing the gene to do the 'job'. Biochem Soc Trans 22: 944-949.
7. Gunning RV, Easton CS, Balfe ME, Ferris IG (1991) Pyrethroid resistance mechanisms inAustralian *Helicoverpa armigera*. Pestic Sci 33: 473-490.
8. Haq SK, Atif SM, Khan RH (2004) Protein proteinase inhibitor genes in combat againstinsects, pests, and pathogens: natural and engineered phytoprotection. Arch BiochemBiophys 431: 145-159.
9. Hilder VA, Gatehouse AM, Sheerman SE, Barker RF, Boulter D (1987) A novelmechanism of insect resistance engineered into tobacco. Nature 300: 160-163.
10. Harsulkar AM, Giri AP, Patankar AG, Gupta VS, Sainani MN, Ranjekar PK, DeshpandeVV (1999) Successive use of non-host plant proteinase inhibitors required for effectiveinhibition of *Helicoverpa armigera* gut proteinases and larval growth. Plant Physiol 121:497-506.
11. Xu D, Xue Q, McElroy D, Mawal Y, Hilder VA, Wu R (1996) Constitutive expression ofa cowpea trypsin inhibitor gene, CpTi, in transgenic rice plants confers resistance to twomajor rice insect pests. Mol Breed 2: 167-173.
12. Bell HA, Fitches EC, Down RE, Ford L, Marris GC, Edwards JP, Gatehouse JA,Gatehouse AM (2001) Effect of dietary cowpea trypsin inhibitor (CpTI) on the growthand development of the tomato moth *Lacanobia oleracea* (Lepidoptera: Noctuidae) andon the success of the

gregarious ectoparasitoid *Eulophus pennicornis* (Hymenoptera:Eulophidae). Pest Manage Sci 57: 57-65.

13. Gatehouse AMR, Davison GM, Newell CA, Merryweather A, Hamilton WDO, BurgessEPJ, Gilbert RJC, Gatehouse JA (1997) Transgenic potato plants with enhancedresistance to the tomato moth, *Lacanobia oleracea*: growth room trials. Mol Breed 3: 49-63.
14. Bravo A, Gill SS, Soberón M (2007) Mode of action of *Bacillus thuringiensis* Cry and Cyttoxins and their potential for insect control. Toxicon 49: 423-435.
15. Pigott CR, Ellar DJ (2007) Role of receptors in *Bacillus thuringiensis* crystal toxin activity.Microbiol Mol Biol Rev 71: 255-+
16. Barton KA, Whiteley HR, Yang NS (1987) *Bacillus thuringiensis* §-Endotoxin Expressedin Transgenic *Nicotiana tabacum* Provides Resistance to Lepidopteran Insects. PlantPhysiol 85: 1103-1109.
17. Fischhoff DA, Bowdish KS, Perlak FJ, Marrone PG, Mccormick SM, Niedermeyer JG,Dean DA, Kusanokretzmer K, Mayer EJ, Rochester DE, Rogers SG, Fraley RT (1987)Insect Tolerant Transgenic Tomato Plants. Bio-Technology 5: 807-813.
18. Vaeck M, Reynaerts A, Höfte H, Jansens S, De Beuckeleer M, Dean C, Zabeau M, VanMontagu M, Leemans J (1987) Transgenic plants protected from insect attack. Nature328: 33-37.
19. James C (2011) Global status of Commercialized Biotech/GM Crops. ISAAA Brief No 43Ithaca, NY, USA.
20. Heckel DG, Gahan LJ, Baxter SW, Zhao JZ, Shelton AM, Gould F, Tabashnik BE (2007)The diversity of Bt resistance genes in species of Lepidoptera. J Invertebr Pathol 95: 192-197.
21. Tabashnik BE, Gassmann AJ, Crowder DW, Carrière Y (2008) Insect resistance to Btcrops: evidence versus theory. Nat Biotechnol 26: 199-202.
22. Storer NP, Babcock JM, Schlenz M, Meade T, Thompson GD, Bing JW, Huckaba RM(2010) Discovery and characterization of field resistance to Bt maize: *Spodopterafrugiperda* (Lepidoptera: Noctuidae) in Puerto Rico. J Econ Entomol 103: 1031-1038.
23. Van Rensburg J (2007) First report of field resistance by stem borer, *Busseola fusca*(Fuller) to Bt-transgenic maize. S Afr J Plant Soil 24: 147-150.
24. Lynch RE, Guo B, Timper P, Wilson JP (2003) United States Department of Agriculture–Agricultural Research Service: Research on improving host-plant resistance to pests.Pest Manage Sci 59: 718-727.
25. Carlini CR, Grossi-de-Sa MF (2002) Plant toxic proteins with insecticidal properties. Areview on their potentialities as bioinsecticides. Toxicon 40: 1515-1539.
26. Franco OL, Rigden DJ, Melo FR, Grossi-de-Sa MF (2002) Plant alpha-amylase inhibitorsand their interaction with insect alpha-amylases. Structure, function and potential forcrop protection. Eur J Biochem 269: 397-412.
27. Bhalla R, Dalal M, Panguluri SK, Jagadish B, Mandaokar AD, Singh AK, Kumar PA(2005) Isolation, characterization and expression of a novel vegetative insecticidalprotein gene of *Bacillus thuringiensis*. FEMS Microbiol Lett 243: 467-472.
28. Fang J, Xu XL, Wang P, Zhao JZ, Shelton AM, Cheng J, Feng MG, Shen ZC (2007)Characterization of chimeric *Bacillus thuringiensis* Vip3 toxins. Appl Environ Microbiol73: 956-961.
29. Kabir KE, Sugimoto H, Tado H, Endo K, Yamanaka A, Tanaka S, Koga D (2006) Effectof *Bombyx mori* chitinase against Japanese pine sawyer (*Monochamus alternatus*) adults asa biopesticide. Biosci Biotech Biochem 70: 219-229.
30. Ferry N, Jouanin L, Ceci LR, Mulligan EA, Emami K, Gatehouse JA, Gatehouse AMR(2005) Impact of oilseed rape expressing the insecticidal serine protease inhibitor,mustard trypsin inhibitor-2 on the beneficial predator *Pterostichus madidus*. Mol Ecol 14:337-349.
31. Maheswaran G, Pridmore L, Franz P, Anderson MA (2007) A proteinase inhibitor from*Nicotiana alata* inhibits the normal development of light-brown apple moth, *Epiphyaspostvittana* in transgenic apple plants. Plant Cell Rep 26: 773-782.
32. Zhu C, Ruan L, Peng D, Yu Z, Sun M (2006) Vegetative insecticidal protein enhancingthe toxicity of *Bacillus thuringiensis* subsp *kurstaki* against *Spodoptera exigua*. Lett ApplMicrobiol 42: 109-114.

33. Estruch JJ, Warren GW, Mullins MA, Nye GJ, Craig JA, Koziel MG (1996) Vip3a, aNovel *Bacillus Thuringiensis* Vegetative Insecticidal Protein with a Wide Spectrum ofActivities against Lepidopteran Insects. Proc Natl Acad Sci U S A 93: 5389-5394.
34. Yu C, Mullins M, Warren G, Koziel M, Estruch J (1997) The *Bacillus thuringiensis*vegetative insecticidal protein Vip3A lyses midgut epithelium cells of susceptibleinsects. Appl Environ Microbiol 63: 532-536.
35. Lee MK, Walters FS, Hart H, Palekar N, Chen JS (2003) Mode of action of the *Bacillusthuringiensis* vegetative insecticidal protein Vip3A differs from that of Cry1Ab deltaendotoxin.Appl Environ Microbiol 69: 4648-4657.
36. Sellami S, Jamoussi K, Dabbeche E, Jaoua S (2011) Increase of the *Bacillus thuringiensis*Secreted Toxicity Against Lepidopteron Larvae by Homologous Expression of thevip3LB Gene During Sporulation Stage. Curr Microbiol 63: 289-294.
37. Markwick N, Christeller J, Dochterty L, Lilley C (2001) Insecticidal activity of avidinand streptavidin against four species of pest Lepidoptera. Entomol Exp Appl 98: 59-66.
38. Burgess EP, Malone LA, Christeller JT, Lester MT, Murray C, Philip BA, Phung MM,Tregidga EL (2002) Avidin expressed in transgenic tobacco leaves confers resistance to two noctuid pests, *Helicoverpa armigera* and *Spodoptera litura*. Transgenic Res 11: 185-198.
39. Markwick NP, Docherty LC, Phung MM, Lester MT, Murray C, Yao JL, Mitra DS,Cohen D, Beuning LL, Kutty-Amma S, Christeller JT (2003) Transgenic tobacco andapple plants expressing biotin-binding proteins are resistant to two cosmopolitan insectpests, potato tuber moth and lightbrown apple moth, respectively. Transgenic Res 12:671-681.
40. Christeller JT, Malone LA, Todd JH, Marshall RM, Burgess EPJ, Philip BA (2005)Distribution and residual activity of two insecticidal proteins, avidin and aprotinin,expressed in transgenic tobacco plants, in the bodies and frass of *Spodoptera litura* larvaefollowing feeding. J Insect Physiol 51: 1117-1126.
41. Murray C, Markwick NP, Kaji R, Poulton J, Martin H, Christeller JT (2010) Expression ofvarious biotin-binding proteins in transgenic tobacco confers resistance to potato tubermoth, *Phthorimaea operculella* (Zeller) (fam. Gelechiidae). Transgenic Res 19: 1041-1051.
42. Ding XF, Gopalakrishnan B, Johnson LB, White FF, Wang XR, Morgan TD, Kramer KJ,Muthukrishnan S (1998) Insect Resistance of Transgenic Tobacco Expressing an InsectChitinase Gene. Transgenic Res 7: 77-84.
43. Fitches E, Wilkinson H, Bell H, Bown DP, Gatehouse JA, Edwards JP (2004) Cloning,expression and functional characterisation of chitinase from larvae of tomato moth(*Lacanobia oleracea*): a demonstration of the insecticidal activity of insect chitinase. InsectBiochem Mol Biol 34: 1037-1050.
44. Wang JX, Chen ZL, Du JZ, Sun Y, Liang AH (2005) Novel insect resistance in *Brassicanapus* developed by transformation of chitinase and scorpion toxin genes. Plant CellRep 24: 549-555.
45. Purcell JP, Greenplate JT, Jennings MG, Ryerse JS, Pershing JC, Sims SR, Prinsen MJ,Corbin DR, Tran M, Sammons RD, Stonard RJ (1993) Cholesterol Oxidase- A PotentInsecticidal Protein Active against Boll-Weevil Larvae. Biochem Biophys Res Commun196: 1406-1413.
46. Corbin DR, Grebenok RJ, Ohnmeiss TE, Greenplate JT, Purcell JP (2001) Expression andchloroplast targeting of cholesterol oxidase in transgenic tobacco plants. Plant Physiol126: 1116-1128.
47. Shukle RH, Murdock LL (1983) Lipoxygenase, trypsin inhibitor and lectin fromsoybeans: effects on larval growth of *Manduca sexta* (Lepidoptera: Sphingidae). EnvironEntomol 12: 787-791.
48. Schroeder HE, Gollasch S, Moore A, Tabe LM, Craig S, Hardie DC, Chrispeels MJ,Spencer D, Higgins TJV (1995) Bean α-Amylase Inhibitor Confers Resistance to the PeaWeevil (*Bruchus pisorum*) in Transgenic Peas (*Pisum sativum* L.). Plant Physiol 107: 1233-1239.
49. Ishimoto M, Chrispeels MJ (1996) Protective mechanism of the Mexican bean weevilagainst high levels of alpha-amylase inhibitor in the common bean. Plant Physiol 111:393-401.

50. Morton RL, Schroeder HE, Bateman KS, Chrispeels MJ, Armstrong E, Higgins TJV(2000) Bean alpha-amylase inhibitor 1 in transgenic peas (*Pisum sativum* ) providescomplete protection from pea weevil (*Bruchus pisorum* ) under field conditions. ProcNatl Acad Sci U S A 97: 3820-3825.
51. Valencia-Jimenez A, Arboleda JW, Avila AL, Grossi-de-Sa MF (2008) Digestive alphaamylasesfrom *Tecia solanivora* larvae (Lepidoptera: Gelechiidae): response to pH,temperature and plant amylase inhibitors. Bull Entomol Res 98: 575-579.
52. Sharma HC, Sharma KK, Seetharama N, Ortiz R (2000) Prospects for using transgenicresistance to insects in crop improvement. Electron J Biotechnol 3: 76-95.
53. Czapla TH, Lang BA (1990) Effects of plant lectins on the larval development ofEuropean corn boror (Lepidoptera: Pyralidae) and Southern corn rootworm(Coleoptera: Chrysomelidae). J Econ Entomol 83: 2480-2485.
54. Hopkins TL, Harper MS (2001) Lepidopteran peritrophic membranes and effects ofdietary wheat germ agglutinin on their formation and structure. Arch Insect BiochemPhysiol 47: 100-109.
55. Fitches E, Gatehouse JA (1998) A comparison of the short and long term effects ofinsecticidal lectins on the activities of soluble and brush border enzymes of tomatomoth larvae (*Lacanobia oleracea*). J Insect Physiol 44: 1213-1224.
56. Down RE, Ford L, Bedford SJ, Gatehouse LN, Newell C, Gatehouse JA, Gatehouse AMR(2001) Influence of plant development and environment on transgene expression inpotato and consequences for insect resistance. Transgenic Res 10: 223-236.
57. Sadeghi A, Smagghe G, Broeders S, Hernalsteens JP, De Greve H, Peumans WJ, VanDamme EJ (2008) Ectopically expressed leaf and bulb lectins from garlic (*Allium sativum*L.) protect transgenic tobacco plants against cotton leafworm (*Spodoptera littoralis*).Transgenic Res 17: 9-18.
58. de Oliveira CFR, Luz LA, Paiva PMG, Coelho L, Marangoni S, Macedo MLR (2011)Evaluation of seed coagulant *Moringa oleifera* lectin (cMoL) as a bioinsecticidal tool withpotential for the control of insects. Process Biochem 46: 498-504.
59. Nauen R, Sorge D, Sterner A, Borovsky D (2001) TMOF-like factor controls thebiosynthesis of serine proteases in the larval gut of *Heliothis virescens*. Arch InsectBiochem Physiol 47: 169-180.
60. Tortiglione C, Fanti P, Pennacchio F, Malva C, Breuer M, De Loof A, Monti LM,Tremblay E, Rao R (2002) The expression in tobacco plants of *Aedes aegypti* TrypsinModulating Oostatic Factor (Aea-TMOF) alters growth and development of the tobaccobudworm, *Heliothis virescens*. Mol Breed 9: 159-169.
61. Schuler TH, Poppy GM, Kerry BR, Denholm I (1998) Insect-resistant transgenic plants.Trends Biotechnol 16: 168-175.
62. Harvey WR (1992) Physiology of V-ATPases. J Exp Biol 172: 1-17.
63. Baum JA, Bogaert T, Clinton W, Heck GR, Feldmann P, Ilagan O, Johnson S, PlaetinckG, Munyikwa T, Pleau M, Vaughn T, Roberts J (2007) Control of coleopteran insectpests through RNA interference. Nat Biotechnol 25: 1322-1326.
64. Mao YB, Cai WJ, Wang JW, Hong GJ, Tao XY, Wang LJ, Huang YP, Chen XY (2007)Silencing a cotton bollworm P450 monooxygenase gene by plant-mediated RNAiimpairs larval tolerance of gossypol. Nat Biotechnol 25: 1307-1313.
65. Bettencourt R, Terenius O, Faye I (2002) Hemolin gene silencing by ds-RNA injecte into *Cecropia pupae* is lethal to next generation embryos. Insect Mol Biol 11: 267-271.
66. Eleftherianos I, Marokhazi J, Millichap PJ, Hodgkinson AJ, Sriboonlert A, ffrench-Constant RH, Reynolds SE (2006) Prior infection of *Manduca sexta* with non-pathogenic*Escherichia coli* elicits immunity to pathogenic *Photorhabdus luminescens*: roles ofimmune-related proteins shown by RNA interference. Insect Biochem Mol Biol 36: 517-525.
67. Fitt GP (1989) The ecology of *Heliothis* species in relation to agroecosystems. Annu RevEntomol 34: 17-52.
68. Trowell SC, Forrester NW, Garsia KA, Lang GA, Bird LJ, Hill AS, Skerritt JH, Daly JC(2000) Rapid antibody-based field test to distinguish between *Helicoverpa armigera* (Lepidoptera: Noctuidae) and *Helicoverpa punctigera* (Lepidoptera: Noctuidae). J EconEntomol 93: 878-891.

69. Sivakumar S, Rajagopal R, Venkatesh GR, Srivastava A, Bhatnagar RK (2007)Knockdown of aminopeptidase-N from *Helicoverpa armigera* larvae and in transfected Sf21 cells by RNA interference reveals its functional interaction with *Bacillusthuringiensis* insecticidal protein Cry1Ac. J Biol Chem 282: 7312-731
70. Estébanez-Perpiñá E, Bayés A, Vendrell J, Jongsma MA, Bown DP, Gatehouse JA, Huber R, Bode W, Avilés FX, Reverter D (2001) Crystal structure of a novel Mid-gut procarboxypeptidase from the cotton pest *Helicoverpa armigera*. J Mol Biol 313: 629-638.
71. Downes S, Mahon R, Olsen K (2007) Monitoring and adaptive resistance managementin Australia for Bt-cotton: Current status and future challenges. J Invertebr Pathol 95:208-213.
72. Patankar AG, Giri AP, Harsulkar AM, Sainani MN, Deshpande VV, Ranjekar PK, GuptaVS (2001) Complexity in specificities and expression of *Helicoverpa armigera* gutproteinases explains polyphagous nature of the insect pest. Insect Biochem Mol Biol 31:453-464.
73. Daly JC, Gregg P (1985) Genetic variation in *Heliothis* in Australia: Species identificationand gene flow in the two pest species *H. armigera* (Hubner) and *H. punctigera*Wallengren (Lepidoptera: Noctuidae). Bull Entomol Res 75: 169-184.
74. Duan X, Li X, Xue Q, Abo-Ei-Saad M, Xu D, Wu R (1996) Transgenic rice plantsharboring an introduced potato proteinase inhibitor II gene are insect resistant. NaBiotech 14: 494-498.
75. Grundy PR (2007) Utilizing the assassin bug, *Pristhesancus plagipennis* (Hemiptera :Reduviidae), as a biological control agent within an integrated pest managementprogramme for *Helicoverpa* spp. (Lepidoptera : Noctuidae) and *Creontiades* spp.(Hemiptera : Miridae) in cotton. Bull Entomol Res 97: 281-290.
76. Xu D, McElroy D, Thornburg RW, Wu R (1993) Systemic induction of a potato pin2promoter by wounding, methyl jasmonate, and abscisic acid in transgenic rice plants.Plant Mol Biol 22: 573.
77. Basinski JMH, Bendena WG, Downe AER (1995) The Effect of CfNPV Infection onSeveral Proteases in the Midgut of the Eastern Spruce Budworm *Choristoneurafumiferana*. J Invertebr Pathol 66: 264-269.
78. Borovsky D, Rabindran S, Dawson WO, Powell CA, Iannotti DA, Morris TJ,Shabanowitz J, Hunt DF, DeBondt HL, DeLoof A (2006) Expression of Aedes trypsinmodulatingoostatic factor on the virion of TMV: A potential larvicide. Proc Natl AcadSci U S A 103: 18963-18968.
79. Murdock LL, Huesing JE, Nielsen SS, Pratt RC, Shade RE (1990) Biological effects ofplant lectins on the cowpea weevil. Phytochemistry 29: 85-89.
80. Eisemann CH, Donaldson RA, Pearson RD, Cadogan LC, Vuocolo T, Tellam RL (1994)Larvicidal Activity of Lectins on *Lucilia Cuprina* - Mechanism of Action. Entomol ExpAppl 72: 1-10.
81. Harper MS, Hopkins TL, Czapla TH (1998) Effect of wheat germ agglutinin onformation and structure of the peritrophic membrane in European corn borer (*Ostrinianubilalis*) larvae. Tissue and Cell 30: 166-176.
82. Williams IS (1999) Slow-growth, high-mortality - a general hypothesis, or is it? EcoEntomol 24: 490-495.
83. Lopes AR, Juliano MA, Juliano L, Terra WR (2004) Coevolution of insect trypsins andinhibitors. Arch Insect Biochem Physiol 55: 140-152.
84. Zavala JA, Baldwin IT (2004) Fitness benefits of trypsin proteinase inhibitor expressionin *Nicotiana attenuata* are greater than their costs when plants are attacked. BMC Ecol 4:11.
85. Silva FC, Alcazar A, Macedo LL, Oliveira AS, Macedo FP, Abreu LR, Santos EA, SalesMP (2006) Digestive enzymes during development of *Ceratitis capitata*(Diptera:Tephritidae) and effects of SBTI on its digestive serine proteinase targets.Insect Biochem Mol Biol 36: 561-569.
86. Room PM (1979) Parasites and predators of *Heliothis* spp. (Lepidoptera: Noctuidae) incotton in the Namoi Valley, New South Wales. J Aust Entomol Soc 18: 223-228.
87. Hellers M, Beck M, Theopold U, Kamei M, Schmidt O (1996) Multiple alleles encoding avirus-like particle protein in the ichneumonid endoparasitoid *Venturia canescens*. InsectMol Biol 5: 239-249.

88. Malone LA, Giacon HA, Burgess EPJ, Maxwell JZ, Christeller JT, Laing WA (1995)Toxicity of Trypsin Endopeptidase Inhibitors to Honey Bees (Hymenoptera, Apidae). JEcon Entomol 88: 46-50.
89. Burgess EPJ, Malone LA, Christeller JT (1996) Effects of two proteinase inhibitors on thedigestive enzymes and survival of honey bees (*Apis mellifera*). J Insect Physiol 42: 823-828.
90. Burgess EPJ, Lovei GL, Malone LA, Nielsen IW, Gatehouse HS, Christeller JT (2002)Prey-mediated effects of the protease inhibitor aprotinin on the predatory carabid beetle*Nebria brevicollis*. J Insect Physiol 48: 1093-1101.
91. Shimizu T, Yoshii M, Wei T, Hirochika H, Omura T (2009) Silencing by RNAi of thegene for Pns12, a viroplasm matrix protein of Rice dwarf virus, results in strongresistance of transgenic rice plants to the virus. Plant Biotechnol J 7: 24-32.
92. Duke SO (2011) Comparing conventional and biotechnology-based pest management. JAgric Food Chem 59: 5793-5798.
93. James C (2009) Global status of commercialized biotech/GM crops: 2009. In: ISAAA:Ithaca N, editor. ISAAA Brief No 41. pp. Ithaca, NY, USA.
94. Brookes G, Barfoot P (2005) GM Crops: The Global Economic and EnvironmentalImpact-The First Nine Years 1996-2004. AgBioForum 8: 187-196.
95. Downes S, Wilson L, Knight K, Kauter G, Leven T (2012) Preamble to the ResistanceManagement Plan (RMP) for Bollgard II 20011/12. Cotton pest management guide 2011-2012: Cotton Catchment Communities. pp. 74-87(http://www. cottoncrc. org.au/industry/Publications/Cotton_Pest_Management_Guide_20 11_2012).
96. Schnepf E, Crickmore N, van Rie J, Lereclus D, Baum J, Feitelson J, Zeigler DR, DeanDH (1998) *Bacillus thuringiensis* and its pesticidal crystal proteins. Microbiol Mol BiolRev 62: 775-806.
97. Höfte H, Whiteley HR (1989) Insecticidal crystal proteins of *Bacillus thuringiensis*. MicroRev 53: 242-255.
98. Crickmore N, Zeigler DR, Feitelson J, Schnepf E, Van Rie J, Lereclus D, Baum J, DeanDH (1998) Revision of the nomenclature for the *Bacillus thuringiensis* pesticidal crystalproteins. Microbiol Mol Biol Rev 62: 807-813.
99. van Frankenhuyzen K (2009) Insecticidal activity of *Bacillus thuringiensis* crystalproteins. J Invertebr Pathol 101: 1-16.
100. Tabashnik BE, Van Rensburg JBJ, Carrière Y (2009) Field-evolved insect resistance toBt crops: definition, theory, and data. J Econ Entomol 102: 2011-2025.
101. Knowles BH, Dow JAT (1993) The crystal δ-endotoxins of *Bacillus thuringiensis*: modelsfor their mechanism of action on the insect gut. Bioessays 15: 469-476.
102. Marvier M, McCreedy C, Regetz J, Kareiva P (2007) A meta-analysis of effects of Btcotton and maize on nontarget invertebrates. Science 316: 1475-1477.
103. Romeis J, Meissle M, Bigler F (2006) Transgenic crops expressing *Bacillus thuringiensis*toxins and biological control. Nat Biotechnol 24: 63-71.
104. Endo Y, Nishiitsutsuji-Uwo (1980) Mode of action of *Bacillus thuringiensis* δ-endotoxin:Histopathological changes in the silkworm midgut. J Invertebr Pathol 36: 90-103.
105. Estela A, Escriche B, Ferré J (2004) Interaction of *Bacillus thuringiensis* toxins with larvalmidgut binding sites of *Helicoverpa armigera* (Lepidoptera : Noctuidae). Appl EnvironMicrobiol 70: 1378-1384.
106. Ryerse JS, Beck JR, Jr., Lavrik PB (1990) Light microscope immunolocation of *Bacillusthuringiensis kurstaki* delta-endotoxin in the midgut and Malpighian tubules of thetobacco budworm, *Heliothis virescens*. J Invertebr Pathol 56: 86-90.
107. MacIntosh SC, Stone TB, Jokerst RS, Fuchs RL (1991) Binding of *Bacillus thuringiensis*proteins to a laboratory-selected line of *Heliothis virescens*. Proc Natl Acad Sci U S A 88:8930-8933.
108. Lane NJ, Harrison JB, Lee WM (1989) Changes in microvilli and Golgi-associatedmembranes of lepidopteran cells induced by an insecticidally active bacterial ?-endotoxin. J Cell Sci 93: 337-347.

109. Knight PJ, Crickmore N, Ellar DJ (1994) The receptor for *Bacillus thuringiensis* CrylA(c)delta-endotoxin in the brush border membrane of the lepidopteran *Manduca sexta* isaminopeptidase N. Mol Microbiol 11: 429-436.
110. Hua G, Masson L, Jurat-Fuentes JL, Schwab G, Adang MJ (2001) Binding analyses of*Bacillus thuringiensis* Cry δ-endotoxins using brush border membrane vesicles of *Ostrinia nubilalis*. Appl Environ Microbiol 67: 872-879.
111. Li HR, Oppert B, Higgins RA, Huang FN, Zhu KY, Buschman LL (2004) Comparativeanalysis of proteinase activities of *Bacillus thuringiensis*-resistant and susceptible *Ostrinianubilalis* (Lepidoptera : Crambidae). Insect Biochem Mol Biol 34: 753-762.
112. Siqueira HAA, Nickerson KW, Moellenbeck D, Siegfried BD (2004) Activity of gutproteinases from Cry1Ab-selected colonies of the European corn borer, *Ostrinia nubilalis*(Lepidoptera: Crambidae). Pest Manage Sci 60: 1189-1196.
113. Tang JD, Shelton AM, Vanrie J, Deroeck S, Moar WJ, Roush RT, Peferoen M (1996)Toxicity of *Bacillus Thuringiensis* Spore and Crystal Protein to Resistant DiamondbackMoth (*Plutella Xylostella*). Appl Environ Microbiol 62: 564-569.
114. Wright DJ, Iqbal M, Granero F, Ferré J (1997) A change in a single midgut receptor inthe diamondback moth (*Plutella xylostella*) is only in part responsible for field resistanceto *Bacillus thuringiensis* subsp. *kurstaki* and *B. thuringiensis* subsp. *aizawai*. Appl EnvironMicrobiol 63: 1814-1819.
115. Gonzalez-Cabrera J, Farinos GP, Caccia S, Diaz-Mendoza M, Castanera P, LeonardiMG, Giordana B, Ferre J (2006) Toxicity and mode of action of *Bacillus thuringiensis* cryproteins in the Mediterranean corn borer, *Sesamia nonagrioides* (Lefebvre). Appl EnvironMicrobiol 72: 2594-2600.
116. Moar WJ, Pusztaicarey M, Vanfaassen H, Bosch D, Frutos R, Rang C, Luo K, Adang MJ(1995) Development of *Bacillus thuringiensis* CryIC resistance by *Spodoptera exigua*(Hubner) (Lepidoptera, Noctuidae). Appl Environ Microbiol 61: 2086-2092.
117. Adamczyk JJ, Holloway JW, Church GE, Leonard BR, Graves JB (1998) Larval Survivaland Development of the Fall Armyworm (Lepidoptera: Noctuidae) on Normal andTransgenic Cotton Expressing the *Bacillus thuringiensis* CryIA (c)-endotoxin. J EconEntomol 91: 539-545.
118. Avisar D, Keller M, Gazit E, Prudovsky E, Sneh B, Zilberstein A (2004) The role of*Bacillus thuringiensis* Cry1C and Cry1E separate structural domains in the interactionwith *Spodoptera littoralis* gut epithelial cells. J Biol Chem 279: 15779-15786.
119. Gill SS, Cowles EA, Pietrantonio PV (1992) The mode of action of *Bacillus thuringiensis*endotoxins. Annu Rev Entomol 37: 615-636.
120. Dow JAT, Evans PD, Wigglesworth VB (1987) Insect Midgut Function. Adv InsectPhysiol: Academic Press. pp. 187-328.
121. Bietlot H, Carey PR, Choma C, Kaplan H, Lessard T, Pozsgay M (1989) Facilepreparation and characterization of the toxin from *Bacillus thuringiensis* var. *kurstaki*.Biochem J 260: 87-91.
122. Nagamatsu Y, Itai Y, Hatanaka C, Funatsu G, Hayashi K (1984) A toxic fragment fromthe entomocidal crystal protein of *Bacillus thuringiensis*. Agric Biol Chem 48: 611-619.
123. Tojo A, Aizawa K (1983) Dissolution and degradation of *Bacillus thuringiensis* δ-endotoxin by gut juice protease of the silkworm *Bombyx mori*. Appl Environ Microbiol45: 576-580.
124. Diaz-Mendoza M, Farinós GP, Castañera P, Hernández-Crespo P, Ortego F (2007)Proteolytic processing of native Cry1Ab toxin by midgut extracts and purified trypsinsfrom the Mediterranean corn borer *Sesamia nonagrioides*. J Insect Physiol 53: 428-435.
125. Rukmini V, Reddy CY, Venkateswerlu G (2000) *Bacillus thuringiensis* crystal δ-endotoxin: role of proteases in the conversion of protoxin to toxin. Biochimie 82: 109-116.
126. Knowles BH, Ellar DJ (1987) Colloid-osmotic lysis is a general feature of themechanism of action of *Bacillus thuringiensis* [delta]-endotoxins with different insecspecificity. BBA-General Subjects 924: 509-518.

127. Haider MZ, Ellar DJ (1989) Mechanism of action of *Bacillus thuringiensis* insecticidal δ-endotoxin: interaction with phospholipid vesicles. Biochimica et Biophysica Acta (BBABiomembranes)
128. Bravo A, Gómez I, Conde J, Muñoz-Garay C, Sánchez J, Miranda R, Zhuang M, Gill SS,Soberón M (2004) Oligomerization triggers binding of a *Bacillus thuringiensis* Cry1Abpore-forming toxin to aminopeptidase N receptor leading to insertion into membranemicrodomains. Biochim Biophys Acta (BBA-Biomembranes), Biomembr 1667: 38-46.
129. Vadlamudi RK, Weber E, Ji I, Ji TH, Bulla LA, Jr. (1995) Cloning and expression of areceptor for an insecticidal toxin of *Bacillus thuringiensis*. J Biol Chem 270: 5490-5494.
*130.* Vadlamudi RK, Ji TH, Bulla LA, Jr. (1993) A specific binding protein from *Manducasexta* for the insecticidal toxin of *Bacillus thuringiensis* subsp. *berliner*. J Biol Chem 268:12334-12340.
131. Jurat-Fuentes JL, Adang MJ (2004) Characterization of a Cry1Ac-receptor alkalinephosphatase in susceptible and resistant *Heliothis virescens* larvae. Eur J Biochem 271:3127-3135.
132. Jurat-Fuentes JL, Gahan LJ, Gould FL, Heckel DG, Adang MJ (2004) The HevCaLPprotein mediates binding specificity of the Cry1A class of *Bacillus thuringiensis* toxins in*Heliothis virescens*. Biochemistry 43: 14299-14305.
133. Nagamatsu Y, Toda S, Yamaguchi F, Ogo M, Kogure M, Nakamura M, Shibata Y,Katsumoto T (1998) Identification of *Bombyx mori* midgut receptor for *Bacillusthuringiensis* insecticidal CryIA(a) toxin. Biosci Biotechnol Biochem 62: 718-726.
134. Gómez I, Sánchez J, Miranda R, Bravo A, Soberón M (2002) Cadherin-like receptorbinding facilitates proteolytic cleavage of helix α-1 in domain I and oligomer pre-poreformation of *Bacillus thuringiensis* Cry1Ab toxin. FEBS lett 513: 242-246.
135. Knight PJK, Crickmore N, Ellar DJ (1994) The receptor for *Bacillus thuringiensis* CryIA(c) delta-endotoxin in the brush border membrane of the lepidopteran *Manduca sexta* isaminopeptidase N. Mol Microbiol 11: 429-436.
136. Knowles BH (1994) Mechanism of action of *Bacillus thuringiensis* insecticidal dendotoxins.Adv Insect Physiol 24: 275-308.
137. Jiménez-Juárez N, Muñoz-Garay C, Gómez I, Saab-Rincon G, Damian-Almazo JY, GillSS, Soberón M, Bravo A (2007) *Bacillus thuringiensis* Cry1Ab mutants affecting oligomerformation are non-toxic to *Manduca sexta* larvae. J Biol Chem 282: 21222-21229.
138. Soberón M, Pardo-López L, López I, Gómez I, Tabashnik BE, Bravo A (2007)Engineering modified Bt toxins to counter insect resistance. Science 318: 1640-1642.
139. Tabashnik BE, Huang F, Ghimire MN, Leonard BR, Siegfried BD, Rangasamy M, Yang Y, Wu Y, Gahan LJ, Heckel DG, Bravo A, Soberón M (2011) Efficacy of geneticallymodified Bt toxins against insects with different genetic mechanisms of resistance. NatBiotechnol 29: 1128-1131.
140. Zhang X, Candas M, Griko NB, Taussig R, Bulla LA, Jr. (2006) A mechanism of celldeath involving an adenylyl cyclase/PKA signaling pathway is induced by the Cry1Abtoxin of *Bacillus thuringiensis*. Proc Natl Acad Sci U S A 103: 9897-9902.
*141.* Zhang X, Candas M, Griko NB, Rose-Young L, Bulla LA (2005) Cytotoxicity of *Bacillus thuringiensis* Cry1Ab toxin depends on specific binding of the toxin to the cadherinreceptor BT-R-1 expressed in insect cells. Cell Death Differ 12: 1407-1416.Bagla P (2010) Hardy Cotton-Munching Pests Are Latest Blow to GM Crops. Science327: 1439.
142. Dhurua S, Gujar GT (2011) Field-evolved resistance to Bt toxin Cry1Ac in the pinkbollworm, *Pectinophora gossypiella* (Saunders) (Lepidoptera: Gelechiidae), from India.Pest Manage Sci 67: 898-903.
143. Luttrell RG, Wan L, Knighten K (1999) Variation in susceptibility of noctuid(Lepidoptera) larvae attacking cotton and soybean to purified endotoxin proteins andcommercial formulations of *Bacillus thuringiensis*. J Econ Entomol 92: 21-32.
*144.* Ali MI, Luttrell RG, Young SY, III (2006) Susceptibilities of *Helicoverpa zea* and *Heliothisvirescens* (Lepidoptera: Noctuidae) populations to Cry1Ac insecticidal protein. J EconEntomol 99: 164-175.

145. Ali MI, Luttrell RG (2007) Susceptibility of bollworm and tobacco budworm(Lepidoptera: Noctuidae) to Cry2Ab2 insecticidal protein. J Econ Entomol 100: 921-931.
146. Carrière Y, Crowder DW, Tabashnik BE (2010) Evolutionary ecology of insectadaptation to Bt crops. Evol Appl 3: 561-573.
147. Gassmann AJ, Petzold-Maxwell JL, Keweshan RS, Dunbar MW (2011) Field-EvolvedResistance to Bt Maize by Western Corn Rootworm. PLoS One 6: e22629.
148. Moar W, Roush R, Shelton A, Ferré J, MacIntosh S, Leonard BR, Abel C (2008) Fieldevolved resistance to Bt toxins. Nat Biotechnol 26: 1072-1074.
149. Tabashnik B, Carriere Y (2009) Insect resistance to genetically modified crops. Environmental impact of genetically modified crops Wallingford: CABI: 74-100.
150. Kruger M, Van Rensburg J, Van den Berg J (2009) Perspective on the development ofstem borer resistance to Bt maize and refuge compliance at the Vaalharts irrigationscheme in South Africa. Crop Protect 28: 684-689.
151. Tiewsiri K, Wang P (2011) Differential alteration of two aminopeptidases N associatedwith resistance to *Bacillus thuringiensis* toxin Cry1Ac in cabbage looper. Proc Natl AcadSci U S A 108: 14037-14042.
152. Jurat-Fuentes JL, Karumbaiah L, Jakka SRK, Ning C, Liu C, Wu K, Jackson J, Gould F,Blanco C, Portilla M, Perera O, Adang MJ (2011) Reduced levels of membrane-boundalkaline phosphatase are common to Lepidopteran strains resistant to Cry toxins from*Bacillus thuringiensis*. PLoS One 6: e17606.
153. Gahan LJ, Gould F, Heckel DG (2001) Identification of a gene associated with Btresistance in *Heliothis virescens*. Science 293: 857-860.
154. Yang Y, Chen H, Wu Y, Wu S (2007) Mutated cadherin alleles from a field populationof *Helicoverpa armigera* confer resistance to *Bacillus thuringiensis* toxin Cry1Ac. AppEnviron Microbiol 73: 6939-6944.
155. Morin S, Biggs RW, Sisterson MS, Shriver L, Ellers-Kirk C, Higginson D, Holley D,Gahan LJ, Heckel DG, Carriere Y, Dennehy TJ, Brown JK, Tabashnik BE (2003) Threecadherin alleles associated with resistance to *Bacillus thuringiensis* in pink bollworm.Proc Natl Acad Sci U S A 100: 5004-5009.
156. Baxter SW, Badenes-Pérez FR, Morrison A, Vogel H, Crickmore N, Kain W, Wang P,Heckel DG, Jiggins CD (2011) Parallel evolution of *Bacillus thuringiensis* toxin resistancein Lepidoptera. Genetics 189: 675-679.
157. Huang FN, Zhu KY, Buschman LL, Higgins RA, Oppert B (1999) Comparison ofmidgut proteinases in *Bacillus thuringiensis*-susceptible and resistant European cornborer, *Ostrinia nubilalis* (Lepidoptera : Pyralidae). Pestic Biochem Physiol 65: 132-139.
158. Li HR, Oppert B, Higgins RA, Huang FN, Buschman LL, Gao JR, Zhu KY (2005)Characterization of cDNAs encoding three trypsin-like proteinases and mRNA quantitative analysis in Bt-resistant and -susceptible strains of *Ostrinia nubilalis*. InsectBiochem Mol Biol 35: 847-860.
159. Tabashnik BE, Croft BA (1982) Managing Pesticide Resistance in Crop-ArthropodComplexes: Interactions Between Biological and Operational Factors. Environ Entomol11: 1137-1144.
160. Gould F (1998) Sustainability of transgenic insecticidal cultivars: integrating pestgenetics and ecology. Annu Rev Entomol 43: 701-726.
161. Tabashnik BE, Carriere Y (2004) Bt transgenic crops do not have favorable effects onresistant insects. J Insect Sci 4: 4.
162. Roush RT (1998) Two-toxin strategies for management of insecticidal transgenic crops:can pyramiding succeed where pesticide mixtures have not? Philos Trans R Soc Lond,Ser B: Biol Sci 353: 1777-1786.
163. Ferré J, Van Rie J (2002) Biochemistry and genetics of insect resistance to *Bacillusthuringiensis*. Annu Rev Entomol 47: 501-533.
164. Moar WJ, Anilkumar KJ (2007) The power of the pyramid. Science 318: 1561-1562.
165. Alcantara EP, Aguda RM, Curtiss A, Dean DH, Cohen MB (2004) *Bacillus thuringiensis*δ-endotoxin binding to brush border membrane vesicles of rice stem borers. Arch InsectBiochem Physiol 55: 169-177.

166. Hernández CS, Ferré J (2005) Common receptor for *Bacillus thuringiensis* toxinsCry1Ac, Cry1Fa, and Cry1Ja in *Helicoverpa armigera*, *Helicoverpa zea*, and *Spodopteraexigua*. Appl Environ Microbiol 71: 5627-5629.
167. Naranjo SE (2011) Impacts of Bt Transgenic Cotton on Integrated Pest Management. JAgric Food Chem 59: 5842-5851.
168. Ryan CA (1990) Protease Inhibitors in Plants: Genes for Improving Defenses AgainstInsects and Pathogens. Annu Rev Phytopathol 28: 425.
169. Jongsma MA, Stiekema WJ, Bosch D (1996) Combatting inhibitor-insensitive proteasesof insect pests. Trends Biotechnol 14: 331-333.
170. Ryan CA (1989) Proteinase inhibitor gene families: strategies for transformation toimprove plant defenses against herbivores. Bioessays 10: 20-22.
171. Tamayo MC, Rufat M, Bravo JM, San Segundo B (2000) Accumulation of a maizeproteinase inhibitor in response to wounding and insect feeding, and characterizationof its activity toward digestive proteinases of *Spodoptera littoralis* larvae. Planta 211: 62-71.
172. Farmer EE, Ryan CA (1990) Interplant communication: airborne methyl jasmonateinduces synthesis of proteinase inhibitors in plant leaves. Proc Natl Acad Sci U S A 87:7713-7716.
173. Gatehouse AM, Norton E, Davison GM, Babbe SM, Newell CA, Gatehouse JA (1999)Digestive proteolytic activity in larvae of tomato moth, *Lacanobia oleracea*; effects of plant protease inhibitors *in vitro* and *in vivo*. J Insect Physiol 45: 545-55
174. Gomes A, Dias SC, Bloch C, Jr., Melo FR, Furtado JR, Jr., Monnerat RG, Grossi-de-SaMF, Franco OL (2005) Toxicity to cotton boll weevil *Anthonomus grandis* of a trypsininhibitor from chickpea seeds. Comp Biochem Physiol B Biochem Mol Biol 140B: 313-319.
175. Gomes CE, Barbosa AE, Macedo LL, Pitanga JC, Moura FT, Oliveira AS, Moura RM,Queiroz AF, Macedo FP, Andrade LB, Vidal MS, Sales MP (2005) Effect of trypsininhibitor from *Crotalaria pallida* seeds on *Callosobruchus maculatus* (cowpea weevil) and*Ceratitis capitata* (fruit fly). Plant Physiol Biochem 43: 1095-1102.
176. Burgess E, Main C, Stevens P, Christeller J, Gatehouse A, Laing W (1994) Effects ofprotease inhibitor concentration and combinations on the survival, growth and enzyme activities of the black field cricket, *Teleogryllus commodus*. J Insect Physiol 40:803-811.
177. De Leo F, Bonadé-Bottino M, Ruggiero Ceci L, Gallerani R, Jouanin L (2001) Effects ofa mustard trypsin inhibitor expressed in different plants on three lepidopteran pests.Insect Biochem Mol Biol 31: 593-602.
178. Tamhane VA, Giri AP, Sainani MN, Gupta VS (2007) Diverse forms of Pin-II familyproteinase inhibitors from *Capsicum annuum* adversely affect the growth anddevelopment of *Helicoverpa armigera*. Gene 403: 29-38.
179. Dunse KM, Anderson MA (2011) Towards the Next Generation of Pest ResistantPlants. ISB News report, Virginia Polytechnic Institute and State University June 2011.
180. Johnston KS, Lee MJ, Gatehouse JA, Anstee JH (1991) The partial purification andcharacterisation of serine protease activity in midgut of larval *Helicoverpa armigera*.Insect Biochem 21: 389-397.
181. Broadway RM (1995) Are insects resistant to plant proteinase inhibitors? J InsectPhysiol 41: 107-116.
182. Bown DP, Wilkinson HS, Gatehouse JA (1997) Differentially Regulated Inhibitor-Sensitive and Insensitive Protease Genes from the Phytophagus Insect Pest, *Helicoverpaarmigera*, are members of Complex Multigene Families. Insect Biochem Mol Biol 27: 625-638.
183. Markwick NP, Laing WA, Christeller JT, McHenry JZ, Newton MR (1998)Overproduction of Digestive Enzymes Compensates for Inhibitory Effects of Proteaseand -Amylase Inhibitors Fed to Three Species of Leafrollers (Lepidoptera: Tortricidae). JEcon Entomol 91: 1265-1276.
184. Zhu-Salzman K, Koiwa H, Salzman RA, Shade RE, Ahn JE (2003) Cowpea bruchid*Callosobruchus maculatus* uses a three-component strategy to overcome a plant defensivecysteine protease inhibitor. Insect Mol Biol 12: 135-145.
185. Broadway RM, Duffey SS (1986) Plant proteinase inhibitors: mechanism of action andeffect on the growth and digestive physiology of larval *Heliothis zea* and *Spodopteraexiqua*. J Insect Physiol 32: 827-833.

186. De Leo F, Bonade-Bottino MA, Ceci LR, Gallerani R, Jouanin L (1998) Opposite effectson *Spodoptera littoralis* larvae of high expression level of a trypsin proteinase inhibitor intransgenic plants. Plant Physiol 118: 997-1004.
187. Alvarez-Alfageme F, Maharramov J, Carrillo L, Vandenabeele S, Vercammen D, VanBreusegem F, Smagghe G (2011) Potential use of a serpin from Arabidopsis for pestcontrol. PLoS One 6: e20278.
188. Altpeter F, Diaz I, McAuslane H, Gaddour K, Carbonero P, Vasil IK (1999) Increasedinsect resistance in transgenic wheat stably expressing trypsin inhibitor CMe. MolBreed 5: 53-63.
189. Lara P, Ortego F, Gonzalez-Hidalgo E, Castanera P, Carbonero P, Diaz I (2000)Adaptation of *Spodoptera exigua* (Lepidoptera: Noctuidae) to barley trypsin inhibitorBTI-CMe expressed in transgenic tobacco. Transgenic Res 9: 169-178.
190. Christy LA, Arvinth S, Saravanakumar M, Kanchana M, Mukunthan N, Srikanth J,Thomas G, Subramonian N (2009) Engineering sugarcane cultivars with bovine pancreatic trypsin inhibitor (aprotinin) gene for protection against top borer(*Scirpophaga excerptalis* Walker). Plant Cell Rep 28: 175-184.
191. Christeller JT, Burgess EPJ, Mett V, Gatehouse HS, Markwick NP, Murray C, MaloneLA, Wright MA, Philip BA, Watt D, Gatehouse LN, Lovei GL, Shannon AL, Phung MM,Watson LM, Laing WA (2002) The expression of a mammalian proteinase inhibitor,bovine spleen trypsin inhibitor in tobacco and its effects on *Helicoverpa armigera* larvae.Transgenic Res 11: 161-173.
192. Hoffmann MP, Zalom FG, Wilson LT, Smilanick JM, Malyj LD, Kiser J, Hilder VA,Barnes WM (1992) Field evaluation of transgenic tobacco containing genes encoding*Bacillus thuringiensis* ∂-endotoxin or cowpea trypsin inhibitor: efficacy against*Helicoverpa zea* (Lepidoptera: Noctuidae). J Econ Entomol 85: 2516-2522.
193. Sane V, Nath P, Aminuddin, Sane P (1997) Development of insect-resistant transgenicplants using plant genes: Expression of cowpea trypsin inhibitor in transgenic tobaccoplants. Curr Sci 72: 741-747.
194. Wu Y, Llewellyn D, Mathews A, Dennis ES (1997) Adaptation of *Helicoverpa armigera*(Lepidoptera: Noctuidae) to a proteinase inhibitor expressed in transgenic tobacco. Mol Breed 3: 371-380.
195. De Leo F, Gallerani R (2002) The mustard trypsin inhibitor 2 affects the fertility of*Spodoptera littoralis* larvae fed on transgenic plants. Insect Biochem Mol Biol 32: 489-496.
196. Heath RL, McDonald G, Christeller JT, Lee M, Bateman K, West J, Van Heeswijck R,Anderson MA (1997) Proteinase inhibitors from *Nicotiana alata* enhance plant resistanceto insect pests. J Insect Physiol 43: 833-842.
197. Charity JA, Anderson MA, Bittisnich DJ, Whitecross M, Higgins TJV (1999) Transgenictobacco and peas expressing a proteinase inhibitor from *Nicotiana alata* have increasedinsect resistance. Mol Breed 5: 357-365.
198. Dunse KM, Stevens JA, Lay FT, Gaspar YM, Heath RL, Anderson MA (2010)Coexpression of potato type I and II proteinase inhibitors gives cotton plants protectionagainst insect damage in the field. Proc Natl Acad Sci U S A 107: 15011-15015.
199. Johnson R, Narvaez J, An G, Ryan C (1989) Expression of proteinase inhibitors I and IIin transgenic tobacco plants: effects on natural defense against Manduca sexta larvae.Proc Natl Acad Sci U S A 86: 9871-9875.
200. McManus MT, White DWR, McGregor PG (1994) Accumulation of a chymotrypsininhibitor in transgenic tobacco can affect the growth of insect pests. Transgenic Res 3:50-58.
201. Jongsma MA, Bakker PL, Peters J, Bosch D, Stiekema WJ (1995) Adaptation of*Spodoptera exigua* larvae to plant proteinase inhibitors by induction of gut proteinaseactivity insensitive to inhibition. Proc Natl Acad Sci U S A 92: 8041-8045.
202. Winterer J, Bergelson J (2001) Diamondback moth compensatory consumption ofprotease inhibitor-transformed plants. Mol Ecol 10: 1069-1074.
203. Abdeen A, Virgós A, Olivella E, Villanueva J, Avilés X, Gabarra R, Prat S (2005)Multiple insect resistance in transgenic tomato plants over-expressing two families ofplant proteinase inhibitors. Plant Mol Biol 57: 189-202.

204. [Luo M, Wang Z, Li H, Xia KF, Cai Y, Xu ZF (2009) Overexpression of a Weed (*Solanumamericanum*) Proteinase Inhibitor in Transgenic Tobacco Results in Increased Glandular Trichome Density and Enhanced Resistance to *Helicoverpa armigera* and *Spodoptera litura*.Int J Mol Sci 10: 1896-1910.
205. Confalonieri M, Allegro G, Balestrazzi A, Fogher C, Delledonne M (1998) Regenerationof *Populus Nigra* Transgenic Plants Expressing a Kunitz Proteinase Inhibitor (Kti3) Gene.Mol Breed 4: 137-145.
206. McManus M, Burgess EPJ, Philip B, Watson L, Laing WA, Voisey C, White D (1999)Expression of the soybean (Kunitz) trypsin inhibitor in transgenic tobacco: Effects onlarval development of *Spodoptera litura*. Transgenic Res 8: 383-395.
207. Nandi AK, Basu D, Das S, Sen SK (1999) High level expression of soybean trypsininhibitor gene in transgenic tobacco plants failed to confer resistance against damage caused by *Helicoverpa armigera*. J Biosci (Bangalore) 24: 445-452.
208. Marchetti S, Delledonne M, Fogher C, Chiaba C, F C, Savazzini F, Giorgadno A (2000)Soybean Kunitz, C-II and PI-IV inhibitor genes confer different levels of insectresistnace to tobacco and potato transgenic plants. Theor Appl Genet 101: 519-526.
209. Falco MC, Silva-Filho MC (2003) Expression of soybean proteinase inhibitors intransgenic sugarcane plants: effects on natural defense against *Diatraea saccharalis*. Plant Physiol Biochem 41: 761-766.
210. Yeh KW, Lin MI, Tuan SJ, Chen YM, Lin CY, Kao SS (1997) Sweet potato (*Ipomoeabatatas*) trypsin inhibitors expressed in transgenic tobacco plants confer resistanceagainst *Spodoptera litura*. Plant Cell Rep 16: 696-699.
211. Senthilkumar R, Cheng CP, Yeh KW (2010) Genetically pyramiding protease-inhibitorgenes for dual broad-spectrum resistance against insect and phytopathogens intransgenic tobacco. Plant Biotechnol J 8: 65-75.
212. Liu HB, Guo X, Naeem MS, Liu D, Xu L, Zhang WF, Tang GX, Zhou WJ (2011)Transgenic *Brassica napus* L. lines carrying a two gene construct demonstrate enhancedresistance against *Plutella xylostella* and *Sclerotinia sclerotiorum*. Plant Cell Tiss Org Cult106: 143-151.
213. Jongsma MA, Bolter C (1997) The adaptation of insects to plant protease inhibitors. JInsect Physiol 43: 885-895.
214. Jouanin L, Bonade-Bottino M, Girard C, Morrot G, Giband M (1998) Transgenic plantsfor insect resistance. Plant Sci 131: 1-11.
215. Girard C, Le Metayer M, Bonade-Bottino M, Pham-Delegue M-H, Jouanin L (1998)High level of resistance to proteinase inhibitors may be conferred by proteolyticcleavage in beetle larvae. Insect Biochem Mol Biol 28: 229-237.
216. Paulillo LC, Lopes AR, Cristofoletti PT, Parra JR, Terra WR, Silva-Filho MC (2000)Changes in midgut endopeptidase activity of *Spodoptera frugiperda* (Lepidoptera:Noctuidae) are responsible for adaptation to soybean proteinase inhibitors. J EconEntomol 93: 892-896.
217. Gatehouse LN, Shannon AL, Burgess EP, Christeller JT (1997) Characterization ofmajor midgut proteinase cDNAs from *Helicoverpa armigera* larvae and changes in gene expression in response to four proteinase inhibitors in the diet. Insect Biochem Mol Biol27: 929-944.
218. Giri AP, Harsulkar AM, Deshpande VV, Sainani MN, Gupta VS, Ranjekar PK (1998)Chickpea Defensive Proteinase Inhibitors Can Be Inactivated by Podborer GutProteinases. Plant Physiol 116: 393-401.
219. Telang MA, Giri AP, Sainani MN, Gupta VS (2005) Characterization of two midgutproteinases of *Helicoverpa armigera* and their interaction with proteinase inhibitors. JInsect Physiol 51: 513-522.
220. Michaud D, Nguyen-Quoc B, Vrain TC, Fong D, Yelle S (1996) Response of digestivecysteine proteinases from the Colorado potato beetle (*Leptinotarsa decemlineata*) and theblack vine weevil (*Otiorynchus sulcatus*) to a recombinant form of human stefin A. ArchInsect Biochem Physiol 31: 451-464.
221. Volpicella M, Ceci LR, Cordewener J, America T, Gallerani R, Bode W, Jongsma MA,Beekwilder J (2003) Properties of purified gut trypsin from *Helicoverpa zea*, adapted toproteinase inhibitors. Eur J Biochem 270: 10-19.

222. Bown DP, Wilkinson HS, Gatehouse JA (2004) Regulation of expression of genesencoding digestive proteases in the gut of a polyphagous lepidopteran larva in response to dietary protease inhibitors. Physiol Entomol 29: 278-290.
223. Chougule NP, Giri AP, Sainani MN, Gupta VS (2005) Gene expression patterns of*Helicoverpa armigera* gut proteases. Insect Biochem Mol Biol 35: 355-367.
224. Girard C, Lemetayer M, Zaccomer B, Bartlet E, Williams I, Bonadebottino M,Phamdelegue MH, Jouanin L (1998) Growth stimulation of beetle larvae reared on atransgenic oilseed rape expressing a cysteine proteinase inhibitor. J Insect Physiol 44:263-270.
225. Cloutier C, Fournier M, Jean C, Yelle S, Michaud D (1999) Growth compensation andfaster development of Colorado potato beetle (Coleoptera : Chrysomelidae) feeding onpotato foliage expressing oryzacystatin I. Arch Insect Biochem Physiol 40: 69-79.
226. Broadway RM (1997) Dietary regulation of serine proteinases that are resistant toserine proteinase inhibitors. J Insect Physiol 43: 855-874.
227. Markwick NP, Laing WA, Christeller JT, McHenry JZ, Newton MR (1998)Overproduction of Digestive Enzymes Compensates for Inhibitory Effects of Proteaseand -Amylase Inhibitors Fed to Three Species of Leafrollers (Lepidoptera: Tortricidae). JEcon Entomol 91: 1265-1276.
228. Mazumdar-Leighton S, Broadway RM (2001) Identification of six chymotrypsincDNAs from larval midguts of *Helicoverpa zea* and *Agrotis ipsilon* feeding on the soybean(Kunitz) trypsin inhibitor. Insect Biochem Mol Biol 31: 633-644.
229. Mazumdar-Leighton S, Ragahavendra Babu C, Bennett J (2000) Identification of novelserine proteinase gene transcripts in the midguts of two tropical insect pests,*Scirpophaga incertulas*(Wk.) and *Helicoverpa armigera* (Hb.). Insect Biochem Mol Biol 30:57-68.
230. Mazumdar-Leighton S, Broadway RM (2001) Transcriptional induction of diversemidgut trypsins in larval *Agrotis ipsilon* and *Helicoverpa zea* feeding on the soybeantrypsin inhibitor. Insect Biochem Mol Biol 31: 645-657.
231. Atkinson AH, Heath RL, Simpson RJ, Clarke AE, Anderson MA (1993) ProteinaseInhibitors in *Nicotiana alata* Stigmas Are Derived from a Precursor Protein Which IsProcessed into Five Homologous Inhibitors. Plant Cell 5: 203-213.
232. Miller EA, Lee MC, Anderson MA (1999) Identification and characterization of aprevacuolar compartment in stigmas of *Nicotiana alata*. Plant Cell 11: 1499-1508.
233. Nielsen KJ, Hill JM, Anderson MA, Craik DJ (1996) Synthesis and structuredetermination by NMR of a putative vacuolar targeting peptide and model of aproteinase inhibitor from *Nicotiana alata*. Biochemistry 35: 369-378.
234. Scanlon MJ, Lee MC, Anderson MA, Craik DJ (1999) Structure of a putative ancestralprotein encoded by a single sequence repeat from a multidomain proteinase inhibitorgene from *Nicotiana alata*. Structure 7: 793-802.
235. Nielsen KJ, Heath RL, Anderson MA, Craik DJ (1994) The three-dimensional solutionstructure by 1H NMR of a 6-kDa proteinase inhibitor isolated from the stigma of*Nicotiana alata*. J Mol Biol 242: 231-243.
236. Lee MC, Scanlon MJ, Craik DJ, Anderson MA (1999) A novel two-chain proteinaseinhibitor generated by circularization of a multidomain precursor protein. Nat StructBiol 6: 526-530.
237. Nielsen KJ, Heath RL, Anderson MA, Craik DJ (1995) Structures of a series of 6-kDtrypsin inhibitors isolated from the stigma of *Nicotiana alata*. Biochemistry 34: 14304-14311.
238. Heath RL, Barton PA, Simpson RJ, Reid GE, Lim G, Anderson MA (1995)Characterization of the protease processing sites in a multidomain proteinase inhibitorprecursor from *Nicotiana alata*. Eur J Biochem 230: 250-257.
239. Lee MCS, Scanlon MJ, Craik DJ, Anderson MA (1999) A novel two-chain proteinaseinhibitor generated by circularization of a multidomain precursor protein. Nat StructMol Biol 6: 526-530.
240. Atkinson A, Lind J, Clarke A, Anderson M (1994) Molecular and structural features ofthe pistil of *Nicotiana alata*. Biochem Soc Symp. pp. 15.
241. Harsulkar AM, Giri AP, Patankar AG, Gupta VS, Sainani MN, Ranjekar PK,Deshpande VV (1999) Successive use of non-host plant proteinase inhibitors requiredfor effective

inhibition of *Helicoverpa armigera* gut proteinases and larval growth. PlantPhysiol 121: 497-506.
242. Murdock LL, Shade RE (2002) Lectins and protease inhibitors as plant defenses againstinsects. J Agric Food Chem 50: 6605-6611.
243. Chiche L, Heitz A, Padilla A, Lenguyen D, Castro B (1993) Solution Conformation of aSynthetic Bis-Headed Inhibitor of Trypsin and Carboxypeptidase A: New StructuralAlignment between the Squash Inhibitors and the Potato Carboxypeptidase InhibitorProtein Eng 6: 675-682.
244. Urwin PE, McPherson MJ, Atkinson HJ (1998) Enhanced transgenic plant resistance tonematodes by dual proteinase inhibitor constructs. Planta 204: 472-479.
245. Oppert B, Morgan TD, Hartzer K, Kramer KJ (2005) Compensatory proteolyticresponses to dietary proteinase inhibitors in the red flour beetle, *Tribolium castaneum*(Coleoptera : Tenebrionidae). Comp Biochem Physiol, C: Toxicol Pharmacol 140: 53-58.
246. Fan X, Shi X, Zhao J, Zhao R, Fan Y (1999) Insecticidal activity of transgenic tobaccoplants expressing both Bt and CpTI genes on cotton bollworm (*Helicoverpa armigera*).Chin J Biotechnol 15: 1.
247. MacIntosh SC, Kishore GM, Perlak FJ, Marrone PG, Stone TB, Sims SR, Fuchs RL(1990) Potentiation of *Bacillus thuringiensis* insecticidal activity by serine protease inhibitors. J Agric Food Chem 38: 1145-1152.
248. Pannetier C, Giband M, Couzi P, Le Tan V, Mazier M, Tourneur J, Hau B (1997)Introduction of new traits into cotton through genetic engineering: insect resistance asexample. Euphytica 96: 163-166.
249. Markwick NP, Reid SJ, Liang WA, Christeller JT (1995) Effects of dietary protein andprotease inhibitors on codling moth (Lepidoptera: Tortricidae). J Econ Entomol 88: 33-39.
250. Tabashnik BE, Gassmann AJ, Crowder DW, Carriere Y (2008) Insect resistance to Btcrops: evidence versus theory. Nat Biotech 26: 199-202

# INDEX